Critical Politica

The politics of environmental science

Tim Forsyth

Routledge
Taylor & Francis Group

LONDON AND NEW YORK

First published 2003
by Routledge
11 New Fetter Lane, London EC4P 4EE

Simultaneously published in the USA and Canada
by Routledge
29 West 35th Street, New York, NY 10001

Reprinted 2004

Routledge is an imprint of the Taylor & Francis Group

© 2003 Tim Forsyth

Typeset in 10/12 Times by Wearset Ltd, Boldon, Tyne and Wear
Printed and bound in Great Britain by Biddles Ltd,
King's Lynn, Norfolk

British Library Cataloguing in Publication Data
A catalogue record for this book is available from the British Library

Library of Congress Cataloguing in Publication Data
A catalog record for this book has been requested

ISBN 0-415-18562-9 (hbk)
ISBN 0-415-18563-7 (pbk)

"A truth is the kind of error without which a certain species of life could not live."

Frederick Nietzsche, *The Will to Power* (1901: 493)

"Call it a lie, if you like, but a lie is a sort of myth and a myth is a sort of truth."

Cyrano de Bergerac, in Edmond Rostand, *Cyrano de Bergerac* (Act 2)

Contents

Illustrations

Preface and acknowledgments

These are controversial times for writing about politics and environmental science. An increasing number of authors are acknowledging the political influences on scientific knowledge and organizations that undertake scientific research. Yet, discussing the link between science and politics commonly leads to accusations of being anti-environmentalist, or epistemologically relativist. There is a need for an approach to environmental politics that acknowledges the social and political framings of environmental science, yet which offers the means to build environmental policy that is both biophysically effective and socially relevant.

This book represents an attempt to rebuild environmental science in a more politicized way. The book is inspired largely by my own research experiences in the developed and developing world, but in part summarizes the concerns of a growing number of researchers about how we understand environmental problems. These concerns do not dismiss the need for environmental protection, or suggest that economic progress will solve all problems. Instead, the concerns are about the grave simplifications and inaccuracies within much environmental debate, often revealing different perceptions between people living in affected regions, and policymakers and concerned public elsewhere. These differences suggest there is a need to rethink explanations of environmental problems in ways that acknowledge the linkages between social factors and the gathering of information about biophysical change.

The book is located within the debate known as "political ecology" because this topic has become associated with assessing the political linkages between society and environmental change. The book, however, seeks to advance this debate by suggesting new ways to integrate political analysis with the formulation and use of "ecology" as the science underlying much environmental debate. It is subtitled *The Politics of Environmental Science* because the book argues that "science" cannot be separated from "politics" but that political factors underlie the formulation, dissemination, and institutionalization of scientific knowledge and networks. Such discussions, however, do not suggest that environmental concern is unwarranted, or that environmental science cannot have predictive success.

The book was written with assistance from the Economic and Social

Research Council of the United Kingdom (award R000 22 2767) held at the Institute of Development Studies, Falmer, and the London School of Economics; and a fellowship to the Global Environmental Assessment Program of the Kennedy School of Government, Harvard University.

I would like to thank Giles Pilbrow for permitting me to reproduce one of his cartoons, originally published in *Private Eye* in Figure 8.1. Figure 4.1 is reproduced with the permission of Pearson Education from Latour, B. (1993) *We Have Never Been Modern*, Hemel Hempstead: Harvester Wheatsheaf. Figure 4.2 is reproduced with the permission of Michael Thompson from Schwarz, M. and Thompson, M. (1990) *Divided We Stand: Redefining Politics, Technology and Social Choice*, Hemel Hempstead: Harvester Wheatsheaf. Figure 9.1 is reproduced with the permission of Elsevier Science from Funtowicz, S. and Ravetz, J. (1993) "Science for the post-normal age," *Futures* 26: September, pp. 739–756. Figure 9.2 is reproduced with the permission of MIT Press from MacKenzie, D. (1990) *Inventing Accuracy: A Historical Sociology of Nuclear Missile Guidance*, Cambridge, MA: MIT Press.

Finally, the following individual scholars may be mentioned specifically for providing advice, conversations, and friendly reproaches over the years that this book has evolved. I am grateful for their contribution, although any errors in this volume remain my responsibility.

Bill Adams, Greg Bankoff, Simon Batterbury, Tony Bebbington, Silke Beck, Raymond Bryant, Dave Cash, Noel Castree, Judy Clark, Alex Farrell, Cathy Fogel, Matthew Gandy, Mike Goldman, Barbara Goldoftas, Hugh Gusterson, Dave Guston, Rom Harré, Alastair Iles, Sheila Jasanoff, Bernd Kasemir, Mojdeh Keykah, Chunglin Kwa, Myannah Lahsen, Melissa Leach, Stuart Leggatt, Diana Liverman, Larry Lohmann, Marybeth Long-Martello, David Lund, Allison MacFarlane, Lyla Mehta, Clark Miller, Dele Ogunseitan, Andy C. Pratt, Paul Robbins, Dianne Rocheleau, Ian Scoones, Paul Sillitoe, K. Sivaramakrishnan, Michael Thompson, Billie Lee Turner II, Damian White, Brian Wynne.

Tim Forsyth

Abbreviations

AIJ	Activities Implemented Jointly
CBA	Cost–Benefit Analysis
CBD	Convention on Biological Diversity
CBNRM	Community Based Natural Resource Management
CCD	Convention to Combat Desertification
CDM	Clean Development Mechanism
COP	Conference of the Parties (to agreements such as the Kyoto Protocol)
CSE	Center for Science and Environment, Delhi
DIPS	Deliberative and Inclusionary Processes in Environmental Policymaking
EIA	Environmental Impact Assessment
EPA	Environmental Protection Agency (of the USA)
GBA	Global Biodiversity Assessment
GCM	General Circulation Model
GIS	Geographical Information System
GM	Genetically Modified
GMOs	Genetically Modified Organisms
IDNDR	International Decade for Natural Disaster Reduction
ICRAF	International Center for Research in Agroforestry
IGBP	International Geosphere–Biosphere Program
I = PAT	(Equation): Environmental Impacts = function of Population growth, Affluence, and Technology
IPCC	Intergovernmental Panel on Climate Change
IUCN	International Union for the Conservation of Nature
JI	Joint Implementation
LRTAP	Long Range Transport of Air Pollution
LTG	Limits to Growth
NASA	National Aeronautics and Space Administration
NGO	Non-Governmental Organization
SBSTA	Subsidiary Body for Scientific and Technical Advice to the UNFCCC
SSK	Sociology of Scientific Knowledge
UNEP	United Nations Environment Program

UNFCCC   United Nations Framework Convention on Climate Change
USCSP   United States Country Studies Program (on climate change)
USIJI   United States Initiative on Joint Implementation
USLE   Universal Soil Loss Equation
WCD   World Commission on Dams
WCED   World Commission on Environment and Development
WCU   World Conservation Union
WED   Women, Environment and Development
WRI   World Resources Institute
WTO   World Trade Organization

1 Political ecology and the politics of environmental science

Abraham Lincoln once remarked that anyone who enjoys eating sausages and using the law should avoid seeing how either is made. The same can be said about many of the scientific "laws" and principles underlying environmental policy and debates today. This book is about why we should treat these apparent environmental "laws" with concern, and instead seek a more accurate and politically aware approach to environmental explanation. The book's key purpose is to show how we need to see the evolution of environmental facts and knowledge as part of the political debate, rather than as a pre-prepared basis from which to start environmental debate.

The time has never been better for reevaluating the political basis of environmental explanations. Few days go by without media reports of environmental crisis. Unusual weather events are taken as evidence of irreversible and catastrophic climate change. Increasingly complex environmental policies and agreements are being agreed, with progressively more control over different aspects of our lives. Inexorably, we seem to slip toward the "Risk Society" of Ulrich Beck (1992), in which lives and politics are organized around the avoidance of risk. Yet, in environmental terms at least, the causal basis of environmental risk, and the implications of proposed solutions to risk, are far from clear.

This book seeks to provide this reevaluation of environmental science by considering the intricate ways in which science and politics are mutually related. This project does not refer to conventional political debates such as public access to scientific information, or the ability to communicate scientific findings to policy. Instead, the project is to develop a political philosophy of environmental science that indicates how social and political framings are woven into both the formulation of scientific explanations of environmental problems, and the solutions proposed to reduce them.

Thus, when Michael Zammit Cutajar, the Executive Secretary of the United Nations Framework Convention on Climate Change commented that: "The science has driven the politics ... if the science is to continue guiding the politics, it is essential to keep the politics out of the science" (2001: 1), he adopted the classic position that environmental science is somehow disconnected from environmental values and politics. This book does not adopt an anti-climate-change position, but seeks to indicate how different political actions and scientific methodologies have led to environmental explanations and solutions that are thoroughly embedded in social and political practices.

So, how does this book proceed? The key objective is to integrate debates in so-called "political ecology" with debates concerning the constructions of science. It is important to note that this approach does not imply rejecting environmental "realism" – or the belief in a biophysically "real world out there." Indeed, the book adopts debates within Philosophy of Science to indicate potential ways to integrate realist biophysical prediction with social and political constructions.

It is also important to note that this book is in no way a supporter of "brownlash," or the criticism of environmental concerns in order to support polluting industries or weaken environmental regulation. Nevertheless, this book does criticize some assertions of environmentalists about the ability of orthodox science to describe environmental change and problems in ways that are politically neutral.

More importantly, though, this book seeks to demonstrate two important and increasingly unavoidable anxieties. First, the adoption of environmental science without acknowledging how it is affected by social and political factors undermines its ability to address the underlying biophysical causes of perceived environmental problems. Second, the adoption of policies based on such unreconstructed science frequently produces environmental policies that unfairly penalize many land users – especially in developing countries – and may even increase environmental degradation and poverty by threatening livelihoods. This book seeks to address these two problems by exploring the links between science and society in order to avoid the replication of inadequate science, and to enable the production of more biophysically accurate, and socially relevant, science.

This initial chapter explains the rationale for this project. The chapter looks specifically at debates in "political ecology" and so-called science studies or science–policy. Readers not familiar with these debates may prefer to turn immediately to Chapter 2.

The separation of science and politics: some past trends in political ecology

It is widely accepted that debates concerning "political ecology" refer to the social and political conditions surrounding the causes, experiences, and management of environmental problems (e.g. Blaikie and Brookfield, 1987; Bryant, 1992; Greenberg and Park, 1994; Zimmerer, 2000). It is, however, remarkable that much writing about political ecology does not define what is meant by "ecology." A variety of authors over the years have revealed different approaches to the meaning of "ecology" in "political ecology."

First, some authors have approached political ecology by explaining environmental problems as the phenomenological interaction of biophysical processes, human needs, and wider political systems. Blaikie and Brookfield wrote:

> The phrase "political ecology" combines the concerns of ecology and a broadly defined political economy. Together this encompasses the constantly shifting dialectic between society and land-based resources, and also within classes and groups within society itself.
>
> (1987: 17)

Second, there is the "politics of ecology" in the sense of political activism in favor of Deep Green environmentalism and its critique of modernity and capitalism. Atkinson wrote: "Political Ecology is both a set of theoretical propositions and ideas on the one hand and on the other a social movement referred to as the 'ecology movement' or, latterly, the Green movement" (1991: 18).

Third, there is the use of "ecology" as a metaphor for the interconnectedness of political relations. This metaphor was adopted by the first book with "political ecology" in its title, *International Regions and the International System: A Study in Political Ecology* by Bruce M. Russett in 1967, even though the book itself had no discussion of biophysical environmental change or conservation. He wrote:

> I have termed this volume "a study in political ecology." As ecology is defined as *the relation of organisms or groups of organisms to their environment,* I have attempted to explore some of the relations between political systems and their social and physical environment.
>
> (Russett, 1967: vii, emphasis in original)

Yet, although this original book did not discuss environmental conservation, later volumes on environment have adopted a similar usage of the term, political ecology. Anderson, for example, wrote:

> Just as environmental ecology refers to interaction and interdependence among soil, air and water, the peasants' political ecology also refers to the interactive interdependence among spheres – the individual, the community, the natural world, and the national society.
>
> (Anderson, 1994: 6)

Fourth, political ecology has been defined as a more specific analysis of Marxist debates about materialism, justice, and nature in capitalist societies, with the view to achieving a fairer distribution of rights and resources:

> Political ecology, like the Marxist-inspired workers' movement, is based on a critique – and thus an analysis, a theorized understanding –

of the "order of existing things." More specifically, Marx and the greens focus on a very precise sector of the real world: the humanity–nature relationship, and, even more precisely, relations among people that pertain to nature (or what Marxists call the "productive forces").

(Lipietz, 2000: 70)

Finally, there is the use of "political ecology" to refer in general terms to the politics of environmental problems without specific discussion of "ecology." Bryant (1992: 13), for example, describes political ecology as an inquiry into "the political forces, conditions and ramifications of environmental change," and may include studies of environmental impacts from different sources; location-specific aspects of ecological change; and the effects of environmental change on socio-economic and political relationships (see also Lowe and Rüdig, 1986). In a later publication, Bryant and Bailey (1997: 190) suggest that "political ecology" as a debate focuses on interactions between the state, non-state actors, and the physical environment, whereas "environmental politics" as a debate concerns the role of the state generally.

This book differs from these approaches by seeking to establish the political forces behind different accounts of "ecology" as a representation of biophysical reality. In this sense, a "critical" political ecology may be seen to be the politics of ecology as a scientific legitimatization of environmental policy. The approach adopted in this book may be seen to differ with the historic approaches to political ecology listed above because these approaches either adopt a priori concepts of environmental science and explanation; or take insufficient steps to avoid the separation of environmental explanation and politics in the analysis of environmental politics. The following discussions describe some themes of these past approaches, and how this book may argue for different approaches. The section after this then discusses how science and politics may be integrated.

Ecology, the subversive science

The first usages of the term "political ecology" in academic publications were made in the late 1960s and 1970s (see Russett, 1967; Wolf, 1972; Miller, 1978; Cockburn and Ridgeway, 1979). Yet, before then, the possibility to integrate political analysis with environmental explanation was widely discussed. The first discussions of ecology as a science with political content emerged in the 1960s during the growing concern about human impacts on the biophysical environment. "Ecology" was seen as both the study of those impacts, but also the new philosophical approach of looking at people–environment interactions as a whole. Indeed, the mood was well represented by Aldous Huxley's paper, "The politics of ecology: the question of survival" (1963).

Rather than simply challenge existing economic development as damaging, the early political ecologists emphasized the philosophical and methodological challenges of "ecology" to existing forms of science. In a collection of papers in the journal, *Bioscience*, in 1964, René Dubos – the future co-author of the companion book to the 1972 World Conference on the Human Environment, *Only One Earth* (Ward and Dubos, 1972) – rejected existing scientific approaches for being reductionist. Instead, Dubos sought a method of seeking "community" or "interrelationships" under ecology as a better basis for understanding environmental change (Dubos, 1964). Similarly, Eugene Odum, the author of *Fundamentals of Ecology* (originally published in 1953), wrote: "*The new ecology is thus a systems ecology* ... [it] deals with the structure and function of levels of organization beyond that of the individual and species" (1964: 15, emphasis in the original).

"Ecology," therefore, was a new science aiming to illustrate the connectiveness of humans and other species. Yet the achievement of ecology, by definition, depended simultaneously upon the development of a new scientific approach highlighting a level of "community" beyond simple individuals, and also the establishment of a new political agenda questioning the destructiveness of human behavior. "Ecology" was therefore inherently "political," and this was expressed most forcefully by Paul Sears in a paper entitled "Ecology – a subversive subject" (1964: 11–12):

> Is ecology a phase of science of limited interest and utility? Or, if taken seriously as an instrument for the long-run welfare of mankind [sic], would it endanger the assumptions and practices accepted by modern societies, whatever their doctrinal commitments? ... By its very nature, ecology affords a continuing critique of man's operations within the ecosystem.

This book shows how this early trend in the politics of ecology is still influential today. Yet, while this initial school is overtly political, the approach does not question how its statements about "community" and "mankind" [sic] might pose problems for establishing universal explanations of environmental problems. Instead, this book discusses newer approaches to less generalized explanations of environmental problems, and the localized and contextual nature of environmental threats.

The domination of nature

Another important theme underlying much debate within political ecology is the preoccupation with what writers have called the "domination" of nature. This theme is also closely linked to the discussion of capitalism as a primary cause of environmental degradation. Such debates have been particularly prevalent among writers influenced by the Frankfurt School of Critical Theory, and particularly Marcuse and Habermas, who described how "human nature" was dominated by the instrumental rationality and exploitation of modern industrial society. The debates also fueled the growth of environmentalism as a "new" social movement in Europe and North America during the 1960s, which was partly premised on concerns about the instrumentality of capitalism, science, and technology. Marcuse, in *One Dimensional Man*, famously wrote:

> Science, by *virtue of its own methods* and concepts, has projected and promoted a universe in which the domination of nature has remained linked to the domination of man [sic] – a link which tends to be fatal to this universe as a whole.
>
> (1964: 166, emphasis in original)

Such statements were also adopted by some of the original writers on ecology and politics. In *The Subversive Science: Essays Towards an Ecology of Man*, Shepard and McKinley wrote:

> To a world which gives grudging admission of the "nature" in human nature, we say that the framework of human life is all life, and that anything adding to its understanding may be ecological. It is life, not man [sic], which is the main contour, and it is ecology in general where human ecology is to be found. Ideas themselves are inseparable from nature and the study of man in nature.
>
> (1969: vii)

And in another early book on political ecology, Miller wrote:

> A primary contribution of a "political ecology" movement should be to demythologize this idealist mystification of the human/nature relationship [as adopted by economic exploitation] and to begin the construction of a new, holistic ethic... This new one-dimensionality of science and of technology, its utilitarianism and devotion to the end goals of the dominant economic class in American society, raises perhaps the ultimate question of environmental values. With nature and people increasingly viewed as having only commodity and exchange value, the acquiescence of science to that same perspective can only lead to a deepening dehumanization within society and to a further exploitation of nature.
>
> (1978: 56, 101)

This book seeks to question how far it is still possible to base explanations of environmental degradation upon such far-reaching criticisms of modern industrial society (see also Castree, 1995; Vogel, 1996; Gandy, 1997). By so doing, this book does not question the need to consider the impacts of modernity on the human condition. But it seeks to explore how far the critique of modernity in the fashion of these critics and the early scholars of the Frankfurt School might also imply forms of environmental explanation that may not acknowledge the complexities of ecological reality in the biophysical world. One important additional aspect of this debate is the link between the "domination of nature" and the ecological science dominated by notions of ecological equilibrium, or a "balance" of nature. André Gorz (in his polemic, *Ecologique et Politique*) wrote:

> Science and technology have ended up making this central discovery: all productive activity depends on borrowing from the finite resources of the planet and on organizing a set of exchanges within a fragile system of multiple equilibriums.
>
> (1975: 12–13)

Such notions of ecological equilibrium are now being questioned among scientific researchers in a variety of disciplines. Instead, a new discourse of "non-equilibrium ecology" is emerging to reflect recent insights into chaos theory, and the problems of inferring ecological explanation across indeterminate time and space scales (e.g. Botkin, 1990; Zimmerer, 2000). Exploring the political origins of different models of environmental explanation therefore requires seeking to integrate the new, non-equilibrium ecologies with social and political understandings of how the two are linked.

Furthermore, this approach also implies questioning how far environmental degradation *per se* may be attributed simply to capitalism, or the exploitation of industry and the state. By questioning the essentialist link between capitalism and environmental degradation, this book challenges virtually all historic approaches to political ecology that have focused on political economy and environment (e.g. Cockburn and Ridgeway, 1979; Atkinson, 1991; Bryant, 1997b; Bryant and Bailey, 1997; Wells and Lynch, 2000). Yet, the aims of questioning the role of capitalism are not to suggest that political ecologists should not be concerned about exploitation of people and resources, but to ask how the opposition to capitalism may have influenced the production of environmental explanations. The aim of a "critical" political ecology is to refocus political ecology from the assessment of capitalism alone as a source of environmental degradation, toward a politicized understanding of environmental explanation beyond the epistemology offered by the critique of capitalism. Indeed, as later chapters describe, the shortcomings of such essential links between capitalism and degradation have been most exposed when they have been applied to societies and environmental problems outside Europe and North America.

Social justice and the developing world

Finally, much debate in political ecology has focused on the social justice of environmental disputes and resource struggles in developing countries (e.g. Watts, 1983; Blaikie, 1985; Escobar, 1995). In part, this is because such environmental conflicts involve the interaction of a variety of actors from state, society, and industry in locations considered to be of global environmental significance (for example, concerning the Amazon, see Bunker, 1985; Hecht and Cockburn, 1989). Yet, in addition, much political ecology within developing countries may be seen to be an extension of so-called cultural ecology, or the research focusing on local environmental practices often in an anthropological fashion (e.g. Conklin, 1954; Geertz, 1963; Rappaport, 1968). The influential cultural ecologist Robert Netting summarized cultural ecology as a focus on the "particular circumstances of geography, demography, technology, and history" that result in a "splendid variety of cultural values, religion, kinship systems, and political structures" in local environmental strategies (Netting, 1993: 1). Indeed, the earliest published journal paper to include "political ecology" in its title (Eric Wolf's 1972 paper, "Ownership and political ecology") draws heavily on Netting's work.

There is much debate concerning the difference between "cultural" and "political" ecology (e.g. see Batterbury and Bebbington, 1999). Generally, political ecology has been seen to focus more on underlying and widespread political explanations for environmental change and degradation, whereas cultural ecology has considered more local and culturally situated practices of land management. Yet within this focus, analysts have identified two broad themes within political ecology in relation to the developing world (see Peet and Watts, 1996; Watts and McCarthy, 1997). The first theme adopts a largely structuralist explanation of land degradation through reference to the forces of capitalism, or oppressive state policies and their impacts on local people and environment (e.g. Blaikie, 1985; Blaikie and Brookfield, 1987; Bryant and Bailey, 1997). The second theme is influenced by poststructuralist approaches to social science that focus instead on the historical and cultural influences on the evolution of concepts of environmental change and degradation as linguistic and political forces in their own right (e.g. Rocheleau, 1995; Leach and Mearns, 1996; Escobar, 1995, 1996, 1998). Yet both are premised upon a sense of social justice for environmental explanation and development.

As Peet and Watts (1996) note, much concern about political ecology in the developing world has reflected the belief that injustices are being committed against both local peoples, and against environmental resources that may be of value to these people or to the world at large. There is a lack of consensus among political ecologists, however, about how to express this concern. On the one hand, some of the orthodox approaches to environmental politics have advocated intervening in environmental struggles in developing countries in order to protect threatened peoples and resources (e.g. Miller, 1978). On the other hand, other critics have suggested that there needs to be more concern about how such struggles are represented. This caution has also been extended to how "indigenous peoples," "threatened ecosystems," or "injustice" themselves are constructed (e.g. Escobar, 1996; Leach and Mearns, 1996). Indeed, according to Hecht and Cockburn (1989), the widespread concern about degradation of the Amazon rainforest has largely been conducted without a chance for the Amazon, or its peoples, to tell its own story. Or, as Yash Tandon rightly commented:

> There is a tendency for movements in the Southern Hemisphere to assume, or to be given, Northern labels.... Environmental movements have a certain "newness" about them, new for the North, not for Africa ... in Africa, the respect (even religious veneration) for land and nature is as old as the hills.
>
> (1995: 172–173)

This book acknowledges these debates about representation of people and environmental problems, and seeks to incorporate them into its discussion of science as the underlying basis through which environmental change is understood. Yet an important challenge in this approach is to integrate the structural focus on state, society, and industry, and the poststructuralist attention to how interactions between such actors co-construct environmental discourses and narratives about the environmental change, and who should be represented as victims and villains. This challenge is also important for the analysis of so-called "local" or indigenous knowledge, which is often considered to be eclipsed by more dominant forms of explanation. Instead of essentializing approaches to "local" knowledge or "local" people, it is more important to ask how, and by whom, each are defined as "local" (or "global"). A "critical" political ecology might contribute to new forms of environmental explanation by providing more inclusive means to acknowledge local environmental concerns, and how such concerns have been addressed under existing environmental science.

The separation of science and politics

A final underlying theme within most historic approaches to political ecology has been the assumption that environmental politics can be separated from the principles and laws of environmental science. This separation may also be observed in many other disciplines and political approaches to environment. In political ecology, it would seem the separation has emerged partly because some researchers have seen further work in biophysical explanation to be unnecessary in essential social science applications.

For example, in a book entitled *The International Politics of the Environment* (Hurrell and Kingsbury, 1992), List and Rittberger argued that social–scientific approaches to environment should not get embroiled in the difficulties of biophysical science. They wrote:

> A more pertinent social scientific analysis, while not denying that ecological problems are at the root of international environmental

issues, would have to take a closer look at the conflicts arising from, or linked to ecological problems.... It is here, in the process of articulating and mediating diverging goals and interests, that the ecological problem gains its political dimension, i.e. that *ecology* becomes *political ecology*.

(1992: 88, emphasis in original)

Similarly, Bryant and Bailey wrote:

Political ecologists tend to favor consideration of the political over the ecological... It is true that political ecologists ought not to ignore advances in the understanding of ecological processes derived from the "new ecology," since, in doing so, they might miss an important part of the explanation of human–environmental interaction ... Yet greater attention by political ecologists to ecological processes does not alter the need for a basic focus on politics as part of the attempt to understand Third World environmental problems.

(1997: 6)

The problem of alleging such a clear separation of science and politics is to avoid the politics in the creation of the science itself. Instead of approaching environmental debates as though the science is already agreed, scholars of environment need to focus more on the mechanisms by which knowledge about environment is produced and labeled, then used to construct "laws," and the practices by which such laws and lawmakers are identified as legitimate in political debate.

The objectives of this approach are multiple. Science is undoubtedly used to legitimize a variety of environmental policies, yet there is often little appreciation of the biophysical uncertainties or political conflicts behind many supposedly well-known problems. The production of scientific knowledge and expertise is also growing as a branch of international development, and financial services. Applying inappropriate environmental policies may lead to social and economic problems for people affected, and fail to address underlying biophysical causes of problems. Yet, as discussed above (p. 8), many explanations of environmental degradation within political ecology have been constructed without the participation of affected peoples, and without acknowledging how explanations may reflect social framings.

The lack of clarity concerning the meaning of "ecology" in "political ecology" has only added to the separation of science and politics. Anderson, for example, demonstrated this by using "ecology" as a metaphor for social connections in Africa, and even as a legitimization for military intervention:

If the awareness of individual/community interdependence found in the environmental movement becomes more generalized and spreads into the political arena, a political ecology founded on environmental-ism may build on itself and spread more widely. For example, the United States and other world powers may become willing to use mili-tary might to serve purposes other than the immediate short-term interests of themselves alone. If this becomes true, then perhaps even the US intervention in Somalia could be seen as an example of an embryonic political ecology at work.

(1994: 171)

The aim of this book is to avoid the presentation of "ecology" – a topic over which there is great biophysical uncertainty and political contestation – into predefined notions of fact, accuracy, and political purpose. This book argues throughout that separating science and politics in environ-mental policy may result in two serious problems: first, many environ-mental policies will not address the underlying biophysical causes of environmental problems; second, many environmental policies will impose unnecessary and unfair restrictions on livelihoods of marginalized people. (Indeed, Vayda and Walters, 1999, and Mukta and Hardiman, 2000, have made similar arguments.) This chapter now summarizes some of the potential ways in which science and politics may be integrated in order to indicate that environmental causality should not be taken as politically neutral.

Integrating environmental science and politics

It is ironic, then, that so much discussion of "political ecology" has pro-ceeded without considering the politics with which "ecology" has evolved as a scientific approach to explaining the biophysical world. This book seeks to achieve this analysis by adopting science studies and science–policy approaches to the evolution of environmental explanation.

The term "science studies" refers to an attempt to integrate a political analysis of environmental conflicts with insights from philosophy and socio-logy of science concerning the nature in which environmental science is made. "Science policy" refers more specifically to the co-evolution (or coproduction) of scientific and political norms within the policy process itself (see Jasanoff *et al.*, 1995; Hess, 1997). As this section of the chapter discusses more fully, these topics are emerging as disciplines within environmental politics, but both have been criticized and misunderstood. Some criticisms of science studies have suggested that discussions of science imply the adoption of an *anti*-science position; or the belief in scientific relativism (the assumption that all truth claims are equally valid), and/or postmodernism (e.g. Gross and Levitt, 1994; Koertge, 1998; Levitt, 1999). All such statements are inaccurate. Instead, it is possible to criticize many statements made by science while still believing in environmental realism (or the existence of a "real world out there"). Similarly, the adop-tion of a critical stance to many environmental discourses does not imply an acceptance of postmodernism or cultural relativism. As this book makes clear, the criticisms of science contained within it are made within the frameworks of debates about science, and the book suggests ways of reforming scientific practice to acknowledge the institutional bases upon which it is conducted.

The following sections discuss different ways in which science and politics may be integrated in order to achieve a more transparent and accountable form of science, which may also be considered biophysically more accurate than many "orthodox" environmental explanations today. The first discussion summarizes the approaches within political ecology itself toward understanding the social influence on ecological science.

Political ecology and the social construction of science

While political ecology was developing in North America as an exploration of holistic links between humans and nature at large, a different approach was adopted in England that focused on the social influences on environmental science as a political tool. The Political Ecology Research Group (PERG) was formed in Oxford, England, in 1976 as an informal association of research scientists and students, and which grew into a research organization focusing on the risks and analysis of new technologies such as nuclear power. Two original associates of the group were Brian Wynne and Peter J. Taylor, who have since published widely on social constructions of science. In a statement foreshadowing many later debates, the group wrote:

> Science is dialectical in nature, i.e. the results of research depend upon the assumptions of the researchers, which depend upon all manner of social factors specific to that researcher or research institution. The current situation, where Government attempts to appoint "impartial" assessors, in a quasi-legal framework, will in our view lead to the increasing dissatisfaction with the inquiry procedure.
>
> (PERG, 1979: 20)

Although the group did not last long into the 1980s, the approach to political ecology as the politics of the application of ecological science continued within British scientific communities, most importantly through the work of Blaikie (1985) and Blaikie and Brookfield (1987). Writing about "ecology" as politics, in *Land Degradation and Society*, they wrote: "It therefore becomes necessary to examine critically the political, social and economic content of seemingly physical and 'apolitical' measures such as the Universal Soil Loss Equation, the 'T' factor and erodibility" (1987: xix).

Such work typified a new trend in social studies of environmental degradation: the analysis of the political and social construction of ecological science that previously had been presented as accurate and politically neutral in assessments of environmental problems. Early examples of this kind of work included the influential *Uncertainty on a Himalayan Scale* (Thompson, Warburton and Hatley, 1986), and *The Himalayan Dilemma* (Ives and Messerli, 1989), although the trend has been continued more recently by volumes such as *Desertification: Exploding the Myth* (Thomas and Middleton, 1994), *People and Environment* (Morse and Stocking, 1995), *The Lie of the Land* (Leach and Mearns, 1996), and *Political Ecology: Science, Myth and Power* (Stott and Sullivan, 2000). Such work has looked into the origin and applicability of various so-called predictions of "crisis" in environmental change, often with reference to the variety of scientific evidence about the presumed ecological changes, and in relation to the political advantages to various parties of portraying an ecological crisis. The themes of such work form a large part of this current book, and will be discussed in more detail in chapters to come.

This body of work presents an important framework through which to analyze the political implications of different approaches to ecological explanation. Perhaps most significantly, it also provides a counterpoint to the definition of political ecology arising from Deep Green critiques of industrialization and modernity (e.g. Atkinson, 1991). But there are still important questions to be answered within the discussion of social constructions of science in political ecology. In particular, there is an underlying tension between so-called "realist" approaches to environmental explanation (relying on the mechanical and universal explanation of biophysical risks and impacts) and more poststructuralist accounts of explanations as historically and culturally situated "storylines" (e.g. Hajer, 1995). Much deconstruction of environmental explanation has referred to the debunking of so-called "myths" of environmental crisis and causality, and instead has preferred to acknowledge the existence of "plural rationalities" about environmental change in their wake (see Thompson *et al.*, 1986; Thomas and Middleton, 1994; Leach and Mearns, 1996). Such statements have often been met with frustration from researchers seeking a more uniform scientific explanation, and who fear that "pluralism" might imply relativism. For example, Blaikie wrote:

> A counterweight to the deconstruction of science must also be provided. A case could be made that the bulk of what is styled as political ecology has been written by social scientists, who have paid little attention to what natural scientists have had to say about their environments, usually with embarrassing results.
>
> (1995: 11)

This book seeks to contribute to these debates by asking how far it is possible to deconstruct scientific "laws" built on orthodox frameworks of science, yet still achieve a biophysically grounded form of explanation that is still socially relevant to the places where such science is applied. But before these themes are discussed further, it is worthwhile to review some basic themes concerning constructivism and realism.

Constructivism and realism

Much discussion about criticisms of science have tended to polarize debates between so-called "constructivists" and "realists," where constructivists may be typified as relativist and postmodern, and realists are empirically grounded yet sociologically unaware. Such a stereotypical representation is, of course, inaccurate, and there are many potential middle positions that may incorporate elements of both constructivist and realist analysis.

One often-criticized aspect of constructivist analysis is a focus upon discourse as a primary tool in discussing environmental science and politics. Dryzek (1997) defines discourse as follows:

> A discourse is a shared way of apprehending the world. Embedded in language, it enables those who subscribe to it to interpret bits of information and put them together into coherent studies or accounts. Each discourse rests on assumptions, judgments, and contentions that provide the basic terms for analysis, debates, agreements and disagreements, in the environmental area no less than elsewhere.
>
> (Dryzek, 1997: 8)

Yet the analysis of discourse has sometimes been accused incorrectly of adopting a relativist, or unempirical, focus on environmental policy. For example, Bryant and Bailey wrote:

> The important role of discourse in conditioning political–ecological conflicts is not to be denied... We are nonetheless concerned that a "turn to discourse" may result in a turn *away* from the material issues that, after all, prompted the birth of Third World political ecology in the first place.
>
> (1997: 192, emphasis in original)

Such comments overlook the relationship between discourse, and the co-generation of so-called "facts" and "norms," which underlies much philosophical analysis of political and scientific debate. A conventional approach to environmental science, as described above, adopts a separation of facts and norms on the basis that science can produce the facts, while politics can establish norms based upon such facts. Alternative approaches focus on the interdependency of facts and norms, and the manner in which "facts" may be identified as meaningful information only in relation to specific predefined discourses. Yet once such "facts" have been identified and recorded, they then support or create further discourses associated with them (see Harré, 1993; Rundle, 1993; Demeritt, 1998; Castree, 2001). "Facts" and "norms" may therefore be seen to be one aspect of the underlying debates concerning epistemology (or the study of knowledge) and ontology (the study of underlying reality). The emergence of a dominant discourse about environmental explanation therefore may be based on historic facts and norms of one society, yet lead onto the construction of scientific knowledge about environment "for other locations or societies" that may not be as "factual" as often assumed (see Box 1.1).

A focus on environmental discourse and constructivism, therefore, does not imply the belief that environmental knowledge is unreal or imagined,

---

Box 1.1   Epistemology and ontology

Epistemology is the theory of knowledge. Debates in epistemology refer to establishing the so-called conditions of knowledge, or the social and philosophical requirements necessary to possess, need, and use knowledge. A crucial problem in epistemology is establishing criteria for defining when we know, and do not know, something. For example, there is much evidence to suggest that anthropogenic climate change ("global warming") is occurring. But accepting such evidence as "proof" requires answering questions about what sort of knowledge allows us to make that conclusion. The sort of criteria used to make that judgment might include philosophical concepts of how far we can make meaningful predictions on the topic; ecological questions of gaining sufficient information to infer change for various time and space scales; and social themes of identifying the legitimacy of which organization or observer makes statements about the nature or meaning of that change. The debate concerning what sort of information is meaningful, who is recognized as speaking with accuracy, and who decides both of these questions, are central epistemological questions relating to the debate concerning anthropogenic climate change.

Ontology is the theory of underlying structures in biophysical or social entities. Ontology aims at discovering a framework for understanding the kinds of things that constitute the world's structure. For example, an ontological approach to anthropogenic climate change would aim to understand the causal mechanisms of climate change, and the accurate apportioning of responsibility to different human causes according to their influence on the biophysical process of warming. Ontology is different to epistemology because it aims to focus on the underlying causes and structures of change. But questions of ontology will inevitably also have to consider questions of epistemology in seeking an explanation of physical changes. Concerning climate change, for instance, one difficult question might be to ask how far seeking the causes of "global warming" might actually lead to the creation of an entity known as "global warming" because of the framing of research in order to assess whether it is occurring (this problem is called reification). Ontology is also closely related to other philosophical debates about "realism," the belief in biophysical reality or important causal social structures, and "truth," the question of how far statements may be considered "true." Ontology, realism, and truth, however, are all subtly different and should not be used interchangeably. (See also definitions of realism and constructivism in Box 3.4.)

Source: Fetzer and Almeder, 1993.

but instead indicates an interest in how statements about the real world have been made, and with which political impacts. As Maarten Hajer, for example, commented:

> The new environmental conflict should not be conceptualized as a conflict over a predefined unequivocal problem with competing actors pro and con, but is to be seen as a complex and continuous struggle over the definition and the meaning of the environmental problem itself.
>
> (1993: 5)

In addition to this focus on discourse and constructivism alone, there are also, of course, different approaches that focus more specifically on environmental realism. Environmental realism in this context should not be confused with the debate in international political theory concerning the study of states' and individuals' political interests in competition with others (e.g. Wiesenthal, 1993), but instead refers to the search for epistemologies (or explanations) that allow an accurate and transferable understanding of underlying ontology (or reality).

One important approach to politicized understandings of scientific realism has come from the school of Critical Realism of Roy Bhaskar, and specifically his book, *A Realist Theory of Science* (1975) in which a rapprochement is attempted between epistemological skepticism and ontological realism (Collier, 1994; Archer *et al.*, 1998; Sayer, 2000). In more simple language, Critical Realism seeks to understand "real" structures of society and the world, while acknowledging that any model or understanding of such structures will reflect only partial experience of them, and social and political framings within the research process. Bhaskar in particular distinguishes between so-called "transitive" explanations (socially constructed and changeable) and "intransitive" knowledge (referring to underlying and unchanging reality). Knowledge may also be classified into three levels of knowledge: empiricism (simple experiences); actualism (experiences, and the events that give rise to experiences); and realism (the underlying ontology and structures that give rise to events and experiences). Under such classifications, much environmental knowledge may refer to short-term indications of long-term transitions; and environmental explanations and models are likely to be transitive structures that reflect partial experience and framings of such complex biophysical events. Some initial writings on Critical Realism and environment have already been achieved in relation to supposed dichotomies between humans and animals (Benton, 1996), and men and women (Jackson, 1997).

Critical Realism, and associated debates such as semantic realism, offers much potential for integrating biophysically grounded explanations of environmental change with political analysis of the social framings of science. Yet the term Critical Realism is also controversial. For example, Peter Dickens' (1996) study of Critical Realism and environment focused exclusively on social ontology, class, and social marginalization under capitalism, and did not engage in any deconstruction of a priori environmental explanations within science. Similarly Hannah (1999) and others have suggested that Critical Realism, as defined by Bhaskar, may be rather too optimistic in its ability to achieve realist explanations based on partial knowledge; and hence has suggested that the term "skeptical realism" may allow a more general approach for the muddied waters of environmental epistemology and ontology.

This book seeks to advance debates about integrating social and biophysical explanations of environment. It draws upon debates in Critical Realism, but also draws upon related debates such as semantic, referential, or institutional realism (e.g. Harré, 1986), in addition to pragmatic analyses of social institutions of science (Rorty, 1989b; Proctor, 1998) and poststructuralist analysis of situated discourses and networks of environmental explanation. In particular, the book draws upon insights within science studies – or so-called science and technology studies (STS) – as means to analyze the drawing of social boundaries around the analysis of complex biophysical processes, and the social networks that support them.

STS and the "science wars"

Finally, it is worth referring to science studies itself as a new and evolving form of analysis. The kinds of debates summarized in this section are all relevant to science and technology studies (STS), which is often assumed to include science studies and science policy – the investigation of science and its relationship to policy formulation. Hess writes:

> Science studies provides a conceptual tool kit for thinking about technical expertise in more sophisticated ways. Science studies tracks the history of disciplines, the dynamics of science as a social institution, and the philosophical basis for scientific knowledge... In short science studies provides a forum where people who are concerned with the place of science and technology in a democratic society can discuss complicated technical issues.
>
> (Hess, 1997: 1)

These objectives are similar to those of the Political Ecology Research Group in the late 1970s, and indeed Irwin and Wynne (one of the original associates of the PERG) wrote: "Science offers a framework which is unavoidably social as well as technical since in public domains scientific knowledge embodies implicit models or assumptions about the social world" (1996: 2–3).

STS, or science studies in general, in brief, aims to question the perceived political neutrality and accurate representative of reality offered by "science," and instead indicate how scientific statements and scientific institutions – such as research institutes, universities, government agencies, and museums – may reflect social and political influences of relevance for how we perceive and manage environment and society.

This book illustrates the relationship of STS to political ecology. Two key themes of STS are worth noting at this stage. First, much STS is concerned with the drawing of boundaries in social discourse in order to indicate domains of explanation or causality. (Some well-known boundaries may exist between "nature" and "society"; "men" and "women"; or "scientists" and "lay" people.) Second, there is also much attention in STS to "hybridity," or the hybrid blending of "facts" and "norms" (e.g. Latour, 1993). Such hybridization may lay down the institutional factors that define many definitions of environmental problems such as desertification or deforestation, which are based on complex interactions of a variety of different biophysical events and processes. (Hybridity has also been used in a variety of other applications of social understandings of biophysical reality, see Braun and Castree, 1998.)

Yet the emergence of STS as an academic debate also triggered intense criticism of its objectives from defenders of positivist science, such as physicists and mathematicians. The launch of this criticism has since been called "the science wars," which took place mainly in the USA in the mid-1990s. In particular, Gerald Holton's (1993) *Science and Anti-Science* warned of the dangers of a new irrationalism in society. Paul Gross and Norman Levitt (1994), in *Higher Superstition: The Academic Left and its Quarrels with Science*, referred to STS as the "cultural left" and criticized a variety of academic themes that had been related to science such as social constructionism, postmodernism, feminism, and environmentalism. (Although it is worth noting that some targets of such criticism, such as Bruno Latour, have commonly criticized postmodernism, see Latour, 1993.)

The main concerns of the science wars critics were to reject what they saw as intellectually lazy, modish trends in social science that avoided the principles of accuracy, honesty, and hard work demonstrated over the centuries within conventional physical science. Perhaps this was illustrated when the physicist, Alan Sokal (1996) persuaded the journal *Social Text* to publish a paper with the suitably deconstructivist title: "Transgressing the boundaries: toward a transformative hermeneutics of quantum gravity." Sokal later revealed this was a hoax intending to indicate sloppy standards in the social science (see Segerstråle, 2000, for a summary of the "science wars"). The science wars represented great bitterness and personal hostility perhaps best indicated by the following quotations from Levitt:

> I think that this current genus of academic nihilism [in STS] is vain, captious, and ultimately unavailing.... To accede to the relativist demand (for that is what it amounts to) that science discard its privileged status as an especially accurate way of learning about reality is not only to defer to questionable philosophy but, as well, to yield the core assumption that drives scientists to endure the considerable pain

and travail of learning their craft and practicing it with rigorous honesty.... Academics who rail or snipe at science are rather like well-brought-up children who have made a deliberate decision to misbehave and outrage their elders on some solemn occasion. They are terribly self-conscious and jittery about the whole business, and gnawed by the suspicion that they might lose their nerve and fail to go through with the thing. When confronted with scientists' hard stares, they fidget and prevaricate and look as though they would really prefer to be elsewhere.

(1999: 10, 23, 302)

Such statements, of course, say more about the writer than the topic under consideration. Yet, many such statements miss the point of science studies, and reduce it – and its critics – to stereotypical positions that avoid many areas of potential overlap. It is hoped that this book will indicate that criticisms of orthodox scientific approaches to environmental problems does not imply cultural relativism, a rejection of epistemological realism, or the rejection of science or scientists.

Some similar concerns to those of Gross and Levitt have also been made specifically in relation to science studies and environmental writing. In particular, concerns have come from defenders of orthodox science who have considered social analysis of environmental risks to be a step toward relativism; environmentalists who seek to defend concepts of an external "nature" against deconstruction; and scientists who see science studies as a tool of industries who would rather weaken environmental regulation. It is important to note that most researchers within STS would consider such criticisms inaccurate.

The resentment of social science approaches to environmental risk was shown in an amusing way at a meeting in 1992 in Britain's most historic research institute, the Royal Society, concerning the publication of a report on risk (Royal Society, 1992). The anthropologist Mary Douglas reported:

Complete decorum reigned until near the end, when a psychologist got up from the floor and reproached the Royal Society report for giving undue space to radical views. When he asked that the term "social construction of risk" be eliminated from the discussion, shouting, clapping and hissing broke out and the meeting was adjourned.

(1993: 122; also in Thompson and Rayner, 1998a: 140)

Concerning the analysis of concepts of "nature," Soulé, for example, wrote:

Why are some social critics in denial about the existence or significance of nature?... The nihilism and relativism of radically constructionist critiques of science and the materiality of nature, while popular

in some academic circles, is sophomoric. Further, it is harmful because
... it undermines efforts to save wilderness and biodiversity.

(1995: 151, 154)

And in *Betrayal of Science and Reason: How Anti-Environmental Rhetoric Threatens our Future*, Paul and Anne Ehrlich (1996) declared:

Brownlash has produced what amounts to a body of anti-science – a twisting of the findings of empirical science – to bolster a predetermined worldview and to support a political agenda. By virtue of relentless repetition, this flood of anti-environmental sentiment has acquired an unfortunate aura of credibility.

(Ehrlich and Ehrlich, 1996: 11)

Under a "critical" political ecology, all statements about ecology have to be assessed for their political construction. There is no agenda to legitimize destructive resource use, or to weaken care for environment where expressed. Yet it is also necessary to reevaluate environmental statements coming from "science" that justify particular policies in preference to others. The overt defense of environmentalism in the statements above, for example, might be subject to criticism for the very reasons such writers use to justify their own positions. The Ehrlichs (1996: 12), for example, suggest that anti-science rhetoric is a "denial of facts and circumstances" that do not fit beliefs, and which may lead to policies that "could lead our society into serious trouble." Yet the same could also be said of their environmental science based on the critique of the domination of nature developed to resist exploitative economics and politics. As discussed in Chapter 2, such scientific explanations can cause immense problems for farmers in developing countries whose activities do not threaten landscape in the ways suggested, but whose livelihoods are restricted by the policies resulting from such science.

The development of an analytical approach that is biophysically grounded yet conscious of social and political constructions is one of the key aims of this book. "Critical Political Ecology" is the attempt to integrate STS into debates about political ecology, but it does not imply relativism, or "brownlash," or the rejection of science altogether. Instead, the aim is to highlight as far as possible the implicit social and political models built into statements of supposedly neutral explanation in order to increase both the social equity of science, and its relevance to environmental problems experienced within diverse social settings.

Building a "critical" political ecology

So, how can we define the objectives of a "critical" political ecology? As the preceding discussions have shown, Critical Political Ecology aims to present debates within political ecology with an approach to environmental politics that allows the successful integration of political analysis with the formation and dissemination of understandings of ecological reality. A key ambition is to avoid the simplistic separation of science and politics (or facts and norms), and the use of a priori notions of ecological causality and meaning, and instead to adopt a more politically aware understanding of the contexts within which environmental explanations emerge, and are seen to be relevant.

This project may legitimately be called "critical" for various reasons. First, the objective to reach an emancipatory form of politics is consistent with the long-term aim of Critical Theory, and its focus on knowledge and science as a product of oppressive regimes (Rasmussen, 1996). Important Frankfurt School scholars such as Marcuse and Habermas discussed how a better, more socially relevant "science" might be developed (Alford, 1985). But such debates have not been integrated fully with new insights from Philosophy of Science or science studies. This book aims to develop such an approach. As Vogel wrote:

> The truth is that, beyond generalized critiques of positivism, little serious consideration has been given to contemporary Philosophy of Science within the postwar tradition of Critical Theory, and this is a significant fault.
>
> (1996: 7)

This book's focus upon environmental science, and contested approaches to explaining the causes of environmental degradation, also means that it discusses functional concepts such as "explanations" and "accuracy." This discussion does not imply that such concepts can exist externally from social contexts (see MacKenzie, 1990; Harré, 1993). But these terms need to be examined in depth because so much environmental policy is based upon the belief that explanations and scientific accuracy have already been established.

Second, the emphasis on science as both a means of explanation, but also rooted in politics reflects the concerns of so-called "critical science," or the reflexive attention of science to the political uses to which it may be put. As noted above, the original movement to link ecology with politics was made by concerned scientists who sought a new methodology for dealing with humans as a "community." As this books shows, however, more recent research has indicated that such a priori decisions about who and what may be considered "community," at both local and global levels, has significant political implications. Furthermore, other authors have acknowledged the apparent contradiction within much current environmentalism by using modern science in order to strengthen its critique of modernization and industrialization (Yearley, 1992, 1996).

In keeping with Critical Science, Critical Political Ecology seeks to engage constructively with the norms of professional science, and to seek influence by discussing possible alternative means to approach environmental explanation. As Nowotny commented:

Coming largely from within [the scientific world], critical science turns against those aspects which it regards as irresponsible on a variety of humanistic, moral and political grounds. Critics oppose a view of omnipotent science which claims legitimacy to indiscriminately bestow upon humanity whatever passes through the heads of scientists regardless of the consequences.

(1979: 21)

Third, Critical Political Ecology also adopts and expands insights from Critical Realism, and associated debates such as semantic and so-called institutional approaches to explanation, that seek to blend epistemological skepticism with ontological realism. These approaches are called "institutional" because, unlike the frameworks of positivist or orthodox science, they acknowledge the institutional bases upon which explanation is contingent (Harré, 1993; Aronson *et al.*, 1994). Yet the book does not restrict itself only to these considerations of scientific practice and realism, and considers more poststructuralist accounts of historical and cultural shaping of environmental narratives and networks (Jasanoff, 1990; Hajer, 1995). Both institutional realism and poststructuralist analyses of environmental "truth" statements with environmental science allow the possibility to achieve a biophysically grounded yet socially relevant form of explanation. By so doing, this book also addresses the acknowledgment of various environmental writers that more attention needs to be given to biophysical agency within political ecology (Grossman, 1998; Woodgate and Redclift, 1998). As Watts and McCarthy commented:

A critical approach to nature has been one of political ecology's weaknesses, and there is surely need for a more social relational understanding of natural science itself (of the institutions of science and scientific regulation) ... including a sensitivity to what one might call nature's agency or causal powers.

(1997: 85)

The attempt to address these questions, however, does not imply that it is possible to explain "reality" in some final, uniform manner. Moreover, this book – unlike *The Skeptical Environmentalist* of Björn Lomborg (2001) – does not dismiss environmental problems in general, but instead questions the manner in which environmental problems are defined, and with which transferability. Environmental problems do exist, and ecologically degrading practices need to be avoided. This book seeks to assist environmental debate and policy. But the book also seeks to reform existing environmental debates to show the problems of current approaches to environmental explanation, and to propose means by which to diversify and localize environmental science, including greater local determination by people not currently represented in science.

Structure of the book

The book is divided into three main sections. Chapters 2–4 summarize the problems of many existing dominant environmental explanations and presents ways of criticizing them from the perspectives of debates in philosophy and sociology of science. One key theme in this section is the distinction commonly made between "science" and "myths," and the different approach to myths as either a falsehood, or a form of truth.

The second section, Chapters 5–7, then looks at the "coproduction" of environmental knowledge and political activism, specifically in relation to environmentalism as a social movement, the globalization of environmental discourse, and the evolution of environmental explanations through the interplay of different scientific, political, and commercial actors. This section also provides some detailed descriptions of environmental problems, and analysis from the perspective of science studies and science policy.

Finally, Chapters 8–10 look at potential solutions to the questions posed in the book. Chapter 8 presents a discussion of potential means of democratizing scientific explanations through scientific practice itself. Chapter 9 reviews debates concerning the political accountability and regulation of environmental science. Chapter 10 provides a conclusion, and a discussion of implications for political ecology in general.

Chapter 2 now starts the analysis by looking at conflicting truth claims about many environmental explanations.

2 Environmental science and myths

This chapter outlines the key problems addressed by this book. The chapter will:

- summarize some of the uncertainties associated with many definitions and explanations of environmental degradation commonly discussed as "fact" by politicians, activists, and in the media. Perhaps surprisingly, the explanations associated with these so-called problems are sometimes highly uncertain and contested by a variety of scientific research and local experience.
- discuss the impacts of such contested explanations on attempts to manage environmental problems, and on the livelihoods of people accused of causing problems. Some environmental policies adopted to address "problems" may actually not address the underlying causes of biophysical changes, and, in some cases, policies may unnecessarily interfere with livelihood strategies. The problems of desertification, soil erosion, and deforestation are summarized as examples.
- introduce the concept of "environmental orthodoxies" to describe common explanations of environmental problems that are considered to be simplistic and inaccurate. Some writers have also called these "myths." The chapter discusses how far such explanations can reasonably be called "science" or "myths" and explains how a focus on these problems does not mean supporting destructive land uses, but a greater attention to how science can engage with environmental problems.

This chapter therefore introduces the book's central theme of showing that many supposedly "factual" explanations of environmental problems are highly problematic and overlook both biophysical uncertainties and how people value environmental changes in various ways. The aim of this discussion is not to deny the existence of environmental problems, nor to legitimize destructive practices. Instead, there is a need to understand the complex social and political influences upon how we explain environmental problems, and then see such explanations as factual. A "critical" political ecology achieves these objectives, and offers the chance to construct more meaningful and effective forms of explaining environmental problems.

Overturning conventional environmental degradation

This chapter describes a problem relating to environmental science and politics that is growing in significance all the time. The problem is that many attempts to find political solutions to environmental problems are based upon well-known, or "orthodox" explanations of how environmental problems occur. Yet, increasingly, people are realizing that many of these orthodox environmental explanations are not as accurate as commonly thought.

It may come as a surprise to many people concerned about environment that some widely known definitions and explanations of environmental degradation are, in actuality, uncertain, highly contested, and misleading. Scientific disagreement about environmental explanations is already well recorded. For example, the media commonly reports on disagreements about whether "global warming" is occurring or not. Yet, in addition to these concerns, there are many other disagreements about topics that are commonly assumed factual and without disagreement. These disagreements can sometimes have serious implications because they can challenge many of our concerns about the impacts of other environmental changes such as global warming.

This chapter starts by analyzing three commonly identified causes of environmental degradation: desertification, soil erosion, and deforestation. These themes are referred to throughout the rest of the book, although other topics may be challenged in similar ways. The purpose of this analysis is to summarize how approaches to environmental degradation relating to these topics may overlook the complexity of changes, and the diversity with which people may view them. As further chapters show, such factors have importance for analyzing the political influence on, and of, environmental science.

Readers should note that the objective of this chapter is not to suggest that environmental problems do not exist, or that desertification, erosion, and deforestation may not, under certain circumstances, present serious problems. The objective, instead, is to show some problems that occur from using these concepts uncritically. Such problems often include the use of common terms such as "deforestation" to denote both environmental changes and degradation at the same time, or the implication that such changes have clearly defined human causes. As later chapters show, these assumptions overlook two key factors: the difficulty of making explanatory statements about long-term and complex biophysical processes; and the social and historical framing of explanations based upon one society's experiences of such changes.

The following discussions are, of necessity, brief, and cannot summarize all debates and uncertainties. The objective is to indicate how these terms have become synonymous with "degradation," yet are rooted in the experiences of particular groups over time, and represent only partial understandings of complex biophysical changes.

Example 1: desertification

Desertification is the concept that refers to land degradation in drylands. It is commonly referred to as an urgent and pernicious process that can lead quickly to associated problems such as drought, agricultural failure, and famine. The co-founder of the Worldwatch Institute, and well-known environmentalist, Lester Brown wrote:

> Easily a third of the world's cropland is losing topsoil at a rate that is undermining its long-term productivity. Fully 50 percent of the world's rangeland is overgrazed and deteriorating into desert... The doubling of the world's herds of cattle and flocks of sheep and goats since 1950 is damaging rangelands, converting them to desert.
>
> (2001: 8, 79)

Such concerns are highly questioned by a variety of scholars. Yet the image of desertification as the dangerous encroachment of deserts remains a popular theme in much environmental rhetoric. In 1975, one report suggested the Sahara might be advancing at the rate of 5.5 km per year (Lamprey, 1975). In a website quoted by Katyal and Vlek, one disaster relief manager wrote: "Like an aggressive cancer, deserts are consuming more and more earth" (2000: 7).

The purpose of this discussion is to show the disparity between such emotive descriptions of environmental degradation, and a wide range of research that questions these statements on grounds of biophysical accuracy, and social relevance to the experiences of local people. These disparities suggest a variety of criticisms concerning how environmental degradation is discussed and explained.

Desertification is sometimes portrayed as an uncontrollable, human-induced phenomenon involving the sudden onset of drought, the death of vegetation, and eventually the transition of fertile land to sandy desert. This image has a long history. Scholars in the eighteenth century, for example, considered the Sahara desert to have been created by the Romans and Phoenicians as the result of deforestation, overgrazing, and overcultivation (Goudie, 1990). Such beliefs were strengthened by the apparent collapse of local empires in North Africa. In 1324, the Emperor of Mali, Mansu Musa, crossed the Sahara to Mecca with 500 slaves and 100 camels laden with gold (Bass, 1990: 13). The caravan's arrival en route in Egypt depreciated the precious metals market there by 12 percent, and spread rumors of the fabulous wealth of the empire's capital in Timbuktu. The empire declined, however, as the result of competition from new Portuguese and Spanish empires, and in 1738 half the population of Timbuktu died of famine. When the city was visited in 1828 by a French traveler, he wrote graphically of his shock at finding apparent evidence of human failure in a barren land:

> I looked around and found that the sight before me did not answer my expectations... [The city] presented, at first view, nothing but a mass of ill-looking houses, built of earth. Nothing was to be seen in all directions but immense plains of quicksand of a yellowish-white color. The sky was a pale red as far as the horizon, all nature wore a dreary aspect; and the most profound silence prevailed; not even the warbling of a bird was to be heard.
>
> (René Caillié, 1828, in Bass, 1990: 13)

Research has since showed, of course, that the Sahara has resulted from the effects of large volumes of rising hot air at the equator, influenced too by the progressive desiccation of northern Africa since the end of the Pleistocene period, 10,000 years before present, when much of northern Europe was under glaciers (Goudie, 1990). Furthermore, other studies have argued conclusively that no threat from expanding deserts existed (Warren and Agnew, 1988). But it is difficult to separate such large-scale biophysical causes of deserts from the effects of apparent land mismanagement on the margins of deserts, such as in the Sahel, south of the Sahara. Paul B. Sears – the author of "Ecology: a subversive subject" (1964) referred to in Chapter 1 – wrote about desertification at the same time as the USA was experiencing the crisis of the Dust Bowl:

> The white man in a few centuries, mostly in one, reversed the slow work of nature that had been going on for millennia. Thus have come deserts, so long checked and held in restraint, to break their bonds. At every step the girdle of green about the inland deserts has been forced to give way and the desert itself has been allowed to expand... If man [sic] destroys the balance and equilibrium demanded by nature, he must take the consequences.
>
> (Sears, 1935: 67, in Worster, 1979: 200)

And Edward Stebbing, a British colonial forester, wrote in a similar vein about the dramatic invasion of deserts:

> Anyone possessing some knowledge of the desert-country types can come and study the stages, quite sufficiently clear-cut once the eye is attuned to discerning them, by which the desert has through the centuries, assisted by man [sic], advanced over rich and fertile regions.
>
> (Stebbing, 1937: 1, in Bass, 1990: 11)

Stebbing's comments also indicate how he considered desertification to result from the actions of irresponsible and misinformed people; and how he considered his own apparently greater knowledge to mark him as an obvious expert.

Such comments today are criticized for a variety of reasons (Correll, 1999). Perhaps most importantly, there is a greater understanding of the underlying biophysical causes of deserts, and particularly the role of climate in controlling relatively wet and dry periods that influence vegetation growth and sand movement in drylands (e.g. Thomas and Middleton, 1994). This research has also questioned the value of some historic approaches to "managing" desert growth by placing fences in the way of sand dunes. Indeed, such fences may even exacerbate the processes of sand movement.

There has also been a much deeper appreciation of adaptive practices adopted by people in drylands in lessening the impacts of drought, and in increasing the efficiency of rangeland management despite uncertainty about rainfall (e.g. Turner, 1993; Scoones, 1994). As such, adaptation strategies may not "prevent" the onset of drought, but they can reduce the immediate economic impacts of drought. Together with the advances in understanding the biophysical causes of "desertification," these responses show that farmers' actions may play only a limited role in causing dryland degradation, and in many cases may actually redress degradation of soils and vegetation (see also Anderson, 1984).

So where does this leave the concept "desertification"? Many writers have now strongly rejected attempts to link so-called desertification with purely social causes. Dregne (1985: 30) wrote "very little land has been irreversibly desertified as a result of man's [sic] activities." And Blaikie commented:

> The case for the globalization of capital being causal in desertification looks rather amateur, since the scientific evidence of permanent damage to the environment points in other directions ... For want of attention to a large and accessible body of climatological and ecological information, the case for adding desertification to the long list of other socially induced woes now looks very thin.
>
> (1995: 12)

Moreover, other writers have called upon the rejection of the term "desertification" itself. Thomas and Middleton (1994: 160), in a book called *Desertification: Exploding the Myth*, identified three commonly held "myths" of desertification: desertification is a voracious process which rapidly degrades productive land; that drylands are fragile ecosystems; and that desertification is a primary cause of human suffering and misery in drylands. In particular, Thomas and Middleton criticize the role of the United Nations Environment Program (UNEP) in prolonging these falsehoods. They wrote:

> The UN has played a major role in conceptualizing desertification since 1977 [the year of the first major UN conference on desertifica-

tion]. It could be considered to have created desertification, the institutional myth. It has been the source of publicity that has frequently had little reliable scientific foundation. The success of UN-derived anti-desertification measures have yet to be reliably demonstrated and, in many cases, appear to have had little relevance to affected peoples. Without the UN, desertification may not be as high on the environmental agenda as it is today.

(Thomas and Middleton, 1994: 161)

Many authors now suggest that the term "desertification" should be avoided as it implodes a variety of different "problems" such as drought, declining soil fertility, or local fuelwood scarcity, into one term that suggests the underlying problem lies in the land (e.g. Biot, 1995; Saberwal, 1997). Instead, critics have suggested official policy and development assistance should seek to provide "drought proofing" or other institutional support to farmers in drylands in order to increase the experience of drought as a life-threatening hazard.

But the old-fashioned images of desertification persist, and they also interfere with programs of social development. Thomas and Middleton (1994) noted, for example, that the government of Chad deferred the implementation of democratization measures during the 1980s because it claimed it needed to maintain control of anti-desertification programs. Many standard proposals for combating desertification, such as destocking, or the reduction of agricultural activities, may actually decrease the economic adaptability of people to drought (Turner, 1993). Some critics have suggested that on-going negotiations for the Convention to Combat Desertification (CCD) need to adopt the new thinking about desertification, and have instigated old divisions between so-called "expert" knowledge from researchers repeating the ideas about ecological fragility, and alternative knowledge relating to local adaptive processes to drought (Correll, 1999). Such criticisms of the CCD do not deny that millions of people face environmental problems in drylands. But evidence is growing that accepting uncritical explanations of desertification may actually impede biophysical understanding, and even inhibit social development.

Example 2: soil erosion

Soil erosion is another common concept of environmental degradation that is usually automatically interpreted as being problematic. Soil erosion refers to the physical removal of soil – primarily by wind or water – and commonly impedes agriculture because it removes nutrients contained in the topsoil. Erosion may also cause further problems in duststorms; unwanted deposition of soil (sedimentation); and in extreme cases, mudslides and landslides, although these may be better understood as a separate but related topic to soil erosion. There is no doubt that soil erosion causes severe problems of decreased agricultural productivity for millions of farmers worldwide. But it is not clear how far addressing "erosion" *per se* can alleviate these problems, or how far the assumptions made about erosion in development projects are applicable to all locations and farmers' practices (Morse and Stocking, 1995; Stocking, 1996).

Perhaps the most graphic illustration of the severe problems caused by erosion was the Dust Bowl in the southern Great Plains of the United States during the 1930s (Worster, 1979; Lookingbill, 2001). John Steinbeck's novel, *The Grapes of Wrath*, vividly captured the tragedy of sudden, apparently unstoppable erosion, and its impact on poor farmers in Oklahoma:

> Every moving thing lifted the dust into the air; a walking man lifted a thin layer as high as his waist, and a wagon lifted the dust as high as the fence tops, and an automobile boiled a cloud behind it ... Men stood by their fences and looked at the ruined corn, dying fast now, only a little green showing through the film of dust. The men were silent and they did not move often. And the women came out of the houses to stand beside their men – to feel whether this time the men would break.
>
> (1939: 1, 3)

Such images and consequences have been replicated in other works on erosion since. One classic example has been Eric Eckholm's (1976) *Losing Ground* that proposed how population growth in many fragile areas of the world would lead to food shortages and crisis.

But despite the obvious problems experienced during the Dust Bowl, the immediate attempt to address soil erosion through research proved exceedingly mixed. Following the erosion in the southern Great Plains area since the 1930s, researchers developed the Universal Soil Loss Equation (USLE) using varied measurements across the USA that intended to predict levels of erosion, and hence allow farmers to keep soil loss to within acceptable levels (USDA, 1961). The equation stated:

$$A = R \times K \times LS \times C \times P$$

Where A = average annual soil loss in tons per acre per year; R = the rainfall and runoff factor by geographic location; K = the soil erodibility factor; LS = the slope length–gradient factor; C = the crop/vegetation and management factor that limit soil loss for crops; and P = the support practice factor, such as contour farming, and other physical management of land locations (Morgan, 1986).

Yet, despite its name, the USLE is far from "universal." Three main problems with the equation have been identified. First, there is a general lack of information concerning the rates of soil formation, and consequently it is difficult to determine acceptable levels of soil loss rather than simply rates of soil loss. Second, the equation uses average rainfall figures rather than referring to the intense storms that cause most erosion in the tropics. Third, no attempt was made in the initial equation to integrate soil erosion research into preexisting practices of soil conservation, or valuations of soil loss (Blaikie, 1985; Hallsworth, 1987).

> While the USLE works excellently across the Great Plains [of the USA], with but little variation from east to west, and sets out clearly the factors that need to be taken into account, the rainfall factor is based on average figures, whereas results from the subtropics have shown that the quantity of soil removed is determined by the occasional highly erosive storm and bear little relation to the average figures. Many attempts have been made to modify the USLE to make it suitable for use in the tropics, but with these two inherent deficiencies the problem is difficult to solve, and the attempts have probably absorbed too much of the relatively slim resources available for conservation work, with the inevitable neglect of work that would have been more relevant.
>
> (Hallsworth, 1987: 145)

Similarly, research has increasingly indicated the role of preexisting biophysical causes of erosion. Carbon dating of soil cores in Australia, for example, revealed that the cycle of erosion starting from the 1850s (when plowed cultivation started) was similar to early cycles of erosion at 390, 3,740 and 29,000 years before present – although these may have been caused by the burning of undergrowth by early human settlers (Walker, 1962). Much research in the Himalayas too has suggested that conventional concerns about soil loss have overlooked the normally high rates of soil movement under tectonic uplift and monsoonal rainfall, and also the roles of naturally occurring gullies on steep slopes in transporting sediment from highlands to lowlands (Höfer, 1993). It has also been shown that only part of erosion occurring on slopes may end up eventually in rivers or deltas (Trimble, 1983). Malin (1946) also argued that drought and dust storms had always existed in the southern Great Plains, and so the Dust Bowl could not always be attributed solely to human action.

Related to these criticisms, it is also clear that "erosion" *per se* need not always be a problem for some farmers because it may also lead to sedimentation of soil on agricultural land that provides nutrients for further agriculture. As Blaikie and Brookfield (1987) wrote, "one farmer's soil erosion is another farmer's soil fertility." Furthermore, in some localities there is evidence to suggest that the perception of sedimentation as a hazard may increase as more and more lowland farmers live in areas close to mountains (Ives and Messerli, 1989). Under such conditions, sedimentation may not have increased in absolute terms over time – or have been caused by upland farmers – but the impression of these may have been given because more lowland farmers experience it as a problem. Such complexity of impacts also suggests that referring to processes of declining soil fertility (or nutrient depletion), plus soil removal (erosion) and deposition (sedimentation) under the general single label of "soil erosion" may be insufficient to appreciate the various physical causes and social implications contained within it.

Yet, perhaps most crucially, research of practices used by farmers in many developing countries has questioned the extent to which erosion may be a "problem" according to both the impact of such erosion on agricultural productivity, and if managed well by local conservation practices. The orthodoxy that erosion is always a problem was shaken by research in Nepal showing that some hill farmers trigger some landslides in order to *improve* soil fertility, and facilitate the construction of terraces (Kienholz *et al.*, 1984). Similarly, other research has revealed that increasing population may also not lead to accelerated erosion. For example, in both the Machakos region of Kenya and in Peru, Tiffen and Mortimore (1994) and Preston *et al.* (1997) argued that careful land management could mean "more people, less erosion" (although these claim have been questioned). In Thailand, research showed that hill farmers deliberately avoided creating erosion on steep slopes (Forsyth, 1996). And in Papua New Guinea, the Wola people have maintained high agricultural productivity despite rising populations by integrating compost into complex soil mounds, and by using crops that do not exhaust nutrients (Sillitoe, 1993, 1998). There are many other examples (Millington, 1986; Richards *et al.*, 1989; Zimmerer, 1996a).

The point of these studies is not to suggest that "erosion" is never a problem, or that the experiences of the Dust Bowl should be discounted. Instead, the implication of this immediate discussion is to question how far the word, "erosion" – with its myriad associations of crisis resulting from the movement of soil by wind or water – is necessarily the best indication of the causes of soil degradation, or the most fitting policies to address it. Some researchers have suggested that it may be more appropriate to assess declining soil fertility as the key problem, rather than erosion (in the same way some have suggested drought is more relevant than desertification) (Reij *et al.*, 1996). Erosion may also preexist human impacts, and not necessarily be enhanced by them.

At present, lumping different experiences of environmental problems under the single category of "soil erosion" may hinder addressing the underlying biophysical causes of soil degradation, and may support proposed solutions that accentuate problems. Where proposals aim to restrict upland agriculture, policies may also impose hardships on agriculturalists when there may be diverse causes of apparent lowland sedimentation. Research of reforestation as a tool to combat erosion, for example, has indicated that many projects have actually increased lowland sedimentation by overlooking the relationship between sheet and gully erosion, and the influence of farmers' activities on reducing runoff (Zimmerer, 1996a, b; Calder, 1999; Driver, 1999). Reforesting land in order to control erosion may therefore have surprisingly counterproductive results.

Example 3: deforestation

Deforestation is probably the most emotive topic of popular environmental debate today. Many people concerned about environment have been persuaded by graphic images of burning forests, or the sight of complex, ancient forests being felled in minutes by loggers who care little for losses to global heritage, biodiversity, and impacts on global climate change. Deforestation has also been linked to causes of desertification and soil erosion too. Such common assumptions were listed by the report of the 1992 Earth Summit:

> The impacts of loss and degradation of forests are in the form of soil erosion, loss of biological diversity, damage to wildlife habitats and degradation of watershed areas, deterioration of the quality of life and reduction of the options for development.
>
> (UNCED, 1992: 233)

Undeniably, forest loss causes a variety of impacts. But again, the key contentions of this statement, and other commonly heard generalizations about deforestation, can be challenged. The commonly ascribed notion that forests – and particularly tropical rainforests – are fragile and pristine ecosystems is highly controversial. Experience of deforestation in the Amazon, for example, has indeed shown that forest regrowth after deforestation may be difficult on account of the lack of nutrients in soils, and the rapid erosion and degradation of soils following deforestation. Yet, new thinking has questioned the permanency of such disturbance; the ability to transfer such experiences to other locations; and the social values that attribute importance to different levels of disturbance.

First, much research has revealed historic rates of change and disturbance in forests. Crapper (1962), for example, estimated that some 90 percent of the forests of Papua New Guinea had been cleared at some point, mostly by fire. Areas now covered with rainforest were also much cooler and drier following the end of the Pleistocene, 10,000 years before present, and so current rainforests are generally newer biomes than sometimes claimed and also have evolved during a variety of changes (Whitmore, 1984).

Second, the role of deforestation in biodiversity loss has also been challenged. It is well reported that forests – again, particularly rainforests – contain significant proportions of the world's species. Early commentaries on rainforest destruction assumed a directly proportional relationship between area of forest lost and species made extinct. Norman Myers (1984), for example, wrote that tropical rainforest destruction represented "the greatest single setback to life's abundance and diversity since the first flickering of life four billion years ago," and estimated that one species was being lost every half hour. Later research has shown, by contrast, that this direct relationship is overstated, and that large numbers of species survive in remaining clumps of forest; that some historic extinctions, such as in the Permian age, were of greater significance; and that other ecosystems such as savanna also have high levels of biodiversity (e.g. Wu and Loucks, 1995).

Such research, of course, is not intended to justify rapid destruction of forests, but they do question the urgent calls of some conservationists that all forests be protected from human impacts. Indeed, other research has shown that forest disturbance itself can provide a boost to certain types of biodiversity. Many studies have indicated wide varieties of species under well-maintained shifting cultivation systems, which often use fire as a way to clear areas of closed forest (Schmidt-Vogt, 1998; Fox *et al.*, 2000). Much biodiversity under shifting cultivation, however, may exclude some "wild" genetic resources and large endangered animals such as tigers and horn-bills that require large areas of forest, and are often incompatible with human land use in the form of settled villages or agriculture. Asserting "deforestation reduces biodiversity" therefore depends in part upon particular definitions of deforestation and biodiversity.

Third, an increasing number of studies question assumed links between deforestation and impacts on climate, hydrology, and erosion. Some of these studies were mentioned above in relation to soil erosion, and the relationships between climate change policy and forests are discussed more in Chapters 6 and 7. But it is now clear that many commonly held assumptions linking deforestation to erosion, water shortages, and even rainfall shortages have been shown to be either poorly supported by data, or contingent upon particular types of measurements (Thompson *et al.*, 1986; Hamilton, 1988; Hamilton and Pearce, 1988; Ives and Messerli, 1989; Calder, 1999). For example, Pereira wrote:

> The worldwide evidence that high hills and mountains usually have more rainfall and more natural forests than do the adjacent lowlands has historically led to confusion of cause and effect. Although the physical explanations have been known for more than 50 years, the idea that forests cause or attract rainfall has persisted. The myth was created more than a century ago by foresters in defense of their trees ... The myth was written into the textbooks and became an article of faith for early generations of foresters.
>
> (1989: 1)

Fourth, much new thinking has also highlighted the importance and diversity of social valuations of different kinds of forest and land uses associated with forest (Barraclough and Ghimire, 1996). It has already been mentioned that some shifting cultivator groups manipulate forest growth to maximize the production of valued species. Such actions may also enhance forest protection. Fairhead and Leach (1996), for example, demonstrated that villagers in Guinea, West Africa, had worked over a period of two or more centuries to produce "islands" of closed forest in the boundary zone between savanna and forest. These "islands" had been created for various reasons, including the facilitation of defense, and the production of forest products. Yet, the finding comes in stark contradiction to official government explanations of forest loss (assisted by historic colonial experts), which alleged such islands were relics of a once larger forest area now lost because of deforestation.

Fifth, partly as a result of preceding challenges, our understandings of deforestation rates are also being challenged. Comparisons of satellite data and ground surveys of forest in many developing countries suggest great statistical ranges in estimates of forest area and quality (Leach and Mearns, 1988; Robbins, 1998). Taking such errors into account, Fairhead and Leach (1998: 183) have estimated that total forest loss in six West African countries since 1900 may reach 9.5–10.5 million hectares, rather than commonly-discussed estimates of 25.5–30.2 million hectares. (Indeed, some agencies, such as the World Conservation Monitoring Center, have placed deforestation in this region even higher, at 48.6 million hectares.) In the Himalayas, a survey of deforestation estimates between 1965 and 1981 revealed a variation in rates by a factor of 67, even after excluding some apparent typing errors (Donovon, 1981; Thompson *et al.*, 1986; Cline-Cole and Madge, 2000). Despite continuing high rates of deforestation in many locations, such statistical uncertainties are often not acknowledged, and as a result, some estimates become seen as factual and unchallenged.

There are clearly many debates about the accuracy of common perceptions of deforestation: this chapter cannot summarize them all. There is no implication in any of the challenges reported here that forest loss should be ignored, or that unregulated destruction of forest ecosystems should be tolerated. Also, it is clear that forests – and other ecosystems – are facing important, and still partially understood, threats from multifarious sources such as from the varied impacts of El Niño, or from projected future changes in climate. But it is clear that many previous accounts of deforestation's impacts have important flaws. Moreover, simply asserting that deforestation is always problematic overlooks both the physical complexity of how deforestation is carried out, and its variety of purposes and impacts:

> The generic term "deforestation" is used so ambiguously that it is virtually meaningless as a description of land-use change ... It is our contention that the use of the term "deforestation" must be discontinued, if scientists, forest land managers, government planners and environmentalists are to have meaningful dialogue on the various human activities that affect forests and the biophysical consequences of those actions.
>
> (Hamilton and Pearce, 1988: 75)

In addition, simply asserting that deforestation should be stopped may both neglect the diverse biophysical causes of supposed impacts such as biodiversity loss and soil erosion, and consequently may not address these problems. It may also impose unwarranted restrictions on agricultural practices used by people in affected zones. These dilemmas may occur in China, for example, where the government imposed a ban on logging in 1998 in order to avoid downstream flooding, and also in other locations where reforestation is now seen to be a panacea for various environmental problems including erosion control, biodiversity conservation, and climate change mitigation.

There is a need to define "deforestation" in more complex ways in order to distinguish between different levels of forest disturbance. Related to this is also the need to identify how and why "forests" may be identified and distinguished from other ecosystems. For example, it is clear that much attention given to tropical rainforests has tended to essentialize various different forest types into one, and also tend to diminish the importance of other forest ecosystems such as savanna (Whitmore, 1984; Solbrig, 1993). But more importantly, there is also a need to understand how such orthodox, and now widely challenged, powerful organizations and campaigners adopt conceptualizations of environmental degradation despite the growing evidence of the inadequacy of such concepts.

> The mindset created by the paradigm which links the absence of forests with "degradation" of water resources, and "more forest" with improved water resources, has not yet been destroyed. Until it is replaced it will continue to cause governments, development agencies and UN organizations to commit and waste funds on afforestation or reforestation programs in the belief that this is the best way to improve water resources.
>
> (Calder, 1999: 37)

Environmental orthodoxies

So, what are the implications of these problems for environmental science and politics? It is important to reiterate that these discussions of desertification, soil erosion, and deforestation do not deny the existence of environmental degradation, but illustrate the inadequacy of the concepts we use to define it. Concepts such as desertification, soil erosion, and deforestation have clearly been associated with severe environmental problems within particular contexts. Yet, used universally and uncritically, these concepts may actually undermine both environmental management and social development by adopting simplistic approaches to the causes of biophysical change, and by encouraging the imposition of land use policies that may only restrict local livelihoods.

Perhaps the most significant feature of such common definitions of environmental degradation is that they continue to be used despite the accumulation of evidence to suggest they are flawed. The continued use of these terms is analyzed in this book, and is seen to be a product of a variety of political influences. Politics underlie the construction of these terms, their continued adoption, and the presentation of them by particular actors as legitimate and accurate representations of reality.

This book uses the term "environmental orthodoxies" to refer to these institutionalized, but highly criticized conceptualizations of environmental degradation. The concept of environmental orthodoxies was used by Leach and Mearns (1996) to describe the persistence of particular explanations of environmental change in policy processes despite the accumulation of evidence to reject or redefine them. Other authors have used similar terms. Calder (1999), for example, uses the term "mother statements," and Adger *et al.* (2001) refer to them as "truth regimes." More generally, these explanations may also be referred to as "environmental narratives" (Roe, 1991, 1995; Harré *et al.*, 1999), and environmental "storylines" (Hajer, 1995). The existence of "myths" or "simplifications" in debates about land-use-cover change have also been noted by a variety of authors in policy debates elsewhere (also see Holling, 1979; Thompson *et al.*, 1986; Batterbury *et al.*, 1997; Adams, 2001; Lambin *et al.*, 2001).

Box 2.1 contains a definition of environmental orthodoxies that is useful for further discussion in this book. Box 2.2 contains some examples of environmental orthodoxies and includes a variety of themes of land-use-cover change. It is also worthwhile defining so-called "environmental adaptations" which are often the examples of local land management that provide exceptions to environmental orthodoxies. Such adaptive practices are discussed further throughout the book.

"Environmental orthodoxies" reveal a variety of characteristics. First, as Boxes 2.1 and 2.2 indicate, orthodoxies are often vague statements or "received wisdom" rather than a narrowly defined scientific theory or hypothesis. Indeed, many physical environmental scientists agree with some of the concerns about vague generalization or biophysical inaccuracies exhibited by orthodoxies (Schumm, 1991; Holton, 1993). Box 2.2 describes some specific orthodoxies relating to topics of land-use-cover change. It is worth noting, however, that similar environmental "myths" or meta-narratives exist in other aspects of environmental debate. For example, the concept of "balance-of-nature" (or non-equilibrium ecology) is examined in Chapter 3; assumptions about environmental impacts concerning gender and other social divisions are discussed in Chapters 4 and 6; debates about environmental "fragility" or "crisis" are covered in Chapter 5; and questions about the supposedly "global" nature of problems are considered in Chapter 7.

Second, the discussion of environmental orthodoxies might appear hostile to many tenets of popular environmentalism because it questions the urgency or role of human action in environmental degradation. This perception may be misplaced, because the purpose of discussing orthodoxies is to improve our understanding of environmental change, and to enhance our means of preventing environmental problems. Furthermore, the discussion of environmental orthodoxies is not necessarily based on a

---

Box 2.1   Environmental orthodoxies and adaptations

Environmental orthodoxies are generalized statements referring to environmental degradation or causes of environmental change that are often accepted as fact, but have been shown by field research to be both biophysically inaccurate and also leading to environmental policies that restrict socio-economic activities of people living in affected zones. Environmental orthodoxies are frequently based upon images of environmental changes as crises brought about by human action, and overlook the role of adaptive practices performed by particular land users in either mitigating or even reversing environmental degradation. They also commonly overlook the role of long-term, complex biophysical factors in causing apparent degradation, such as non-anthropogenic climate change; tectonic uplift; or the historic frequency of events such as floods or fires. Research on environmental orthodoxies has been associated with, yet is not necessarily part of, the discussion of "non-equilibrium" (or non-linear) ecology that emphasizes the prevalence of disturbance and change within ecological systems, and the social influences on the identification of time and space scales.

Environmental adaptations are practices adopted by people to mitigate the environmental impacts of resource scarcity or environmental change on local resources. Adaptations may be divided into adaptive *strategies* and adaptive *processes*. Adaptive strategies are practical decisions by an individual to change productive activities, such as selling livestock during drought years, or building small-scale soil conservation measures such as mounds or *diguettes* (stone lines) to prevent declining soil fertility. Adaptive processes are more long-term decisions that create socio-economic trends, such as the decision to undertake long-distance migration, or the building of terraces on agricultural land. Usually, the adoption of environmental adaptations may be associated with actions that contradict the predictions of environmental degradation resulting from environmental orthodoxies. Moreover, environmental adaptations may also be seen as the opposite to environmental orthodoxies, as orthodoxies represent generalized expectations based on prior assumptions about population growth and ecological fragility, whereas environmental adaptations illustrate local instances where the negative impacts of degradation have been avoided.

Sources: Leach and Mearns, 1996; Batterbury *et al.*, 1997;
Batterbury and Forsyth, 1999.

---

statement that environmental problems do not exist, but instead that the terms used to describe them are inaccurate and unhelpful. In this sense, discussing environmental orthodoxies is different to some attempts to dismiss environmental concerns on grounds of optimism about economic growth (such as Björn Lomborg's *The Skeptical Environmentalist,* 2001). (The potential clashes between environmental orthodoxies and environmentalism are discussed later in this chapter.)

Third, engaging in debates about environmental orthodoxies also implies raising questions about scientific realism. By their very nature,

Box 2.2   Examples of environmental orthodoxies

Orthodoxy and new findings (simplified)	*Sample references*

Desertification

Pro-orthodoxy
Sears, 1935;
Stebbing, 1937;
Lamprey, 1975;
Brown, 2001

- Orthodoxy: the belief that population growth, deforestation, and intensive agriculture on the margins of desert areas is leading to irreversible increase in desert areas, decline in rainfall, and associated famine. (Such beliefs have often led to policies that seek to restrict livestock and agricultural holdings in drylands; or strategies to "prevent" desertification by planting trees or building fences to prevent the spread of sand dunes.)

- New findings: researchers now understand the greater significance of long-term fluctuations in rainfall and climate in drylands, and that efforts to prevent movement of sand by placing barriers to sand dunes may make problems worse. Farmers may not be culpable for causing desertification, as there are ways in which they reduce impacts on soils, and the diversity of causal factors is high. "Desertification" has often been confused with "famine" and "drought," but "drought" may be a more effective means of assessing livelihood concerns than "desertification."

Anti-orthodoxy
Dregne, 1985;
Biot, 1995;
Thomas and Middleton, 1994;
Blaikie, 1995;
Hoben, 1995;
Saberwal, 1997;
Rasmussen *et al.*, 2001

Tropical deforestation

Pro-orthodoxy
Richards, 1952;
Myers, 1984;
Mather, 1992;
Mather and Needle, 2000;
Brown, 2001

- Orthodoxy: a variety of beliefs referring to the fragility of tropical (often rain) forests; the role of forests in maintaining biodiversity; and the pressures upon forests from rising populations, especially of local agriculturalists such as shifting cultivators or poor people in search of fuelwood. Disturbances such as deforestation and fire may cause severe and long-lasting damage to forests and biodiversity. (Such beliefs have led to a variety of policies that seek to protect forests from interference from local people.) (See also "Shifting cultivation.")

- New findings: research has questioned many aspects of orthodox concepts of deforestation. While not denying a role for population growth or poverty, movements of people who undertake deforestation are more likely to be

Anti-orthodoxy
Leach and Mearns, 1988; Agarwal and Narain, 1991;
Rocheleau and

affected by government policies that encourage migrants, or loss of political stability in frontier regions. Similarly, "deforestation" need not signify clearfelling, or complete loss of land cover, but instead a variety of impacts, sometimes minor. Some farming communities may even contribute to the growth and protection of forests. The role of disturbance, such as by fire, is acknowledged as a source of change and development of biodiversity within certain forest ecosystems. Biodiversity also need not be maintained only through preserving forest areas, as neighboring grasslands or savanna systems may also have high biodiversity. Impacts of population growth on rural energy requirements need not necessarily lead to uncontrolled deforestation, and instead need to be understood alongside other sources of energy.

Ross, 1995; Barraclough and Ghimire, 1996; Fairhead and Leach, 1996, 1998; Cullet and Kameri-Mbote, 1998; Robbins, 1998; Angelsen and Kaimowitz, 1999; Cline-Cole and Madge, 2000; Kull, 2000; Lambin *et al.*, 2001

Shifting cultivation

- Orthodoxy: the belief that shifting cultivation, or "slash and burn" agriculture, is of necessity destructive of forests; has low agricultural productivity; and causes a variety of lowland impacts such as water shortages and sedimentation. (These beliefs have led to policies that identify shifting cultivators as responsible for various forms of environmental degradation, and, consequently, efforts to resettle them, or restrict upland agriculture through re/afforestation.) (See also "Himalayan degradation" and "Watershed degradation.")

Pro-orthodoxy
Myers, 1984; Mather and Needle, 2000

- New findings: research has indicated that there are many different forms of shifting cultivation, and that environmental impacts depend on the length of tenure at specific sites by settlers: some cultivators adopt semi-sedentary practices such as terracing, soil conservation, or coppicing of forests. Shifting cultivation in general may not cause "loss" of forest, but instead may encourage development of specific types of forest and biodiversity. Many supposed impacts of upland agriculture may be caused by preexisting and long-term biophysical processes such as gullying or factors leading to low levels of water retention in highland zones.

Anti-orthodoxy
Conklin, 1954; Geertz, 1963; Angelsen, 1995; Fairhead and Leach, 1996; Sillitoe, 1993, 1998; Schmidt-Vogt, 1998; Fox *et al.*, 2000

Rangeland degradation

- Orthodoxy: the belief that rangelands (or grasslands) are natural "climax" vegetation systems that are determined by edaphic factors such as soil or climate. Rangelands may also therefore have natural "carrying capacities" for people and livestock. (Such beliefs have led to policy proposals to limit numbers of livestock or restrict agriculture.)
- New findings: research has indicated that large areas of rangelands are maintained by interactions of human impacts on longer-term biophysical changes. Restricting human activities may therefore lead to rapid changes. Multiple states of stability may be experienced with different forms or stages of vegetation growth. Grazing may be necessary to maintain such states.

Pro-orthodoxy
Harris, 1980

Anti-orthodoxy
Solbrig, 1993;
Turner, 1993;
Scoones, 1994;
Bassett and Zuéli, 2000; Oba *et al.*, 2000

Agricultural intensification

- Orthodoxy: the belief that population growth is leading smallholders, especially in developing countries, to increase agricultural intensification toward unsustainable levels. High levels of agricultural intensification may lead to erosion, or exhaustion of land and water resources. (These beliefs have, in part, led to policies that seek to rationalize agriculture in many developing countries.) (See also "Shifting cultivation.")
- New findings: research has indicated that methods of agricultural intensification are complex, and may involve a variety of livelihood strategies including income diversification (perhaps involving part-time migration or non-agricultural income); or intensified methods of increasing production without environmental degradation.

Pro-orthodoxy
Eckholm, 1976;
Ehrlich and
Ehrlich, 1991

Anti-orthodoxy
Netting, 1993;
Tiffen and
Mortimore, 1994;
Mortimore and
Adams, 1999;
Bebbington and
Batterbury, 2001

Watershed degradation and water resources

- Orthodoxy: a series of inter-connected beliefs relating to the degradation of soils and forests on watershed areas (or zones, commonly mountainous, that are seen to supply water to other areas, often in lowlands). Beliefs may include: that forests increase rainfall; forests increase runoff; or that forests reduce erosion and floods. (These beliefs have often led to policies that seek to relocate farmers from watershed zones; to reforest watersheds,

Pro-orthodoxy
Wittfogel, 1956;
Openshaw, 1974;
Postel, 1993;
Revenga *et al.*, 1998

often with plantation forestry; or to achieve all of these by converting watersheds into national parks or other protected lands.) (See also "Himalayan degradation.")

- New findings: a wide variety of research has questioned either the scale or the uniformity of orthodox beliefs. For example, the effects of forests on rainfall are small, but cannot be totally dismissed. Similarly, the impact of forests on erosion is highly variable, depending on types of forest and types of erosion (plantation forests may increase sheet erosion; much gully erosion may be greater under "natural" forests than on cultivated slopes). The belief that forests increase runoff, however, has been widely dismissed (although there are commonly changes to the speed and seasonality of discharge, although evidence linking floods to deforestation is highly variable). The influence of lowland increase in demand for water in causing apparent water shortages also needs to be acknowledged.

Anti-orthodoxy
Hamilton, 1987, 1988; Hamilton and Pearce, 1988; Pereira, 1989; Alford, 1992; Chapman and Thompson, 1995; Chomitz and Kumari, 1996; Calder, 1999; Custodio, 2000; Gyawali, 2000; Calder and Aylward, 2002

Theory of Himalayan environmental degradation

- Orthodoxy: the belief that increasing population and agricultural intensification in the Middle Hills of the Himalayas (and similar regions) is leading to a downward cycle of deforestation, erosion, landslides, and lowland sedimentation. (Beliefs have supported policies seeking to restrict highland land use, resettle villages, or reforest large areas of hillslopes.)

Pro-orthodoxy
Eckholm, 1976; Cronin, 1979

- New findings: research has since shown that much erosion is caused by processes other than agriculture (such as gullying or the effects of tectonic uplift); that farmers may adopt practices to mitigate erosion and land failure; that much degradation of agricultural land has been related to historic large-scale land clearance; and that lowland floods have diverse causes. Increasing population is more likely to decrease soil fertility on gentle slopes where fallow periods decline, rather than lead to cultivation on steeper slopes, as many farmers appreciate that this is where erosion, and hence declining soil fertility, is highest.

Anti-orthodoxy
Thompson *et al.*, 1986; Hamilton, 1987; Ives and Messerli, 1989; Metz, 1991; Forsyth, 1996; Gyawali, 2000; Calder and Aylward, 2002

environmental orthodoxies are explanations that have questionable accuracy and relevance. Seeking more accurate, and more relevant, explanations must therefore require examining questions of epistemology and ontology concerning environmental science and biophysical change (see Chapter 1). This kind of analysis may be different to many other debates in environmental sociology or politics that focuses on contested environmental values (e.g. McNaughten and Urry, 1998) because it also considers how far a "real" biophysical world may exist alongside the biophysical explanations of it. Such analysis, therefore, needs to incorporate debates about science studies and biophysical epistemology in ways that environmental sociology or politics commonly do not do.

Fourth, the ability to learn about environmental orthodoxies has usually come when existing conceptualizations of environmental degradation have been shown to be deficient. Deficiencies may be in terms of biophysical environmental management, such as in the case of fences to stop desertification, or when policies have caused widespread local resentment. These factors have significance for debates about how we learn about the inaccuracies of environmental science, and are discussed more in Chapter 8.

Finally, it is important not to underplay the potential impacts of environmental orthodoxies on affected peoples. Some proposed "solutions" to problems of desertification, soil erosion, and deforestation, for example, have included placing restrictions on livestock numbers or planting practices of poor people living in zones considered to be at risk from degradation. Other forms of control, such as taxation, fines, and even imprisonment have been applied to practices that may be claimed to be not degrading. Fairhead and Leach described such social injustices in relation to the Kissidougou region of Guinea:

> It is hard to underestimate the importance of the degradation discourse's instrumental effects on many aspects of Kissidougou's life. These have impoverished people through taxes and fines, reduced people's ability to benefit from their resources, and diverted funds from more pressing needs. They have accused people of wanton destruction, criminalized many of their everyday activities, denied the technical validity of their ecological knowledge and research into developing it, denied value and credibility to their cultural forms, expressions, and basis of morality, and at times even decried people's consciousness and intelligence. The discourse has been instrumental in accentuating a gulf in perspectives between urban and rural; in undermining the credibility of outside experts in villagers' eyes; in provoking mutual disdain between villages and authority, and in imposing on the farmer images of social malaise and incapacity to respond to modernity.
>
> (1996: 295)

Challenging the I = PAT equation

The preceding discussion of environmental orthodoxies highlighted the problems of environmental explanations on specific themes. Yet these specific explanations also reflect some broader debates that underlie much general environmental concern. One of these frameworks is the so-called I = PAT equation.

The I = PAT equation has been employed – often implicitly – as the basis for the study of environmental degradation since the early 1970s (Ehrlich and Holdren, 1974; Kates, 2000a). The equation states, simply, that environmental impacts (I), are a function of population growth (P); the affluence, or rate of consumption of particular societies (A); and the technological innovations that may either enhance rates of consumption, or allow societies to reduce impacts on resources through greater efficiency or by the management of degrading influences (T). The equation is closely linked to the long-running Limits to Growth debate, in which Malthusian notions of environmental change (accentuating the adverse effects of population increase on limited resources) may be offset by more optimistic Boserupian thinking (that stresses the ability for technological innovation and adaptation to allow apparent limits to be exceeded). It is also linked to the "tragedy of the commons" model that proposes environmental collapse will result following unrestricted access of private actors upon public resources (Hardin, 1968).

The I = PAT equation has also been linked to many "orthodox" conceptions of the role of poverty (or lack of affluence) in environmental degradation. Some statements reflecting the equation were made in the 1987 Brundtland Commission (WCED, 1987), for example:

> Poverty is a major cause and effect of global environmental problems. It is therefore futile to attempt to deal with environmental problems without a broader perspective that encompasses the factors underlying world poverty and international inequality.
>
> (1987: 59)

Or:

> Many parts of the world today are caught in a vicious downwards spiral: poor people are forced to overuse environmental resources to survive from day to day, and their impoverishment of their environment further impoverishes them, making their survival ever more difficult and uncertain.
>
> (ibid.: 27)

Increasingly, however, there are important reasons to question the uniformity of these statements. Indeed, some observers have called these statements a further set of "myths" (see Box 2.3). These claims reiterate the

---

Box 2.3   Myths and oversimplifications concerning poverty and environment

"Myth"	*"New thinking"*
1. The poor cause most environmental degradation.	In general, the rich use more resources and have greater environmental impact than the poor. Poverty, however, often forces people to use resources unsustainably.
2. Economic growth inevitably leads to environmental degradation.	Economic growth can help pay for a better environment, and improved environmental management enhances and sustains growth.
3. The poor don't care about the environment.	The poor are acutely aware of the negative effects of a poor environment on their lives, particularly as they often depend directly on the environment for survival.
4. The poor lack the knowledge and resources to improve their environment.	The poor can and do invest in better environmental management, particularly where incentives and information are available. Their traditional knowledge is often undervalued or ignored.

Source: DFID, 2002; also see Forsyth, Leach and Scoones, 1998;
Leach and Mearns, 1991.

---

importance of so-called "environmental adaptations" as means of establishing environmental protection and livelihoods.

There is no suggestion in the environmental orthodoxy debate that population, affluence/poverty, or technology play no role in environmental degradation, or that we should not seek to alleviate world poverty. But the implications of environmental orthodoxies are that the assumptions underlying much thinking influenced by the $I = PAT$ equation are simplistic for two key reasons. First, the equation overlooks the diverse ways in which environmental changes and impacts may (or may not) be experienced as degradation. Second, it fails to acknowledge how poor people do not necessarily cause environmental degradation through the adoption of environmental adaptations or practices that conserve environmental resources, even in the presence of population growth and supposed ecological fragility. These flaws can be attributed to the failure of the $I = PAT$ equation to acknowledge the role of social norms and organization on both sides of the equation, concerning how "population growth, affluence, and technology" (PAT) may be managed, or in relation to the definition and meaning of "impacts" (I).

Much research within cultural ecology has acknowledged the role of local adaptive processes in influencing how population, affluence, and technology may influence environmental impacts. For example, the soil mounds of the Wola in Papua New Guinea mentioned above may be considered a "technology," but the training, and integration of soil mounds into other forms of livelihood are all functions of social organization (Sillitoe, 1998). These factors suggest that it is difficult to assess the impacts of population, affluence, and technology without acknowledging the social setting.

Furthermore, environmental "impacts" may also be contextualized. As discussed above, a variety of changes in environment may be seen alternatively as positive or negative depending on the objectives of different land users. Such alternative objectives might include the vision of forest as a source of nutrients for soil, and a barrier to agriculture (as some shifting cultivators might perceive some areas of forest); or the appreciation of forests as aesthetically pleasing and endangered forms of landscape. The dilemma for the $I = PAT$ equation is that, clearly, the discussion of "impacts" are dependent upon such valuations, yet the equation does not acknowledge how, or by whom, such valuations are made (see Hynes, 1993).

This book builds on the criticisms of the $I = PAT$ equation by presenting a variety of analyses of how both "I" and "PAT" may be affected by social norms and organization. Again, this critique does not imply that population, affluence, or technology need never contribute toward environmental degradation (see also Kasperson *et al.*, 1995; Batterbury *et al.*, 1997; DeHart and Soulé, 2000; Lambin *et al.*, 2001). Instead, the objective is to ensure that environmental explanations are not made uncritically and universally in ways that overlook the biophysical complexities of how environmental degradation occurs, or that the policies linked to such explanations do not restrict local livelihoods.

Science or myths?

This book, therefore, examines the means by which different environmental explanations become dominant; the political implications of such different explanations; and the ways such dominant explanations may be democratized in order to make environmental science more accurate and relevant to a wider number of people. This task, however, requires rethinking approaches in both environmental science and politics.

It is tempting, for example, to refer to environmental orthodoxies as "myths," in the sense of "falsehoods," because they refer to statements that are commonly taken as "fact," but which have been shown to be highly flawed in practice. Thomas and Middleton (1994), for example, adopt this approach in their book, *Desertification: Exploding the Myth*. Consistent with orthodox science, this approach assumes that the problem of environmental orthodoxies can be overcome by improving the flow of information to policy debates and agencies in order to correct the false-hood.

Yet the word "myth" need not only refer to information that is "false," but also to systems of knowledge and belief that are seen essentially as "true." (For example, see the quotation from *Cyrano de Bergerac* repeated at the front of this book: "Call it a lie, if you like, but a lie is a sort of myth and a myth is a sort of truth.") Influenced by Roland Barthes, Rangan wrote: "Myths are produced through narratives that render particular social events significant by transporting them from their geographical and historical contexts into the realm of pure nature" (2000: 1).

Such "truthful" forms of myth may take various forms. On one hand, much "local" knowledge or cultural practices such as environmental adap-tations may be referred to as mythology or "lore," because they represent embedded trusted knowledge (Johnson, 1992). On the other hand, environmental orthodoxies, or dominant scientific explanations from out-side, may also be considered "mythical" if they form a source of concep-tual organization and authority from which to approach environmental management. Indeed, Karl Popper, the great defender of the scientific method, wrote that much of the popular power of science lay in its "poetic inventiveness, that is, story-telling or myth making: the invention of stories about the world" (Popper, 1994: 40). The evolution of such orthodoxies from conventional "science" may therefore not diminish their mythic stature (see the debate between Metz, 1989 and Thompson, 1989; Forsyth, 1998a).

Instead of seeking a once-and-for-all definition of what may be con-sidered true or false about environmental explanations, perhaps it is more constructive to examine how, and under which conditions, statements about environmental causality may be considered true. This book there-fore aims for a different approach to that commonly adopted within ortho-dox science sometimes known as "synoptic rationality" in which decisions are made based on first collating "all the facts" (Collingridge and Reeve, 1986: 63). Synoptic rationality has often been applied to environmental science, such as through Baarschers' (1996) book, *Eco-facts and Eco-fiction*. In contrast, this current book questions the very meaning of the word "fact," although this does not mean that accuracy or realism are impossible.

Such an approach to ecological reality, however, commonly attracts two kinds of criticism. First, it is often thought (incorrectly) that the deconstruction of scientific discourse in the manner of the environmental orthodoxy debate is a movement toward cultural relativism – or the belief that social factors have more relevance to the dominance of particular scientific explanations than any resemblance to the "real world." Contrary to expectations, the environmental orthodoxy debate does not suggest that *any* scientific statement may be considered truthful, or that there is no "real world" about which to build explanations. The objective, rather, is to examine how explanations of biophysical events and processes may emerge as the result of different social and political experiences, and to analyze their political implications. This objective is discussed in more detail throughout the book.

Second, some observers have claimed that criticisms of dominant environmental science might also imply a rejection of environmentalism. Indeed, as noted in Chapter 1, Paul and Anne Ehrlich (1996) published a book on this subject entitled *A Betrayal of Science and Reason.* In particular, this book described "brownlash" as a form of environmental research that deliberately undermines environmental concern. Brownlash is commonly sponsored by large industries that seek to avoid environmental regulation such as research publicized by the Global Climate Coalition (http://www.globalclimate.org/). Indeed, some similar concern has been raised in Great Britain by the publication of some monographs about environmental orthodoxies by the British pro-market think tank, the Institute of Economic Affairs, even though these monographs do not explicitly discuss pro-market ideas (see Morris, 1995; Stott, 1999).

It is important to note that the debate about environmental orthodoxies is not a form of brownlash. There are many differences between brownlash and research focusing on environmental orthodoxies. First, most research on environmental orthodoxies has been unrelated to any work conducted on behalf of large industries. As discussed above, many studies highlighting environmental orthodoxies has come from cultural ecology, or work conducted by researchers working in regions where such orthodoxies are clearly inaccurate. Second, research on orthodoxies has often revealed that dominant scientific explanations *get in the way of* achieving environmentalist objectives. For example, research on water shortages in watershed regions has often indicated that plantation reforestation will reduce rather than improve supply of water to the lowlands. Indeed, research has also shown that some orthodoxies may result in insufficient regulation of other, more environmentally damaging activities, such as high water demand outside watershed areas (Forsyth, 1996; Calder, 1999). Third, much research on orthodoxies has been conducted within the frameworks of orthodox science – for example, by using detailed empiricism and a critical engagement with hypotheses – rather than an outright rejection of scientific practice. And fourth, many studies have sought to demonstrate the negative impacts of hegemonic environmental explanations on poor people who have often protected resources from degradation.

But while there are many ways in which the environmental orthodoxy debate should not be seen as brownlash, there are also ways in which this debate can still be critical of some environmentalist statements. As discussed above, such statements may include simplified explanations that overlook the complexity of biophysical changes; or those values or policies that restrict local livelihoods.

One possible example of this kind of explanation could come from the Ehrlichs themselves. Writing about a visit to Rwanda in central Africa, they stated:

> Going around the world in search of butterflies also gave us a personal view of then little-recognized signs of environmental deterioration. . . . We would have been hard-pressed to find relatively undisturbed habitat at many of our stops in what we had imagined to be an "unspoiled" tropical paradise. . . . In the early 1980s we traveled through Rwanda to the Parc National des Volcans, home of the rare mountain gorilla. The nation presented a classic picture of overpopulation and environmental deterioration: steep hillsides farmed to the tops with little or no erosion control, patches of exotic (non-native) eucalyptus trees being heavily coppiced for firewood, and rivers running red with eroded soil.
>
> (1996: 5–6)

The problem with this kind of statement is that it ascribes a notion of "unspoiled paradise" to many locations of the developing world that experience rapid processes of rainfall, soil movement, and vegetation change regardless of human activities. Furthermore, while it is clear that human settlement does impact on ecosystems, in many locations such settlement (and agriculture) interacts with local ecosystems to produce different, yet no less viable, biogeographic systems. The quotation's romantic image of "rivers running red with eroded soil" – apparently because of human mismanagement – is misplaced because there is no other evidence (in this quotation at least) that erosion did not predate agriculture, or that it causes severe problems for the people in this village. Finally, many people in developing countries might object to the primacy afforded in this quotation to butterflies and the image of an unspoiled paradise when the villagers at this site are engaged in building livelihoods through agriculture. (One could ask whether the cities of North America and Europe also reflect forms of ecological sustainability and irresponsibility.) Many people living in such regions may be struggling with short-term survival against a range of social, economic, and political problems, and consequently may value butterflies and wildlife less.

The point of this discussion is not to denigrate the environmental concern shown by the Ehrlichs, or to suggest that brownlash should not be criticized. Furthermore, there is no intention to suggest that we have to choose between economic livelihoods and wildlife such as butterflies, or that economic growth should be tolerated whatever its costs.

Instead, the aim is to indicate that many discussions of what should count as "science and reason" under popular environmentalism reflect many tacit assumptions about environmental values and science that can be challenged on many grounds. Indeed, some of these themes can be described as mythical, either in terms of myths as falsehoods (such as the automatic assumption that erosion is degrading or human-induced), or myths as guiding principles about how things should be (such as in the vision of an "unspoiled paradise").

It is therefore difficult to distinguish between "myths" and "science," even though the stated intention of science is to achieve a privileged form of knowledge different from opinions and folklore. "Science" itself is subject to social influence, either in the formulation of objectives that reflect social agendas, or in its rhetorical use to legitimize particular conceptualizations of environmental explanation against others.

This book seeks to overcome some of these dilemmas by looking more closely at the social and political factors that influence the constitution and use of environmental science. Under a "critical" political ecology, there can be no unpoliticized use of the word "ecology," and every statement about the nature or causes of ecological degradation is examined to reveal how this link was established, and how far it may hide political assumptions and implications. This approach may challenge some commonly held beliefs about environmental degradation. But it may eventually create a more accurate and relevant form of environmental explanation.

Summary

This chapter has summarized some of the book's central questions that will form the basis for discussion in later chapters.

Many popular and political debates about environment are based upon conventional beliefs, or "received wisdom" about environmental degradation that are highly challenged and uncertain. Indeed, some observers have called these explanations "myths." The chapter summarized examples of such contested science in relation to desertification, soil erosion, and deforestation. Many conventional approaches to these problems have resulted in land-use policies that have either simplified the underlying biophysical causes of apparent problems, or even imposed restrictions on the livelihoods of local people.

These conventional – yet questionable – explanations are referred to as "environmental orthodoxies." Yet, such orthodox thinking may also include simplistic generalizations about the role of population, affluence, and technology in environmental degradation (the $I = PAT$ equation), or the view that "nature" should be in balance. Discussing the problems of such explanations does not deny the existence of environmental degradation, but rather criticizes the concepts and approaches we have used to define it.

This book seeks to explain how such environmental orthodoxies have emerged, and how they may be challenged with more relevant approaches to environmental science. Yet, rather than simply suggesting that environmental orthodoxies are "myths" in the sense of falsehoods, it may be more constructive to see how orthodox explanations are seen to be true. Dominating visions of environmental explanation and science may continue to exist because they are seen by many to be fair and accurate, and because they may uphold visions of how the world should be. The following chapters consider both the "false" and "true" aspects of environmental myths.

3 Environmental "laws" and generalizations

The preceding chapter spelt out the main problem addressed by this book: many widely held principles and understandings of environmental degradation are commonly accepted as "fact" within popular and political debates. Yet an increasing amount of research has indicated that many of these explanations are biophysically inaccurate or lead to policies that are socially unjust. There is consequently a need to understand how explanations of environmental degradation evolve in order to make environmental science more meaningful to people who experience environmental problems, and to avoid the inaccuracies and injustices of many "orthodox" environmental explanations.

This chapter starts the discussion by examining some of the underlying problems of making scientific statements about complex environmental processes and events, and how these relate to social and political debates.

In particular, this chapter will:

- introduce debates from the Philosophy of Science about the problems of so-called "laws of nature" and the social basis of generalized statements about environmental change under orthodox frameworks of science;
- discuss the significance of "non-equilibrium ecology" as an alternative to historic approaches to environmental change based on equilibrium, evolution, and a "balance of nature";
- outline differences between "realist" and "constructivist" approaches to environmental explanation; and
- suggest ways in which scientific inquiry can acknowledge diversity and non-equilibrium in ecology, yet still allow explanations about a "real" biophysical world.

This chapter sets the tone for much of the book. It describes how much environmental science has been based on historic practices of sampling and inference that may not fully acknowledge the social and political contexts in which environmental problems are experienced. Furthermore, the chapter discusses how many practices of regulating scientific findings – through peer review or "conjecture and refutation" of ideas – have been criticized by many scholars in science studies for failing to challenge powerful interests within science.

A "critical" political ecology is an ability to reveal the hidden politics within supposedly neutral statements about ecological causality. This chapter discusses many of the hidden social biases within the frameworks of "orthodox" ecological science. These problems are then discussed further in following chapters, which outline how such orthodox science has been co-produced with social trends and political activism.

The frameworks of orthodox science

Chapter 2 summarized some of the problems resulting from the uncritical adoption of environmental explanations sometimes known as "environmental orthodoxies." These problems include biophysical simplicity – or a failure to acknowledge complex, long-term biophysical changes such as tectonic uplift or non-anthropogenic climate change that may predate many human impacts. Similarly, problems may also include a failure to acknowledge how "environmental adaptations" adopted by people in affected zones may lessen the impacts of population growth or economic activities on environmental degradation. As a result, many land-use policies based upon environmental orthodoxies may end up not addressing the underlying biophysical causes of environmental degradation, and may even unfairly restrict local livelihood strategies of poor people.

How did such unhelpful and inaccurate explanations come into being? This chapter examines how such "orthodox" explanations of environmental degradation have evolved, and in particular focusing on the frameworks of "orthodox" science that may have caused these problems.

The term "orthodox science," however, needs some definition. Science is not a uniform, or unchanging, institution, and many professional scientists are themselves critical of some dominant trends in scientific practice (Nowotny and Rose, 1979). "Orthodox science" in this book is taken to mean scientific practices that characterize much basic research in ecology involving principles loosely labeled as "positivism." These practices may be called "orthodox" because they date from early thinking about science and scientific practice that sought to establish scientific knowledge as a privileged source of accuracy and political neutrality. Orthodox science, in this context, may be characterized by three key tenets: the mechanism of inference from samples; the self-regulation of findings and research by scientists; and an underlying belief in the ability of science to indicate reality in an accurate and unbiased way.

It is important to note, however, that there are still important debates and uncertainties about these tenets – including from professional scientists. Many "orthodox" scientists are also concerned at simplistic generalizations about environmental causality, and would like to challenge some of the environmental policies proposed by environmentalists (Whitmore, 1984; Levitt, 1999). The discussion of "orthodox science" in this chapter, therefore, is not meant to be a criticism of science, or scientists in general. Instead, this chapter is a summary of how the tenets of "orthodox science" may be insufficiently sensitive to appreciate the social and political meanings inherent in the identification of environmental problems, or of how environmental science is both influenced by, and employed within, different political debates.

Positivism and its critics

"Positivism" is most commonly the starting point for discussions about the inadequacies of standard scientific practice, and it is usual for such discussions to criticize positivism strongly. In practice, such criticisms of positivism often conflate a variety of philosophical positions relating to science and realism, and overlook that it is often through positivism that science has made its greatest contributions to society. It is perhaps more useful to assess the various positions associated with positivism, and to assess how far each position raises questions for the development of accurate generalizations about environment.

The popular definition of "positivism" is an approach to science that adopts the principle of the scientific method, or the use of carefully selected and examined samples to infer properties to the bodies from which the samples were selected. The basis of inference under "positivism," however, is not always clear. Early positivists such as Mach (1838–1916) sought to infer generalizations about reality by summarizing apparent trends in existing datasets (indeed, this is the group that philosophers of science refer to exclusively as "positivists"). This approach was later developed by the so-called Vienna School of Logical Positivism in the 1920s to focus on *verification* of such patterns as the key process of inference. Under verification, patterns or trends were seen to be accurate if further sampling revealed similar patterns.

The framework of logical positivism was also closely linked to so-called logical empiricism, which was an adaptation of the early empiricism of the British empiricists Locke and Hume to a more powerful inferential method. The practice of logical empiricism reflected a belief that knowledge existed only through experience, and the recording of such knowledge could lead to logical analysis of the observed patterns. Yet, in turn, such beliefs depended in part upon further assumptions about the ability to understand reality that are now generally the cause of most criticisms of positivism as a research technique today. In particular, logical empiricism was closely related to foundationalism, or the belief that existing knowledge could indicate underlying reality in a clear and unproblematized way (or, that there were absolute epistemological foundations for knowledge). Similarly, logical empiricism also supposed naturalism – or the application of similar research techniques to subjects generally known as natural and social sciences, without consideration to how far such unitary scientific practice may be problematic. Both foundationalism and naturalism may be criticized under both social theory and debates concerning scientific realism for overlooking how far the systems we used to create knowledge may reflect social agendas, or also overlook complex processes beyond the ability of techniques used (see Box 3.1). Yet, although logical empiricists, early positivists, and the Vienna School all approached foundationalism and naturalism in generally unproblematized ways, many later advocates of the scientific method do not necessarily accept these principles.

Logical empiricism was most famously challenged by Karl Popper in the 1960s, who argued that verification should be replaced by *falsification*

Box 3.1 Philosophies of orthodox science

Foundationalism is an approach to knowledge dating from Ancient Greece that holds there are basically known propositions that do not derive from other known positions or beliefs. Early examples include the notion of pure ideas of Plato, or the rationalism and logic of mathematics associated with Descartes. The essence of a foundationalist approach to science is the belief that there are inherent assumptions that cannot be challenged.

Logical empiricism is associated with the Vienna School of Logical Positivism of the 1920s, and asserted that knowledge only exists through experience. Logical empiricism is therefore an attack on more metaphysical positions, and sought to deny the existence of anything that could not be measured and verified. Such a position therefore assumed that events described by data were physical in origin, and consequently that a unified approach to social and natural sciences could be achieved (see "naturalism").

Logical positivism is also associated with the Vienna School, and was the scientific method linked to logical empiricism. The key purpose of logical positivism was the search for verification of patterns observed in datasets, and the assumption that such verification indicated invariant laws that determined the relations between observable facts.

Naturalism is the approach to scientific discovery that seeks to identify prospects pertaining to the innate "nature" of subjects. An implication of this approach is the assumption that the scientific method applied to "natural" sciences (those generally looking at biophysical entities) could also apply to "social" subjects such as social behavior.

Critical rationalism is the name often given to the approach of Karl Popper in the 1950s and 1960s as a critique of logical positivism. Critical rationalism sought to replace verification with falsification as the basis of inference. Falsificationism, as it is sometimes also known, was argued to be a more powerful basis for science because it allowed the testing of generalized *theories* about reality, rather than simple observed trends, and consequently allowed the replacement of apparent descriptions of reality with more complex propositions about the structuring and causes of reality. Although Popper's approach is quite different to the early approaches to "positivism," it is still popularly referred to as "positivist."

Sources: Fetzer and Almeder, 1993; Morrow, 1994.

as the means of inferring statements about reality. Falsification was stronger as a means of inference than verification because it did not seek to summarize perceived trends in existing measurements but instead allowed the proposition of a *theory* about how reality is structured, and which might apply to all potential datasets, and also infer reasons (or causes) for the existence of such trends. The purpose of the scientific method under Popper's system of so-called critical rationalism was therefore to test hypotheses, or theoretical propositions about reality, which may be falsified by each new sample. Failure to find sufficient evidence to reject the hypothesis therefore allowed the theory to continue unchallenged until new data provided new opportunities for showing the inadequacies of one theory, and therefore allowing alternative explanations of phenomena (Morrow, 1994: 69–71). Although the falsificationism of Popper is essentially a criticism of both early positivism and the logical positivists, many popular debates still refer to Popper's technique as another form of "positivism."

There is, of course, insufficient space in this study to review all the different arguments relating to positivism, whether defined in terms of the Vienna School or Popper (see Harré, 1986; Lynch and Woolgar, 1990). But it is important to acknowledge that the methods of explanation achieved under either approach indicated a belief that "laws," or apparent causal connections between different objects, could be found from such research, and that such "laws" were considered both truthful and universal until further evidence challenged them.

Orthodox science, therefore, is highly ambitious about its ability to make generalized explanations, but the criticisms of Popper also began to indicate an awareness of the influence of social influences on such explanations. Under falsificationism, Popper acknowledged that new empiricism would take place from the perspectives of existing theories, and accordingly that scientific inquiry took place in a process of "conjectures and refutations" (Popper, 1962), rather than simply through "blind" empiricism, or the collection of "facts" as they emerge. Furthermore, the scientific practice of Popper also questioned the ability to make statements about reality that were absolute, eternal truths because each theory held as the best explanation at any one time could be falsified quickly in accordance with new data and test results. Such practice is far from the early foundationalism of early empiricists, and instead highlights the social dynamics underlying empirical inquiry, and the decisions by which new results are seen to falsify old theories. The assumption that science can produce politically neutral generalizations about environmental explanation, despite the acknowledged importance of such social direction in its practice, is a further tenet of "orthodox" science, and needs further examination.

The social regulation of science

One of the key tenets of "orthodox" science, therefore, is the belief that the scientific method – or hypothesis testing involving careful sampling – can ensure accuracy in understanding reality. Yet this approach too is shaped firstly by a set of social codes adopted by scientists, and, second, on the broader social and political influences under which science is conducted.

Much orthodox science is assumed to proceed under certain assumptions of logic, consistency, fairness, and impersonality that reflect science's perceived position as a politically neutral mechanism for understanding reality (see Hess, 1997: 53–60). In particular, the writings of the American sociologist Robert Merton (1973) in the essay, "The normative structure of science," originally published in 1942 as "Science and technology in a democratic order," described science as an adequate, reliable, and progressive social institution. The prescriptions for scientific practice contained in these writings became known as "Mertonian norms" and included four key elements. First, science is universal: truth claims can be proposed regardless of the scientist's race, nationality, and religion. Second, science is communal: science's findings should be seen as common heritage to be shared with all (without disregarding the scientist's claim to intellectual property rights). Third, science is disinterested: scientists must subject their work to the rigorous scrutiny of fellow experts (rather than lay people), and must adjust writings accordingly. Fourth, science is organized skepticism: science should always seek to challenge institutionalized beliefs such as emanating from religion or politics (ibid.: 55–56).

Such "norms" have, of course, been the source of much bitter disagreement. Many orthodox scientists have sought to defend these norms as either adhered to rigorously by career scientists, or at least as underlying the principle of conducting scientific research. Other researchers, notably sociologists of scientific knowledge, have argued that these norms are regularly broken. For example, many scientists have experienced intense trouble in getting work published or funded because of the institutional political interests from competitors who hold influential positions in academic journals or funding organizations. Also, much science can barely be said to be "universal" because of the different relevance to, or participation from, different social groups in the formulation of theories and apparent "laws." Indeed, the discussion of environmental orthodoxies in Chapter 2 demonstrated that many statements coming from orthodox science, such as the supposedly universal Universal Soil Loss Equation, are indeed anything but "universal" in spatial terms.

There is again insufficient space in this study to list all the various debates that have considered the pros and conservation of Mertonian norms (see Knorr-Cetina and Mulkay, 1983; Latour, 1987; Morrow, 1994; Hess, 1997). Indeed, the principles of supposedly free debate and criticism of science are also apparent in the "Open Society" discussed by Popper

(1945), or the "Republic of Science" of Polanyi (1962), which were conceptualizations of public debate and scientific progress based upon the belief in conjecture and refutation of ideas and theories. (Indeed, such approaches may be considered to be consistent with general political trends during the Cold War in which free-market ideas and politics were seen counter-posed to the central planning of Socialist economies.)

It is worth mentioning, however, that the debate about the relationship of practices of peer review and internal regulation of science among scientists has been a major part of the bitter controversies known as the "science wars" (see Chapters 1 and 9). Many defenders of so-called "orthodox" science have stringently criticized attempts to highlight variances from Mertonian norms. Some have developed further guidelines for how "science" should proceed (e.g. Bunge, 1991; see Box 3.2). The objectives of such guidelines have been to ensure that knowledge generated by "science" can be seen to be more accurate and trustworthy than many "pseudo-sciences," or social sciences that do not adopt a regularized knowledge-gathering system similar to the "scientific method." But it is not always clear how far such rigorous definitions of science may either exist, or be feasible, particularly when dealing with environmental "problems," which, by definition, involve human processes of identification and experience.

For example, the renowned philosopher of science, and defender of (orthodox) science against "pseudo-sciences," Mario Bunge, wrote this, which few people would find disagreeable:

> It is foolish, imprudent, and morally wrong to announce, practice or preach important ideas or practices that have not been put to the test or, worse, that have been shown in a conclusive manner to be utterly false, inefficient, or harmful.
>
> (1991: 256)

But, as listed in Box 3.2, Bunge then went on to list ten criteria for the achievement of successful and accurate science that may seem difficult, if not impossible, to achieve. The ten criteria include, first, the existence of "a social system composed of persons who have received a specialized training, hold strong communication links amongst them, and initiate or continue a tradition of inquiry (not just of belief)" (ibid.: 246). The following requirements then include a society that supports such a network of scientists; the concern with real entities rather than "floating ideas"; the adoption of up-to-date and transparent theories and methods; and the long-term aim to use research and testing as a means to achieve generalizable theories and hypotheses. Bunge also describes science as: "the ethos of the free search for truth, depth and system (rather than, say, the ethos of faith or that of the bound quest for utility, profit, power or consensus)" (ibid.: 247).

Critics of Mertonian norms claim such assertions about the rigor and

Box 3.2   A definition of "science," after Bunge (1991)

Despite the common usage of the word "science" in popular debates, there is still great uncertainty concerning what this term actually means. Some theorists have strongly argued that "science" needs to be defined very strictly in order to avoid its misuse by other forms of inquiry. These other forms of inquiry could include social "science," or "pseudo-sciences" such as astrology.

The eminent Argentine-born philosopher of science, Mario Bunge (1991), used the following definition to explain what may be considered "factual science," or a form of science that can be considered a rigorous and analytical form of inquiry. Indeed, defenders of orthodox science have since referred to, and supported, this definition during the so-called "science wars" (e.g. Levitt, 1999: 357). Many scholars of science studies, on the other hand, would consider this definition unrealistic and typical of how "orthodox" science fails to recognize the social conditions that underlie scientific inquiry.

The definition is as follows.

According to Bunge's definition, "R" may be considered to be "a family of factual scientific research fields" representable by the following conditions:

$$R = <C, S, D, G, F, B, P, K, A, M>,$$

where, at any given moment in history,

i   "C," the research community of "R" is a social system composed of persons who have received specialized training, hold strong communication links amongst themselves, or continue a tradition of inquiry (not just of belief);

ii   "S" is the society (complete with culture, economy, and polity) that hosts C;

iii   the *domain* "D" of "R" is composed of real entities, rather than freely floating ideas;

iv   the *general outlook* "G" of "R" consists of (a) a realist ontology (rather than ghostly or miraculous things); (b) a realist epistemology; (c) the "ethos of the free search for truth, depth, and system (rather than, say, the ethos of faith or that of the bound quest for utility, profit, power or consensus)";

v   the *formal background* "F" of "R" is a collection of up-to-date mathematical and logical theories;

vi   the *specific background* "B" of "R" is a collection of up-to-date and reasonably well-confirmed data, hypotheses, and theories, plus research methods;

vii   the *problematics* "P" of "R" consists of cognitive problems concerning the nature (i.e. laws) of the members at least of "D";

viii   the *fund of knowledge* "K" of "R" is a collection of up-to-date and testable (although not final) theories, hypotheses, and data compatible with those in "B" and obtained by members of "C";

ix the *aims* "A" of the members of "C" include *discovering or using the laws*, trends and circumstances of the "D"s, systematizing into theories about "D"s;

x the *methodics* "M" of "R" consists exclusively of *scrutable* (checkable, criticizable) and *justifiable* (explainable) procedures.

Bunge also noted that the scientific research field represented by "R" should have associated research fields that may be seen as complementary. Furthermore, the last eight components of "R" listed above should be expected to evolve as scientific inquiry proceeds.

This rather rigid list of requirements for so-called "factual science," of course, indicates how difficult it is to achieve scientific findings that may be considered consistent with the expected frameworks of this approach. Furthermore, the list calls for a strong disciplinary following with regularized training and objectives, and for the rejection of all research ambitions that might compromise the "free search for truth" such as "profit, power, or consensus." Debates in science studies have argued that many of these assumptions and prescriptions are unrealistic and inoperable.

Source: Bunge, 1991: 246.

moral strength of scientific inquiry simply cannot be expected or upheld. These criticisms refer to the way in which scientific knowledge is either produced or regulated within social systems of scientists each with their own personal and institutional differences and objectives. In addition, and probably more important, much debate has also pointed to the broader social and political influences on science, in enhancing paradigmatic stages in the perceived purpose of science, and hence the empiricism and resulting "laws" that result. According to Popper, the transition from verifying perceived "laws" (under logical positivism) to the testing of hypotheses (under falsificationism) implied that the statements within each theory would shape scientific research. The development of a strong theory, which other researchers considered important, would therefore result in further research to test and expand this theory.

Later sociologists and philosophers of science have, of course, elaborated such directional influences in science. Most notably, Thomas Kuhn (1962) introduced the term "paradigm" to indicate a generalized trend of research that followed the investigation of a particular theme or theory. Scientists working within this paradigm could be described as following "normal science" because this was consistent with pre-identified objectives. Imré Lakatos (1978) later refined this analysis of paradigms to include research programs, as a more politically focused examination of scientists as actors in ensuring the focus of science on particular objectives. Such writings have clear parallels with the stated problem of environmental orthodoxies, in how certain simplistic conceptualizations of environmental causality can continue despite the growing evidence to discount them. This chapter has only introduced such themes. The further implications of these debates for environmental explanation and a "critical" political ecology are discussed more fully in Chapters 7 and 8.

Building "laws" of nature

One of the main products of so-called orthodox science is the apparent ability to make generalizations, or universalistic statements, about either the nature of biophysical or social entities, or the causal links that exist between them. Indeed, science is at its most powerful – as either a tool of explanation, or as an influential institution – when it summarizes apparently complex relations to a succinct statement or formula such as $E = mc^2$. Some other, less elegant formulas may also result, such as the Universal Soil Loss Equation, or some other, less formulaic expressions of knowledge such as the statement that "water freezes at zero degrees Centigrade," or that "deforestation causes erosion."

There are clearly many differences in the accuracy and universality of such statements, either in terms of how they were originally formulated, or in how they are generally meant. Yet such statements share a common impact of being used to describe some generalized statement of causality that are often the basis of policy, and which many people, if challenged, would find hard to deny. As such, they may be considered effective "laws" that have been produced by scientific research, and which may be used uncritically. In principle, "laws of nature" are considered universally true, not just occasionally or generally (Harré, 1993). Yet, of course, there are great difficulties in establishing a "law" of nature, and in applying apparent "laws" to different circumstances.

First, there are many different types of apparent "law," which may not qualify as universally true. For example, over the years, "laws" have been interpreted variously as summaries of apparent trends in datasets; statements about the dispositions of things; or also the prepositional character of a theory, which, of course, may be falsified. All of these interpretations may be adopted and used as apparent "laws," even though they may not be considered as universally and totally acceptable.

One of the most powerful tendencies that influence the identification of apparent laws, and then their adoption within societies as "laws" is the assumption of repeated patterns. Hume (1711–1776), one of the most influential empiricists, wrote in 1739:

> Regularities impress themselves upon us to such an extent that we come to expect that events will continue to follow the patterns that they have already displayed. Thus we come to expect an event of the same type as the effect on the occurrence of an event of the same type as the cause.
>
> (Hume, 1739, in Harré, 1993: 81)

Yet this tendency also raised concerns of later scientists, even the early positivists. In 1894, Mach wrote:

The communication of scientific knowledge always involves description, that is, a mimetic reproduction of facts in thought, the object of
which is to replace and save the trouble of new experience. Again, to
save the labor of instruction and acquisition, concise abridged description is sought. This is really all that natural laws are.

(Mach, 1894, in Harré, 1993: 40)

The approach to both the early positivists and the Vienna School,
however, was actualism, or the perception of trends in datasets and the
inference (through verification, under logical positivism) of such trends
elsewhere. Falsificationists, under Popper, sought similar inference, but
through the more imaginative, and prepositional use of theories about
causality, rather than by simply using perceived trends in measurement.
Yet both mechanisms of inference involve statements about reality that
are based in existing knowledge sources or datasets. Moreover, they
assume such datasets are both representative of the wider reality not yet
measured, and free of social influences that reduce their representativeness. Such assumptions, as discussed above, may be difficult to uphold,
particularly for environmental "problems" that include social framings and
the experience and needs of diverse people.

Research by historians of science has also indicated further social influence on apparent laws of nature. Some well-known laws, such as Boyle's
law stating the proportionality of volume to pressure, were constructed
with a surprising level of social influence on experimental technique. Such
influence affected the selection and manipulation of equipment used
during measurement; the noting of results under experimental conditions;
and the negotiation and politicization of results in the scientific societies of
seventeenth-century London (see Shapin and Shaeffer, 1985). Modern-day
anthropological work by Latour and Woolgar (1979) also showed the practice of science in laboratories and the translation of "laws" observed under
experimental conditions to the outside world, based on the assumption
that the same conditions exist there as in the laboratories. A further level
of socialization of science may also lie in how scientific techniques of an
age may also reflect the underlying social and political norms of the time.
Critics have suggested, for example, that the rigid rejection of metaphysics
under logical empiricism of the Vienna School may have been typical of
events in Germany and Austria between the two world wars and the contemporaneous rise of Nazism, in the same way that Popper's "Open
Society" and "conjecture and refutation" reflected free-market idealism
after the Second World War.

The implication of such historical, and often ethnographic, analysis is
that apparently politically neutral and universal statements from science
reflect a variety of social influences. The importance of such influences is,
of course, still controversial. In the case of Boyle's law, for example, the
relationship between pressure and volume remains plausible today, mostly
because no alternative dataset or theoretical challenge has been experi-

enced, even though the directly proportional relationship proposed by the "law" is not always experienced because of a wide variety of apparently circumstantial influences. Yet some other apparent "laws" – such as the workings of the Universal Soil Loss Equation, or the beliefs in some basic environmental orthodoxies such as "deforestation causes soil erosion" – are more obviously inaccurate in a wider variety of contexts. There are, of course, clear conceptual and semantic differences between "laws" that relate to generic physical properties such as "pressure" and "volume" on one hand, and entities such as "deforestation" and "erosion" that are afforded many more types of local context and meaning. Yet it is not clear if such differences have any impact on the practical application of such generalized statements in environmental management, when in practice such statements are taken as "laws."

The question about the evolution and adoption of generic "laws" – or environmental orthodoxies – in environmental management are discussed throughout the book, and especially in Chapters 4 and 8 (where some of these dilemmas are addressed). But the point of the initial discussion in this chapter is to highlight the basis of many environmental orthodoxies upon the principles of "orthodox" science, and especially through the tendency for orthodox science to seek to produce generalized "laws" of nature. In particular, environmental generalizations have been based upon the initial use of datasets about environmental problems that have only reflected partial information in both biophysical terms and in regard to how such problems have been evaluated by different social groups. Such datasets, however, have been the basis for further sampling and inference using the scientific method, and the eventual production of generalized statements about environmental degradation that might be applicable to all people or all locations. As Castree wrote:

> While we should applaud the Realist belief in real social and natural events beyond the immediate horizon of epistemology, Realists ... do not *problematize* the assumption for that belief nor do they seem alive to the consequences of its unreflexive invocation. Rather, they work *within* it, worrying more about the problems of *accessing* the real as if the real is just there.
>
> (1995: 39, emphasis in original)

The rest of this chapter now assesses the problems of this framework for environmental explanation according to new thinking about the non-equilibrium nature of ecological change, and also seeks new directions for building a less universalistic basis for generalization.

The challenge from non-equilibrium ecology

The preceding section summarized how so-called "orthodox" science has been based on the search for universally applicable "laws" of nature, using methods that give the appearance of objectivity and accuracy. These object-ives can be criticized, however, for failing to recognize how such "laws" may reflect the norms and experiences of the people who make them. Yet, at the same time, there are also debates in ecology that challenge the use of such universally applicable "laws" and generalizations. These debates have been called "non-equilibrium" (or non-linear) ecology, and they have many implications for political ecology and environmental explanation.

Non-equilibrium ecology may be defined as an approach to ecological explanation that emphasizes the variable, and often chaotic, nature of change within ecological systems, at a series of spatial and temporal scales. Such variability means that some commonly held notions of ecological stability, gradual evolution, or a "balance of nature" are no longer tenable as accurate representations of environmental change and ecosystems (see Holling, 1973, 1979; Wiens, 1977; Botkin, 1990; Zimmerer, 2000).

Since the emergence of ecology as a topic for debate in the nineteenth century, much analytical discussion has assumed that ecosystems may be inherently stable, or exhibit homeostasis, or the self-regulation of ecosys-tems. George Perkins Marsh made one classic statement of early ecology in his groundbreaking work, *Man and Nature* in 1864:

> Nature, left undisturbed, so fashions her [sic] territory as to give it almost unchanging permanence of form, outline, and proportion, except when shattered by geologic convulsions; and in these compara-tively rare cases of derangement, she sets herself at once to repair the superficial damage, and to restore, as nearly as possible, the former aspect of her dominion.
>
> (Marsh, 1864: 29)

Later writings in ecology were also clearly influenced by Darwinian notions of evolution as well as the concepts of balance and equilibrium (Simpson, 1960). William Morris Davis' geomorphological model of land-scape evolution and "stages of maturity" (dating from the 1880s) reflected influences from both evolution and stability. Frederic Clements' concept of plant succession (first published in 1916) also explained vegetative change as a gradual progression toward some pre-defined "climax" vegeta-tion, or equilibrium point. As Odum wrote: "[Ecological succession is] an orderly process of community development that is reasonably directional and, therefore, predictable ... [succession] culminates in a stabilized ecosystem" (1969: 262, in Botkin, 1990: 54).

Such concepts of gradual transition and equilibrium have, of course, played an important role in the formation of much environmentalist, or conservationist, ideology. Odum and other writers emphasized these con-cepts in the early political writings on ecology as a "subversive science" (see Chapter 1). Many other environmentalists still use these concepts today. Lester Brown, the co-founder of the World Watch Institute, for example, refers to "the fragile balances of nature" (Brown, 2001: 79).

Yet, despite such widespread repetition of notions of ecological equilibrium, much research and debate relating to non-equilibrium ecology has questioned these assumptions. First, research on ecological *processes* has indicated a greater agency to changes in underlying inputs to ecosystems than previously thought (e.g. Holling, 1973; Wiens, 1977). As a result, ecosystems may change more rapidly because of changes in processes considered (at least partially) external to the physical landscape or form known as the ecosystem.

Second, the flux *within* ecosystems has become more widely understood, most obviously as a result of research on disturbance (or patch dynamics) in forests (e.g. Wu and Loucks, 1995), or changes in vegetation or soils occurring in pastoral systems in drylands (e.g. Turner, 1993; Scoones, 1994; Dougill *et al.*, 1999). The implications of such flux are greater apparent variability in ecosystems in terms of spatial variations of species, soil fertility, or vegetation structure; and also in terms of temporal variations of these characteristics. Episodes of ecosystem disturbance resulting from, for example, fire, might have been identified as degradation under equilibrium-based ecology. Under non-equilibrium ecology, such disturbances may be seen as a longer-term or spatially wider form of change within the ecosystem, even though such changes may be considered to look like forms of degradation to many observers. Ecosystems may indicate a number of stable states, identifiable at a variety of spatial and temporal scales. Furthermore, the significance of different disturbances (such as incidences of fires, storms, floods, or disease epidemics) in explaining ecological change will vary according to the length of time and the area over which change is analyzed (May, 1974; Pickett and White, 1985; Adams, 1997).

Third, many scholars have also questioned how far the notions of ecological "balance" and "equilibrium" might result from *social norms* about how "nature" should be. For example, Denevan (1989) has questioned the "myth" that the Americas were "pristine" before the arrival of Columbus in 1492. Concepts such as "wilderness" or "untouched nature" may also be inaccurate because they suggest a state of unchanging equilibrium that may not exist. Indeed, according to Cronon (1996: 47), "Wilderness ... is quite profoundly a human creation." These comments have had implications for debates about environmentalism (see below, pp. 108–109). But they have also suggested that many time or space scales that have been identified as exhibiting "equilibrium" in the past have been identified because they are of social significance, rather than representative of how "nature" really works. Such factors may also influence the identification of specific "regions" or "eco-zones" where ecological change is analyzed, including terms such as "watersheds," "forests," or similar units where policies have been directed. Similarly, they may also indicate specific zones identified as "unspoilt" by powerful or colonial landowners in contrast to the experiences of other residents (e.g. Neumann, 1996, 1998).

These debates within non-equilibrium ecology have had immense implications for environmental explanation and orthodox science aiming for "laws" of nature. Most importantly, these themes indicate, again, that the underlying events and processes occurring within ecological systems are far more complex than commonly suggested by initial explanations offered through the construction of observable "laws." They also show how attempts to build laws about ecological change have reflected social and political factors of people constructing the laws. Such criticisms are similar to the general criticisms of the frameworks of orthodox science listed above (pp. 57–61).

But, non-equilibrium ecology also weakens some of the common beliefs within popular environmentalism about ecological fragility or the irreversible impacts of disturbance on ecosystems. Botkin (1990: 156) noted that non-equilibrium ecology was a "Pandora's box for environmentalists," and Adams wrote:

> Gone, therefore, are the days when conservationists could conceive of "nature" in equilibrium and hence portray human-induced changes in those ecosystems as somehow "unnatural." Gone too are comfortable certainties about naturalness and the management regime needed to sustain it.
>
> (1997: 286)

Box 3.3 summarizes some of the implications of non-equilibrium ecology for conservation practice, and some of the dilemmas for agencies that were established because of beliefs about ecological stability and fragility. Of course, these criticisms do not imply that there is no need to be concerned about environmental degradation. But they do raise questions of how supposed "laws" of environmental degradation (such as environmental orthodoxies) have been constructed without sufficient understanding of factors influencing ecosystems, or of how social norms may influence such laws. Second, they also raise questions of how such apparent criticisms of existing explanations have not been adopted by many governments, environmental agencies, or academic disciplines.

Zimmerer raises a significant political question for the continued adoption of many environmental policies based upon notions of ecological equilibrium: "Many abuses that have stemmed from conservation policies are rooted in the belief, held by policymakers, politicians, scientists, and administrators, of a balance or equilibrium-tending stability of nature" (2000: 357).

Furthermore, he notes that such changes have also been resisted in certain academic disciplines or professional sciences such as forestry:

> Oddly, the proliferating mass of land-use assessments [of forest patch dynamics] has remained slow to account for the important role of such disturbances. This halting awareness may be due to the especially strong legacy of concepts from equilibrium ecology or system ecology

Box 3.3 Rethinking ecological equilibrium in British conservation

The insights of non-equilibrium ecology have profound implications for the activities of environmental conservation and attempts to manage landscapes. Since the establishment of formal institutions for nature conservation in Great Britain and North America in the late nineteenth century, much practical conservation action has been associated with a conservation ideology seeking to protect "natural" places from human action. One early writer described the movement as for "all lovers of the wilderness, all worshippers of uncontaminated nature" (Lankester, 1914: 35). Such statements, of course, do not lie easily with insights from non-equilibrium ecology that highlight the prevalence of disturbance, change, and chaos in landscapes. As Botkin (1990: 10) asked: "how do you manage something that is always changing?"

In Great Britain, the advice of ecologists such as Julian Huxley (1947) about the potential risk to natural landscapes led to an act of parliament in 1949 that created the new organizations of Nature Conservancy and the National Parks Commission. In the words of W.M. Adams (1997: 279), "these debates left British conservation an idiosyncratic institutional framework, with a somewhat perverse organizational divide between the conservation of species and ecosystems ... and the conservation of natural beauty and open countryside." This divide encouraged the use of ecological scientific discourse to legitimize the protection of specific landscapes, or the perception that "ecology" was the science of "conservation," or the protection of landscapes considered beautiful. Indeed, under the 1949 National Parks and Access to the Countryside Act, the Nature Conservancy could establish National Nature Reserves, and "Sites of Special Scientific Importance" (later "Interest"). The assumption that "scientific interest" could justify the value of ecosystems and species persisted in later legislation, such as the Wildlife and Countryside Act of 1981. Yet the apparent unity between "ecology" and "conservation" did not always seem logical to everyone. One critic of conservation policy in Britain wrote, "ecological research is evidently to be merely the handmaiden of conservation" (Williams, 1958: 87).

The implications of this history are that views about "balance" in landscapes, and the science of ecology, co-evolved to be mutually supportive.

> Using ecology, conservationists have diagnosed the pathologies of nature and prescribed remedies to make it regain its rightful form ... the words and concepts borrowed so blithely by ecologists from engineering to describe nature, such as ideas of thermodynamics, energetics, equilibrium and control, were absorbed uncritically by conservationists.
> (Adams, 1997: 284, 285)

A further implication is that conservation now needs to reconsider its historic alliance to notions of order and stability, and instead find ways to integrate ecological change with more flexible forms of environmental protection. Such "soft engineering" may already be occurring in practices such as allowing limited river restoration (including flooding, where feasible); or in the managed retreat of coasts under erosion. These approaches require rethinking the purposes of conservation among the public, and within official conservationist organizations. Source: Adams, 1997.

qua systems. Influential models of livestock-carrying capacity and explanations for the biodiversity of economic plants, for example, persist in overlooking the prevalence of natural ecological disturbances in the rangeland and agricultural land of poor countries,

(Zimmerer and Young, 1998: 15)

Why do concepts of equilibrium ecology still persist so strongly in many academic and policy circles? In many ways, this is the same question posed about so-called environmental orthodoxies – or apparently inaccurate explanations of environmental degradation in general – discussed in Chapter 2. The following chapters in this book discuss these themes further. In particular, Chapters 5, 6, and 7 focus on the use of science as a further means of politics by a variety of political actors. The rest of this chapter, however, considers alternatives to the universalizing "laws" of nature that have been so problematic in explaining environmental problems at different times and places. The following section may be seen, alongside discussions in Chapter 4, to outline various ways in which environmental explanations may be made more diversified and flexible in order to reflect social and ecological diversity – and hence avoid the problems of environmental orthodoxies.

Diversifying "laws" of nature

The preceding discussions showed that environmental explanations based on universalistic statements of causality and concepts of equilibrium and a teleological progression to climax are highly problematic. The explanations are problematic because they impose visions of order and predictability upon complex biophysical processes and diverse evaluations of environment that say more about the social practices of making science than the ability to know and explain biophysical reality in its totality. Yet despite the growing consideration of non-equilibrium ecology principles among ecologists, and the increasing evidence that much existing environmental explanation is either wrong or socially unjust, these principles of environmental explanation are still deeply entrenched within many environmental organizations and political debates. Despite the evidence against them, the principles of universality and equilibrium (seen in statements such as "deforestation causes erosion" or "desertification is a dangerous threat resulting from poor land management") in environmental debate are still generally treated as "laws," and much environmental research still sees the identification of such guiding "scientific" principles as a key objective.

There is consequently a need to go beyond these kind of inadequate generalizations and instead develop bases for explanation that are both biophysically more accurate and socially more just. Trudgill and Richards (1997) and Massey (1999) describe this as a search for greater "subsidiarity" in environmental generalization. Yet achieving this greater diversity is a difficult task. On one hand, diversification requires engaging with the philosophical principles upon which "laws" and generalizations are made. On the other, it also implies seeking greater political representation of different social viewpoints and experiences of environmental change. In essence, both of these requirements indicate a more relativist stance to scientific generalization, but one that seeks to diversify the basis for making realist statements about environmental causality, rather than dissolve any attempt to make causal statements simply to the different viewpoints adopted.

"Relativism," like "positivism," is often taken as an extreme position in the debate about science, and similarly is often used as a term with which to dismiss opponents (Proctor, 1998). The term is usually meant to imply the philosophical position in which each truth claim is reduced simply to different viewpoints, and consequently that any attempt to construct more generalized "laws" or explanations is futile. Harré and Krausz define relativism in this strong form:

> Relativism takes its strongest form when the natural and social worlds and their phenomenal appearances are themselves claimed to be products of the language and other culturally distinctive practices of the persons who inhabit them.
>
> (1996: 9–10)

(A further discussion of relativism, and its usual counterpart, realism, is contained in Box 3.4.)

This particular form of relativism, however, is somewhat stereotypical and unusual. Harré and Krausz (ibid.: 23–24) list four different varieties of relativism. Semantic relativism is relativity according to the meaning of language (for example, the different interpretation of similar words). Ontological relativism is the relative existence of conceptual systems (such as the varied belief in things such as witches). Moral relativism is the adoption of different ethical or moral guidelines. Aesthetic relativism is the adoption of varying values and appreciations of similar objects. These varieties can then be evaluated according to the degree of relativism adopted. At the weakest level is a denial of universalism – or that the "laws" of explanation cannot be applied universally. Next is a denial of objectivism – or a belief in no objective viewpoint. Finally, there is a denial of foundationalism – or that there is only one, permanent foundation for assessments of meaningfulness. Relativism is at its weakest when universalism alone is denied. Relativism becomes stronger when objectivism and foundationalism are also denied.

These different levels of relativism are important when seeking to diversify the universalistic "laws" of nature that have underlain much environmental explanation and environmental orthodoxies. The discussion of environmental explanation to date has shown that apparent "laws" result from both social framings, and a partial experience of biophysical

Box 3.4 Realism, relativism, and constructivism

Realism is the belief that science can reflect underlying structures of causality or ontology. It is also the belief of *necessity* in nature, or the ability to make theoretical propositions similar to: "if *x* then *y*." Realism is an assumption underlying all orthodox philosophies of science, and sometimes (under the "strong" school of Realism) has been associated with the assumption that scientific statements *are* accurate representations of the "real" world. More recent debates (such as in Critical Realism) have adopted a less orthodox approach to realism, and instead have defined realism as the belief in underlying causal structures, without the assumption that such causal structures can be identified through science. (Often orthodox, so-called "naïve" approaches to science are called Realist with a capital "R," and alternative less orthodox approaches are spelt with a small "r," or referred to as the "weak" school of realism.)

Relativism is the belief that statements about reality are not universal, and are influenced by the viewpoint, assumptions, and ambitions of the individual or organization making statements. Relativism is often described simplistically as the abandonment of all hope of achieving any form of scientific explanation or causal statements beyond the level of individual perspectives. But relativism can also imply a variety of different positions referring, least controversially, to the rejection of universal truth statements, and, more extremely, to the denial of objectivism or any unitary purpose of explanation.

Constructivism is the belief that statements about biophysical reality are shaped at least in part by human influences. Constructivism is often referred to stereotypically (and inaccurately) as equivalent to the "imagining" of things, with no attention to whether such things really exist. Instead, it is more accurate to refer to constructivism as the appreciation of a wide variety of social influences on how reality is presented. Such influences may include references to language (the semantic and semiotic influence of language on the representation of physical objects); the historical influence of past societies or individuals on our comprehension of biophysical objects (such as the statement that our understanding of soil erosion is in part constructed by the experience of the Dust Bowl and Universal Soil Loss Equation); or the current cultural context within which physical concepts are presented and consumed (for example, the perceived importance of environmental "crisis" as an alleged indication of the failings of economic growth and modernity). Adopting a constructivist position usually implies agreeing with principles of relativism. But, ironically, neither relativism nor constructivism necessarily implies a rejection of realism (although they do imply a rejection of orthodox Realism), and so it is possible to be realist, relativist, and constructivist at the same time.

Source: Harré, 1986; Castree, 1995; Harré and Krausz, 1996;
Sayer, 1997, 2000; Proctor, 1998.

reality, or sampling, that have been conducted within these social framings. Seeking a more diversified approach to "laws of nature" may therefore engage with both the biophysical complexity as revealed by past scientific methods, or also with the role of social framings in shaping that science. The interface between social trends and constructions of biophysical reality is a vast topic (see Archer *et al.*, 1998; Gieryn, 1999), and it is dealt with throughout this book. But three broad themes in approaching this dilemma may be identified at this stage, and which indicate a transition from a generally realist to generally relativist position in terms of placing different emphasis initially on the biophysical ordering of nature, and then later on social framings of how nature is seen.

Stratification and emergence

The most realist approach to diversifying "laws" of nature is the process known as stratification and emergence. The approach is based on the Critical Realist writings of Roy Bhaskar, which identify nature as stratified according to various levels of meaning and causality, and the belief that these structures "emerge" to human observers as the result of scientific inquiry.

Bhaskar argued there are different types of structuring of nature. At the most basic, our experience of environmental change can be differentiated into three main domains. The "empirical" domain is the simplest, and refers to those experiences of underlying reality that we can observe and measure. The domain of the "actual" contains more than just measurements of change, but also the events by which environmental change may take place. The most fundamental domain is of the "real," and includes causal mechanisms as well as events and experiences (Bhaskar, 1975: 13). The crucial lesson of Critical Realism in this respect is to appreciate that existing explanations of nature are unlikely to refer immediately to the domain of the real, and consequently it is important to question constantly how far existing explanations may only be based on empirical or actual domains. As Collier wrote:

> There is a tendency in empiricist Philosophy of Science (unavoidable given its actualistic assumptions) to deny the status of explanation to any but the most basic explanatory stratum. Explanations in terms of higher-level mechanisms are seen as mere "explanation-sketches," standing in for explanations not yet achieved. But this misrepresents the development of science.
>
> (1994: 110)

The beauty of this approach is that it advances a powerful framework for understanding how environmental explanations may emerge as the result of partial empirical research and lead to fixed models of causality that do not reflect more complex underlying causes of change. Under stratification and emergence, environmental explanations may be seen as a series of interconnected black-box statements, in which one level of apparent causality may be replaced with a more complex level of understanding, similar to the act of peeling an onion layer by layer. Stratification and emergence also allows a more complex appreciation of time and space scales in environmental explanation, and potentially allows for the reconciliation of apparent paradoxes in environmental observations. For example, the very fundamental statement "water always flows downhill" may be contradicted by the so-called partial area runoff model of stream formation (Dunne and Black, 1970; Dunne and Leopold, 1978) that shows how saturated slopes may allow streams to progress upwards from the base to the top of a slope. Both observations – of water either flowing up or downhill – can be said to be either within the domain of the empirical or actual, and can lead to the development of explanations that summarize these events. But neither causal statement engages sufficiently with the "real" mechanisms of how water forms into rivulets or moves over space. Indeed, both statements, referring to water moving up or downhill, may be used in effect as "laws" in similar circumstances to which they were observed. But neither is designed to look into mechanisms deeper than the superficial perceived events of water movement as it affected humans at the time each statement was developed.

As a consequence of such different levels of explanation, Bhaskar also argued that the stratification of nature could also refer to a hierarchy of academic sciences. Physics, for example, could be seen to be more basic than chemistry, which is more basic than biology, which is more basic than human sciences, and so on (Collier, 1994: 107). The structuring of explanation under Critical Realism (as discussed by Bhaskar) consequently acknowledges the subjugation of existing explanations to human perspectives. But the underlying implication of stratification and emergence is that biophysical reality is still essentially knowable through a long-term process of uncovering different mechanisms.

Semantic realism

Semantic realism represents a more socially constructed view of generalization and truth than that taken under stratification and emergence. Semantic realism can refer to a variety of theoretical positions, but is based upon the belief that it is more useful to analyze the social orderings and basis for truth statements to be made, rather than in assuming that the truth values are there to be discovered. Semantic realism does consider that causal statements can be made about complex biophysical processes and objects. But the construction of such causal statements depend on the ordering of complex events and experiences ("facts") into units of meaning that have a clearly defined starting place, process, and end result. Indeed, such structures are often called sentences, and it is a tenet of semantic realism that truth claims can only be made in sentence form. The individuals or organizations that create such sentences therefore have greater power over the construction of supposed truth. As Bertrand Russell wrote:

> On what may be called the realist view of truth, there are "facts," and there are sentences related to these facts in ways which make the sentences true or false, quite independently of any way of deciding the alternative. The difficulty is to define the relation which constitutes truth if this view is adopted.
>
> (1940: 245)

Under this approach, it is possible to understand apparent contradictions revealed by research on environmental orthodoxies as the conflict between different truth sentences (see also Tennant, 1997). For example, Blaikie and Brookfield's statement (1987) that "one farmer's soil erosion is another's soil fertility" indicates different evaluations of the process of soil movement according to its effect on soil fertility. Similarly, the role of many historic shifting cultivators in increasing biodiversity by introducing regular forest disturbance through fire can also lead to different truth sentences. Sentences could refer to either a positive or destructive influence of shifting cultivation on "environment" depending on the value attributed to forest disturbance. For example, if "forest" is defined as a dynamic repository of biodiversity (if affected by regular disturbance), then some forms of cultivation may be seen as positive. If forests are seen as ecosystems that need to be preserved in its current state, with particular emphasis on large trees and key indicator species, then the kind of disturbance offered by shifting cultivators may be seen as destructive.

The approach to explanation adopted by semantic realists has been tailored further by Searle (1995) into a distinction between so-called "brute facts" and "institutional facts." Under this conceptualization, "facts" may be classified according to the meaning attributed to their social functions, and the extent to which such functions are shared by all society. For example, a pen may be described in brute fact terms as a piece of plastic because this is the most basic unfunctional description of it. Calling it a pen implies that it has a specific function of writing. But the pen could equally well be used as a backscratcher or gambling chip. Such descriptions that imply function are considered "institutional facts" as they imply a shared specific meaning. Again, this approach can be used to translate the brute fact of "soil movement," for example, into either "soil fertility" or "soil degradation" as discussed by Blaikie and Brookfield (1987). Yet "erosion" (most simply meaning the weathering and removal of matter) is still generally interpreted as a negative environmental experience.

The power of different words to imply particular meanings that may not be shared by all has also been discussed under the context of semantic realism. Earlier work by Poincaré (1958) (also in Harré, 1993: 49) distinguished between "brute facts" and "scientific facts" as a way to re-describe (or re-label) observed events (or actuality) into a more formal statement of mechanism or theory. As a result of this co-evolution of scientific language with scientific understandings, the language used to express a particular "problem" or "cause" are often taken to be synonyms with the causes and problems themselves, even when the initial attempts at explanation were superficial. The example of "erosion" as a word describing a process, which is also taken to imply a "problem," and the assumption that it is always a problem, is an example of this semantic influence on environmental understanding.

Transcendental realism

A further, and even more socially oriented, approach toward environmental generalization refers to transcendental realism. Transcendental realism again comes from the writings of Bhaskar, and can be summarized simply as the social basis under which empiricism is undertaken. For example, much environmental research to date has identified soil fertility and soil erosion as key topics, and many statements of causality have been established. The transcendental approach to this scientific inquiry would be to point out that such empiricism and analysis can only proceed under the necessary precondition that such measurements are meaningful, and that soil fertility is something that needs to be measured. Such presuppositions, of course, reflect a variety of social and political factors that are not apparent when the debate about soil, or the findings of research, are stated as "fact." The approach of transcendental realism is therefore a step toward relativism and phenomenology because it challenges the foundationalist basis of existing systems of scientific inquiry.

Bhaskar's use of the term "transcendental realism" is a careful distinction from older and decidedly more constructivist uses of the term dating from Kant's *Critique of Pure Reason*. In this work, Kant discussed transcendental "realism" and "idealism" as an investigation into the boundaries of possible knowledge or experience. Since then, so-called neo-Kantian approaches to science have generally been taken to imply an extreme version of constructivism in which society alone was seen to control what was thought of as nature (see Sismondo, 1996; Demeritt, 1998). Bhaskar's use of transcendental realism, however, implies a greater engagement with the resulting empiricism and scientific statements that result from social processes, rather than the neo-Kantian objective of demonstrating social control alone. The assessment of transcendental realism under Bhaskar, therefore, may seek to identify how the search for "necessary truths" (or universalistic statements of causality) may also imply a tendency to let the conclusion ("the world must be thus") shape the search for activities or events that support this conclusion, or at least show alternative accounts to be implausible (Bhaskar, 1975, 1986; Collier, 1994: 25).

A focus on transcendental realism therefore implies looking for the social assumptions that lead to empiricism and scientific laws, rather than assuming that empiricism and science are automatically indicative of neutral and universally applicable reality. In environmental terms, questioning the transcendental basis of environmental causality might include adopting an approach to environmental management that allows the maintenance of food supply and access to resources for poor people to be of higher significance than assuming the purpose of environmental management is to keep ecosystems free of damage from human influence. Indeed, the term "damage" has transcendental implications as it may be taken automatically to indicate that human influence on ecosystems is particularly degrading, but disturbances from other factors (such as from fire, long-term climate change, etc.) may be seen differently. The diversification of environmental "laws" by transcendental means consequently implies considering alternative presuppositions for meaning, and accepting the associated changes in empiricism and analysis such changes would bring. Most commonly in environmental terms, it means seeking to include the transcendental structures of social groups not previously included in the formulation of environmental explanation, and who are currently penalized under existing alternative forms of explanation. (The practical applications of such new science are explored in Chapters 8 and 9.)

These three categories of diversification, stratification and emergence, semantic realism, and transcendental realism, are an initial classification of alternatives to orthodox scientific generalization, and can be distinguished further. The key point of this discussion, however, is to indicate that much environmental debate and explanation in political debates today is conducted on the basis of alleged frameworks of environmental causality and structure that have been widely criticized from debates in philosophy and sociology of science; from within new approaches to ecology; and from research that focuses on the effectiveness and justice of environmental policy in various locations around the world. Seeking alternative forms of generalization can be achieved by questioning both the universalistic assumptions of orthodox "law" making and the social directions for inference and meaning in environmental terms. The next chapter now engages more fully with the social and political basis for the selection of different bases for inference and generalization, and seeks to illustrate the various means by which environmental generalizations have emerged.

Summary

The last chapter described some of the main problems experienced in environmental science: many well-known and unquestioned explanations of environmental degradation (so-called "environmental orthodoxies") are increasingly considered to be inaccurate, or to lead to environmental policies that are unfair. This current chapter has begun the process of highlighting the inherent politics in such explanations by summarizing debates in Philosophy of Science concerning the problems of scientific inference, and then applying these to debates in ecology.

The chapter highlighted how approaches under "orthodox," or positivist scientific method may have contributed to inaccurate explanations by building "laws" of nature without recognizing how such laws reflect the experiences of the people who make them. Moreover, such frameworks of science claim to produce politically neutral findings in either the short-term dissemination of results, or in long-term directions, or paradigms, of scientific inquiry. Such claims are criticized within science studies, which seek instead to demonstrate how such apparent "laws" may reflect the experiences or actions of specific social groups rather than be universally applicable to all.

The discussion also highlighted the impacts of new thinking concerning non-equilibrium ecology upon orthodox environmental explanations or notions of stability, or "balance" in nature. Non-equilibrium ecology has demonstrated that disturbance and flux are prevalent in ecosystems, and that many supposed zones or periods of stability may be identified largely through social expectations of what "nature" is supposed to be. A more politicized analysis of environmental science may therefore seek to indicate how notions of equilibrium or "wilderness" have reflected particular groups' viewpoints, or represented some other groups (such as shifting cultivators) as disrupters of ecological balance. A further politicized analysis might seek to show how new insights from non-equilibrium ecology have been resisted by different political actors. These themes are addressed in more detail in later chapters.

The chapter concluded by summarizing further debates from Philosophy of Science that propose more localized and diversified alternatives to universal (and inaccurate) "laws" of nature. These debates – such as semantic and Critical Realism – offer possibilities for the determination of statements of environmental causality to be determined more locally and more relevantly than environmental orthodoxies or "laws" of nature. Under a "critical" political ecology, discussion focuses on how supposedly neutral and unchallengeable environmental science may reflect the perspectives of particular groups; and on how such science may be made more politically transparent and reflective of more people. The next chapter now introduces insights from Sociology of Scientific Knowledge that complement this chapter in showing how environmental science is inherently political.

4 Social framings of environmental science

The objectives of this chapter are to summarize further problems with environmental science resulting from debates about language and social divisions. The chapter will:

- introduce debates from Sociology of Scientific Knowledge concerning the role of framings and language in influencing how environmental information is collected and presented;
- discuss the drawing of "boundaries" – between different social groups; scientists and "lay" people; or human and non-human objects – and their importance for the politics of environmental science; and
- outline different approaches to acknowledge the influences of different social framings on the evolution of environmental knowledge and explanations, with implications for making current environmental science more transparent and representative of different social groups.

This chapter builds on the discussion of Philosophy of Science in Chapter 3, and forms a further illustration of how social norms and experiences have become absorbed into scientific statements commonly considered as both factual and universal. Understanding how such social framings influence science is a crucial component of a "critical" political ecology. Later chapters discuss how we can change existing science toward more transparent and socially representative outcomes.

Social framings of science and knowledge

The preceding chapter discussed the importance of perspective, or local experience, in how scientific inquiry proceeds. Indeed, according to debates such as semantic or transcendental realism, the local perception and evaluation of different biophysical processes can be crucial in determining how environmental "changes" are considered to be environmental "problems."

The local perception or evaluation of environmental changes may be referred to as "framings." This term refers to the principles and assumptions underlying political debate and action. An environmental debate, for example, may consider whether to instigate a logging ban, a national park, or a tax on timber operations as alternative ways to reduce deforestation in one locality. The frame of such a debate, however, would be the assumption that deforestation is a degrading and uncontrolled practice, and needs effective action to address it. The analysis of underlying frames and assumptions in political debate is also an important aspect of discourse analysis and psychoanalytical research (see Silverman, 1993). Peet and Watts (1996: 37), for example, use the concept of framings when they refer to "environmental imaginaries," or the frameworks through which different individuals or societies perceive and evaluate aspects of environmental change.

The identification of frames, however, is not always easy. Frames are generally implicit rather than explicit, and a distinction has to be made between an explicit policy position or choice, and the more tacit frames that give rise to explicit positions.

> The frames that shape policies are usually tacit, which means that we tend to argue *from* our tacit frames *to* our explicit policy positions. Although frames exert a powerful influence on what we see and how we interpret what we see, they belong to the taken-for-granted world of policy making, and we are usually unaware of their role in organizing our actions, thoughts, and perceptions.
>
> (Schön and Rein, 1994: 34)

In order to reflect on how frames influence politics, we have to become more aware of them, and how (and from whose actions) they were constructed. Indeed, to acknowledge the importance of frames is also to accept the importance of constructivism in general in political analysis, and in relation to how notions of ecological reality have evolved.

> Constructivist policy analysis recognizes not only that issue framings do not flow deterministically from problems fixed by nature, but also that particular framings of environmental problems build upon specific models of agency, causality and responsibility. These frames in turn are intellectually constraining in that they delimit the universe of further scientific inquiry, political discourse, and possible policy options.
>
> (Jasanoff and Wynne, 1998: 5)

Frames therefore have influence in defining the basis of transcendental realism in justifying empirical projects on specific themes, and in shaping the nature of knowledge production in general. But assessment of frames should not just be limited to those that are labeled as important at present, but also seek to consider alternative framings that may not currently be considered important in political debates.

There are a variety of approaches to framing. The objective of this section is to outline some key types of framing, and the mechanisms in which they can influence the gathering and ordering of environmental knowledge.

Problem closure

The most direct way in which social factors may influence the nature and purpose of empiricism is through problem closure. Problem closure is the pre-definition of the purpose of inquiry, and is consequently effectively the transcendental structures that establish the basis of empiricism (see Chapter 3).

One typical example of problem closure in environmental terms may be the approach to deforestation that identifies the purpose of forest conservation to be the preservation of wilderness for aesthetic reasons, rather than the potential loss of resources available to local settlers. Policies that may result from such problem closure could include the proposal of parks or exclusionary land-use measures that would prevent local access to timber or non-timber products (indeed, this was observed in Guinea by Fairhead and Leach, 1996). Similarly, a dominant approach to climate change policy is to define the problem in terms of reducing atmospheric concentrations of greenhouse gases, rather than in preparing different societies to reduce vulnerability and exposure to the effects of climate change (see Shackley, 1997; Demeritt, 1998). Because of this problem closure, proposed policies have tended to identify the reduction of greenhouse gas concentrations as their primary purpose, rather than increase the ability of different societies to adapt to climate change (see Chapters 6 and 7 for fuller discussions).

According to Habermas (1974), much problem closure occurred through the production of knowledge for "technical cognitive interest" alone – or the economic exploitation, or mechanical control of objects. In this sense, the oppression of modernity had its own implicit problem closure of instrumental control of nature and society. The purpose of alternative, or liberatory, politics therefore was to provide different purposes for knowledge relating to other, non-exploitative uses.

Semiotics and metaphors

Predefined economic objectives are, of course, important in framing the nature and purpose of data collection. But problem closure and framing is also performed by and through language. Indeed, linguistic approaches to political frames are more encompassing than criticisms of economic exploitation because all communication and description is performed through language. When Mary Douglas, for example, defined "pollution" simply as "matter out of place" (1978: 35) she was indicating that there is nothing essentially bad about materials that constitute pollution, but instead that the definition of pollution depended on social regulations of where, when, and how much of such materials exist. A similar expression is "wild flowers in the wrong fields are weeds" (McHenry, 2000). In both cases, the words "pollution" and "weeds" imply an environmental problem, but the items constituting such problems are not always considered problematic.

The use of such words to indicate problems beyond the constitution of their parts can be seen as an application of Searle's (1995) distinction between "brute" facts and "institutional" facts (see Chapter 3). Words such as "pollution" and "weeds" imply a negative function and meaning to particular constituents (such as emissions or plants) that in other locations, quantities or periods may not be considered problematic (the emissions and plants on their own may be considered, according to Searle, to be unproblematized, or "brute" facts). The use of such words can be explained under semiotics as signifiers of environmental processes considered inherently bad, when there are, in actuality, wide disagreements about the universality of such statements.

For example, as discussed in Chapter 2, deforestation and erosion are still considered by various scientific or popular environmental debates to be equivalent to "degradation," when this equivalence is challenged by a variety of land users and scientists worldwide. (In some locations too, such as in Thailand, the act of deforestation is also identified as a "crime" under domestic law.) Part of this confusion is also due to the use of clumsy terms such as deforestation and erosion to refer to a variety of assumed environmental changes and impacts that may not always occur as the result of cutting down trees or the movement of soil. "Degradation," "deforestation," and "erosion" are all inelegant summaries of various constituent elements that may themselves all be challenged by identifying disagreements about the meaning and impact of what these words are supposed to mean.

The problem experienced in this example is the implicit assumption that language conveys an accurate and generally agreed representation of reality. In practice, however, language has evolved over time as a result of successive episodes of problem closure and the development of specific terms to capture the simultaneous occurrence and evaluation of physical events (Castree, 1995). Such co-evolution of language with social evaluations and political objectives of development implies, as Haraway (1991: 3) wrote, "grammar is politics by other means."

The politics of language in shaping environmental science has been illustrated in regard to the use of metaphors in both shaping empirical research, and then in replicating social perceptions in powerful ways. Writing on Critical Realism, Bhaskar (1991) highlighted that metaphors allow the rapid reference to a presumed common public sphere of presumed "fact" (see also Ortony, 1993; Lewis, 1996). According to Bhaskar, metaphor could be used as either "conversational references," denoting a term for a hypothetical activity; or "practical references," involving the physical measurement or fixing of hypothesis. In this case, the conversational application of metaphor created the transcendental structure necessary to presuppose the gathering of empirical data.

Metaphors can also reinforce social perceptions when used as a condensed expression of ecological reality. For example, referring to the Earth as a "lifeboat"; a rainforest as a "living fossil"; or human impacts on the environment as a "time bomb" clearly suggest a particular meaning with political purpose. Yet even referring to biophysical reality as an "ecosystem" or "forest" is also metaphorical because, as Demeritt (1994: 177) wrote, "human knowledge of nature comes to us already socially constructed in powerful and productive ways ... ecology is a discourse, not the living world itself." Using metaphors – or indeed all language – uncritically is to risk reification, or the presumption that the concept used to discuss an item is the item itself. Smith (1988) used the term "cerning" to refer to the act of circling or enclosing supposed units of society in such a way as to suggest the subjects are exactly as referred to by language, and that there is no political disagreement about such definitions. The term is used as an alternative to "discernment," which implies a more critical analysis of the concepts used to describe reality (see also Castree, 1995; Shapin, 2001).

The importance of such debates for framing is the need to explore how far there may be political disagreement about certain terms used as accurate representations of reality, and to ask how far proposed policy may have implications for different sectors of society. In effect, exploring framings means questioning how, when, and by whom such terms were developed as a substitute for reality. For example, the term "desertification" emerged within scientific communities as an indication of rapid degradation of fragile dryland ecosystems (see Chapter 2). This term was developed initially by visitors to drylands with tacit assumptions of land being used for agricultural or livestock production, and also the view that local settlers were in some way unable to perform adequate land management (Stebbing, 1937). The term "desertification" implies a sense of permanent despoliation, of land suddenly covered in sand. Research conducted more recently suggests greater agency of factors beyond social control in determining the non-equilibrium ecology of drylands (e.g. Dougill *et al.*, 1999), and accordingly, a greater self-determination of the needs of people living in such a zone. Replacing one metaphor ("desertification") with others (such as "dryland degradation," or "problems of drought") may be ways to indicate greater engagement with what we now know to cause environmental problems in drylands, and the problems as actually experienced by people living there (see also Biot, 1995; Mortimore and Adams, 1999). Box 4.1 describes a further linguistic analysis of the term "tropical rainforest," which has been claimed to represent similar inaccurate and unhelpful images of ecology.

Box 4.1   Tropical rainforests and language: one radical view

The influence of language on the framing of environmental problems has been discussed in a deliberately confrontational manner by the British bio-geographer, Philip Stott (1999). Stott argued that the complexity of tropical ecosystems cannot be adequately captured by the simple phrase "tropical rainforest." Moreover, this phrase can easily be manipulated by different political actors to indicate a variety of other themes that are not necessarily applicable to the ecosystem(s) known as "tropical rainforest." The word(s) "tropical rainforest" (or *tropische Regenwald*) are generally considered to have been used first by the German biologist, Andreas Franz Wilhem Schimper (1856–1901). Yet the concept has since been used to indicate various themes of forested wilderness; ever-wet tropics; and diverse fecundity since. Such underlying themes, however, do not necessarily indicate the understandings of biologists working in rainforests, or the relationship with other (if related) tropical ecosystems such as savanna. Indeed, all tropical forests have advanced in extent notably since the end of the last Ice Age some 16,000 years ago. "Whereas 'cup' and 'cupness' clearly relate to an object in which the boundaries are not noticeably fluid, when exactly, by contrast is a savanna woody species a tree and not a shrub?" (Stott, 1999: 9).

Stott analyzed the linguistic content of ten documents about rainforests published on a variety of websites from NGOs or centers offering information about rainforests. Such discourse analysis revealed a number of inherent assumptions about the overall value of rainforests to human society that may suggest first that "one" clear ecosystem of the "tropical rainforest" may exist, and second that different organizations may be placing particular normative values on "rainforests" that come more from social discussions about how forests should be seen. For example, one statement from the Rainforest Information Center (Australia) stated: "rainforests have been called the womb of life..." One Fundamentalist Christian organization in Arizona, USA (Kid's Quest) stated, "we are reminded that trees are created to be pleasant..." Such comments clearly reflect human valuations of forests, and overlook the importance of other ecosystems (such as savanna) in upholding biodiversity or in providing other uses to humans. Another common concept that rainforests are the "lungs of the earth" is particularly confusing because lungs take in oxygen and emit carbon dioxide, whereas this statement is meant to imply the opposite (although frequently forests actually *do* emit carbon dioxide).

The objective of this research is not to suggest that "tropical rainforests" should not be valued or protected, but to make people more aware of the ways in which scientific discourse is used to carry a variety of values that are not necessarily accurate in relation to the biophysical entit(ies) known as rainforests. Indeed, such uses of scientific discourse may present a simplified account of rainforest biology, such as overlooking the diversity of rainforests, or the complex relationship between forest disturbance and biodiversity (as discussed in Chapter 2, some forms of disturbance may enhance biodiversity). Moreover, such assumptions may encourage policies that penalize farmers engaging in limited forms of cultivation and forest use. According to Stott, such policies are unjust because they are based on outsiders' mythical beliefs about the importance and fragility of rainforests. Stott argues there are few material reasons for needing rainforests, and calls rainforest conservationism a "New Age form of colonialism." Other biologists, however, might consider this to be an extreme view, and would seek ways to allow rainforest conservation with the protection of local livelihoods.

Source: Stott, 1999; www.ecotrop.org.

Social divisions: gender, class, and race

Social divides such as gender, class, and race may also impact both on the transcendental structures that guide the collection of empiricism, and on the role of science in reinforcing such divides. The investigation of the relationship of social divides and epistemology, however, is controversial. On one hand, it seems reasonable to expect that different social groupings have different environmental perceptions and framings; indeed, much research on environmental orthodoxies has revealed this (Jewitt, 1995, 2000; Rocheleau *et al.*, 1996; Rocheleau and Edmunds, 1997; Jackson, 1997). Yet on the other hand, the repeated and uncritical use of existing classifications of society may enact another form of cerning, and hence reify groups, or the generalized association of particular behavior or environmental framings associated with such groups. One important aspect of such cerning is not to assume, despite the importance of differences, that such categories can be neatly divided and assessed independently, without acknowledging their mutual embeddedness (see Leach *et al.*, 1997).

There are many reasons to indicate the importance of social differentiation in explaining the social basis of environmental explanation. For example, writing about feminist political ecology, Rocheleau *et al.* wrote:

> Environmental science and "the international environmental movement" have been largely cast as the domain of men. In fact, while the dominant and most visible structures of both science and environmentalism may indeed be dominated by men, mostly from wealthier nations, the women of the world – and many men and children with them – have been hard at work maintaining and developing a multiplicity of environmental sciences as well as grassroots environmental movements.
>
> (1996: 6)

Furthermore, debates in environmental racism show how the location of waste dumps, highways, or similarly undesirable structures have been linked to localities where inhabitants are people of color or recent migrants (Westra and Lawson, 2001). Similarly, social class – whether defined in conventional terms of working class, or socially less powerful groups in general – is a crucial element in defining which environmental framings may become dominant, or who may receive a higher than average distribution of environmental services or potential hazards. It is highly unlikely, for example, that a wealthy suburban dweller would frame environmental needs in similar terms to poor inner-city squatters.

It is not the intention of the book at this stage to discuss the implications of such framings (these are discussed in more detail in Chapters 5, 6, and 7). The purpose of this initial discussion is to highlight the crucial importance of such social divisions to the achievement of a "critical" political ecology, or a more diversified and meaningful environmental explanation. But in addition, there are severe problems of establishing such a more democratic framing for environmental science based upon social divisions. Three main types of problem may be identified.

First, it is difficult to achieve a more balanced and representative analysis of less powerful groups in society because the language, science, and assumptions we use to do so are all imbued with the historic social evaluations that helped create the marginalization of such groups (Longino, 1990). The Royal Society of London, for example, excluded women from membership when it was established in the seventeenth century. In the eighteenth century, literature was excluded from "science" because it was considered "feminine." Goethe's reputation as a poet was said to ruin his reputation as a scientist (Barr and Birke, 1998: 112). Feminist critiques of science have argued that women have effectively been written out of history, or excluded from creative input into explanation (Harding, 1986).

The exclusion of women may have led to the reflection of stereotypical gender roles in much scientific discourse. For example, Martin (1991) noted that accounts of sexual reproduction portrayed sperm in active terms such as "active," "forceful," and "self-propelled," whereas eggs were described more passively as "swept," "transported," or "drifting down" fallopian tubes. Early scientific accounts of reproduction in plants also reflected metaphors from human society. Linnaeus, for example, writing in the eighteenth century, used words such as "nuptials" to denote fertilization of plants; "wedding gowns" to describe the blooming of trees and shrubs prior to pollination; and "bridal beds" to signify flower petals (Schiebinger, 1993: 23). It is therefore important not to see science and research as neutral tools to redress the balance of explanation in favor of marginalized groups, but instead to appreciate there is also the need to redress the tools themselves.

The second main problem is that attempts to focus on particular groups may tend to reduce the representation of social diversity to reified and stereotypical categories that only add to the need to express social differentiation. For example, Haraway (1991: 243) argued that the research seeking to acknowledge the importance of "gender" by looking only at women is to avoid the dynamics that lead to the cerning and marginalization of women in the first place. Similarly, it is also misleading to equate "people of color" with "race." It is important to note that each social differentiation such as gender, race, and class are not static and may vary within themselves and over time. Such groupings also reflect power relations and marginalization of society in general. Indeed, in *Death of Nature*, Carolyn Merchant (1980) used nature as a metaphor for women.

This desire to represent marginalized sectors of society also raises a third main problem of redressing social divisions in science. The problem relates to the epistemological impacts of undertaking such a task, or the desire to redress environmental priorities *on behalf of* groups considered to be marginalized or disempowered. Social concerns seeking to redress social imbalances along lines of gender, race, sexuality, and also environmentalism are generally associated with the so-called "new" social movements that emerged initially in Europe and North America in the 1960s. Such new social movements differed from the old, class-based movements because they were considered to be either classless, or the activities of one class (usually a "middle" class) on behalf of all society (see Habermas, 1981; Offe, 1985). This theme is discussed in more detail in Chapters 5, 6, 7, and 9. But, as noted in these chapters, it is not always clear if such well-meaning attempts to improve conditions for marginalized groups by more powerful sectors of society may not replicate pre-conceived identifications of such groups. Indeed, the priority should not be to "get" women, or other apparently marginalized groups, "into" science and its institutions, but instead to see how science and institutions may be reformed in order to understand more effectively how social exclusions have been created. Indeed, as Box 4.2 shows, there are many forms of inequality and exclusion with environmental science along the lines of gender.

A "critical" political ecology seeks to indicate how far explanations of environmental problems reflect – or fail to reflect – the perspectives of different social groups. Yet, successfully democratizing environmental science involves questioning existing definitions of environmental problems (problem closure); the linguistic basis of science and reference; and the problems in identifying, communicating, or empowering the perspectives arising from different social groupings. The next section now considers the implications of a politicized approach to framings for political approaches to understanding how environmental science reflects social norms and divisions.

Contested boundaries and hybrids

The preceding discussion outlined how the process of "framing" – through complex processes of problem closure, language, and social participation – may lead to the reification of particular viewpoints as scientific "fact." For example, it is sometimes difficult to use terms such as "desertification" or "tropical rainforest" without also adopting many of the inherent valuations associated with them. This section now discusses how sociologists of scientific knowledge have approached this problem, in order to make environmental science more transparent. Three concepts are worthy of

Box 4.2   Gender differences within academic research on environment

As with any organization or network of people, the composition of people who conduct academic research on environment may reflect inequalities in social groupings such as class, gender, and race. For example, one study in 1996 of social scientists in the USA reported that "geography" (as one discipline focusing on environment) contained the lowest percentage of women Ph.D.s (24 percent of all PhDs were women) when compared to clinical/counseling psychology (65 percent); anthropology (54 percent); sociology (62 percent); political science (29 percent); and economics (25 percent) (Holden, 1996, in Luzzarder-Beach and MacFarlane, 2000: 408).

A more specific survey of physical geographers in 1995 and 1996 (Luzzarder-Beach and MacFarlane, 2000) further indicated other divisions: many more women were at the (more junior) level of "assistant professor" than men, and some 29 percent of women had achieved tenure status, compared with 59 percent of men. More women researchers were seen to be prominent in "biogeography" than other fields of physical geography, reflecting claims made elsewhere that biology represented the most likely route of entry for women into science (The *Economist*, 1996). Men, by comparison, were more prominent in the fields of geomorphology and climatology. Concerning research techniques, 9 percent of women questioned saw laboratory work to be prominent in their research, compared with 30 percent of men.

It is, of course, difficult to draw clear conclusions from these kinds of inequalities observed between women and men in the field of physical geography. It is unclear, for example, how far such trends reflect the generally fewer applications by women to physical geography PhD programs, or whether there are other factors restricting female advancement. It is interesting to note, however, that women physical geographers complained of more frustrations in their career than men. For example, 73 percent of women PhDs questioned described frustration at the competition for research funding (compared with 50 percent of male PhDs); 50 percent (versus 18 percent) complained at the lack of peer communication; and 39 percent (versus 16 percent) protested at apparent barriers in the tenure process. Whatever the reason for poor female recruitment, it does seem that women see more barriers to advancement in the career of physical geography than men.

Source: Luzzarder-Beach and MacFarlane, 2000.

noting as ways to describe the problems of language, or discourse, in attempting to explain complex reality.

First, concepts or explanations such as "desertification" or "deforestation" may also be referred to as "black-boxed" (after Bruno Latour, 1987). According to Latour, a concept or term can be said to be black-boxed when their internal nature is taken to be objectively established, immutable, or beyond the possibility for human action to reshape it. The concepts of "shifting cultivation," or "pollution," for example, may be seen to be black-boxed within many popular debates of environmentalism when they are framed as automatically degrading, or when people see no need to discuss what these terms might mean or why they are seen to be damaging.

Second, the process by which such framing has influence is through the imposition of boundaries. Boundaries may be drawn at specific times and places to make the frame relevant to the creation of knowledge or policy. The drawing of a boundary around social groups, biophysical entities, or their interactions, is, in effect, to establish an ordered vision of events. The resulting structure therefore reflects the viewpoint of the boundary creator, and provides a precedent for explanation that may eventually be accepted as "fact" (see Kukla, 1993; Barnes *et al.*, 1996; Gieryn, 1999).

Third, Bruno Latour (1993) again captured some epistemological effects of such boundary closure through the concept of "hybrids." Hybrid objects are commonplace objects or "things" that appear to be unitary, real, and uncontroversial, but in practice reflect a variety of historic framings and experience specific to certain actors or societies in the past. A hybrid may be compared with the "institutional facts" of Searle (1995), the "cyborgs" of Haraway (1991), and with "environmental orthodoxies" that signify commonly accepted explanations of environmental degradation based upon only partial experiences and values. The evolution of hybrids, through the interplay of framing and boundary drawing, has profound implications for what we understand as "ecology," and who participated in creating this concept.

According to Latour, the ability to draw boundaries between "nature" and "society" is dependent on a dichotomy that can only be maintained by so-called "purification," or the separation of two distinct ontological zones of human beings and non-humans. Yet Latour's argument is that such purification can only take place superficially on the basis of concepts that are hybrid blends of human and non-human. As a result, the objective of purification – or the establishment of rational, neat explanatory devices for causal relations between nature and society – are doomed to failure because they overlook the interrelated experience of nature and society, and the inaccurate simplicity of the nature–society dualism. As an alternative, it is necessary to look more at the process of "translation" – or the creation of networks between social and natural objects – as the means to identify how we have experienced "nature" in specific ways. Figure 4.1 illustrates what is meant by "purification" and "translation." In this diagram, the first social dichotomy is between objects that are supposedly "purified" between "nonhuman" or "nature" and those that are "humans" or "culture." The second dichotomy is between these two groups of objects and those that are not, or not yet, categorized into these groups. Latour's point is that both dichotomies are false, and that any attempt at purification is likely only to reflect questions of social choice.

By focusing on translation instead of purification, research looks more at the experiential and cumulative construction of apparent "facts" over time as the result of particular actors' boundary decisions, rather than

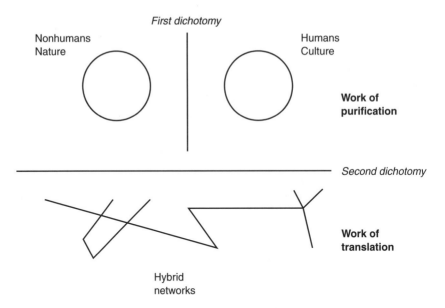

Figure 4.1 Purification and translation
Source: Latour, 1993: 11.

the belief that such facts are objective and universal representations of reality, from which further objective causal links may be identified. Latour wrote:

> Maybe social scientists have simply forgotten that before projecting itself on to things society has to be made, built, constructed?... Dualism may be a poor solution, but it provided 99 per cent of the social sciences' critical repertoire, and nothing would have disturbed its blissful asymmetry if science studies had not upset the applecart.... By trying the impossible task of providing social explanations for hard scientific facts – after generations of social scientists had tried either to denounce "soft" facts or to use hard sciences uncritically – science studies have forced everyone to rethink anew the role of objects in the construction of collectives, thus challenging philosophy.
>
> (1993: 54–55)

A practical manifestation of hybridization, or the cumulative construction of a perceived "fact" out of diverse experiences and framings, is a "quasi-object" (ibid.: 51). This term is related to Haraway's own metaphor of "cyborg" for so-called artefactual nature, in which the preexisting biophysical reality may not be represented totally by the concepts such as "plants" or "animals" that have emerged in a similarly situated way to metaphors and quasi-objects. Haraway wrote:

Organisms are *biological* embodiments; as natural-technical entities, they are not preexisting plants, animals, protistes, etc., with boundaries already established and awaiting the right kind of instrument to note them accurately. Biology is a discourse, not the living world itself. But humans are not the only actors in the construction of the entities of any scientific discourse; machines (delegates that can produce surprises) and other partners (not the "pre- or extra-discursive objects," but partners) are active constructors of natural scientific objects.

(1991: 298, also in Demeritt, 1994: 181)

The drawing of boundaries between "nature" and "society" is therefore a key way in which social framings, either coupled with, or influenced by actual experiences of environmental change, have led to the establishment of "factual" or universal statements based upon those combined experiences and valuations. Yet there are also less obvious ways in which such "boundary-work" (Gieryn, 1999) can be emplaced. Usually, the adoption of historic boundaries presented as "fact" may have the effect of replicating past dichotomies in the analysis of new problems and the drawing of new boundaries. Much criticism of exploitative economic development, for example, has highlighted the apparent adoption of an Enlightenment-based dualism between "people" and "nature" as innately damaging to "environment" and the interconnectedness of humans and other species (e.g. Worster, 1979; Atkinson, 1991). Yet, in addition, boundary-work may also be used to criticize the assumptions contained in any statement that does not examine critically the implicit boundaries contained within it. For example, in a linguistically based critique of William Cronon's (1991) examination of the impacts of development on "nature" in the USA West (*Nature's Metropolis: Chicago and the Great West*), Demeritt drew reference to the uncritical adoption of equilibrium-based notions of ecological science as a basis for the narrative.

In categorical statements such as, "In nature's economy, all organisms, including human beings, consume high grade forms of the sun's energy," Cronon (1991: 150) matter of factly states what nature is. This certainty, however, is dependent upon the silent appropriation of ecological science and the trophic-dynamic ecosystem models pioneered by Eugene Odum. Ecology is a discourse, not the living world itself. By conflating the two in categorical statements about first nature, Cronon (1991: xvii) fixes the very "boundary between human and nonhuman, natural and unnatural," that his book so brilliantly shows to be "profoundly problematic."

(1994: 177)

The drawing of boundaries, therefore, effectively allows the establishment of an order and framework from which to proceed. Yet the construction of items as a result of drawing boundaries can always be done alternatively, given different problem closure, language, and social divisions. The decision to place boundaries in particular forms around different problems, or in favor of particular groups, therefore facilitates the achievement of political objectives of those who draw the boundaries. The replication of boundaries may simply follow from a lack of awareness of potential alternative framings; or may result from an intention to enforce the political objectives associated with the boundaries; or because it may also support new and different political objectives in a further debate. As Demeritt (ibid.: 174) commented: "environmental historians rely upon ecological science for explaining concepts like ecosystem and equilibrium that organize their narratives."

The implicit acceptance of proposed boundaries is therefore a key way in which either propositional truth claims about reality, or historical observations of pattern, may become accepted as "fact." Yet "boundary-work" may also be applied to the social regulation of science and policymaking. As discussed in Chapter 3, one tenet of so-called science is the belief in self-regulation of results by peer review and criticism. Such regulation also has a strong disciplining effect on both scientists (or knowledge producers in general) and on who can participate in both the production of knowledge, and the regulation of findings. According to Gieryn:

> Epistemic authority does not exist as an omnipresent ether, but rather is enacted as people debate (and ultimately decide) where to locate the legitimate jurisdiction over natural facts. . . . Real science is demarcated from several categories of posers: pseudoscience, amateur science, deviant or fraudulent science, bad science, junk science, popular science. Boundary-work becomes a means of social control: as the borders get placed and policed, "scientists" learn where they may not roam without transgressing the boundaries of legitimacy, and "science" displays its ability to maintain monopoly over preferred norms of conduct.
>
> (1999: 15–16)

Indeed, one might wonder how responses to this book may demonstrate elements of the above statement. It has already been noted, for example, that raising concerns about environmental orthodoxies in conferences full of orthodox development practitioners such as watershed scientists and foresters leads to much resentment. According to Ian Calder (whose book, *The Blue Revolution*, 1999, questions much orthodox thinking on watershed management): "Sometimes it feels like you need to have a motorbike waiting outside with its engine running after you have given a paper" (Calder, pers. comm. 2000).

The implication of such social regulation of science is that only particular forms of knowledge or explanatory frameworks may be accepted as "science" or "legitimate" by particular organizations and actors. Usually this screening and disciplining is defended in terms of protecting society from ideology, or non-rigorous "pseudoscience" that may wrongly claim to have the privileged status of "scientific" knowledge (e.g. Bunge, 1991). But the impacts of such screening go further, in blurring the lines between scientists as knowledge producers, to scientists as policy advisors, or custodians of public debate about topics of environmental concern. In effect, this may mean the institutionalization of existing concepts and explanations (hybrids or orthodoxies) as non-negotiable "fact," and the use of these hybrids for further explanation: the process of "purification" so criticized by Latour. Furthermore, they also construct a definition of science that is used as a means to show authority in political debate about "nature" or "environment," that may both strengthen their political position, and weaken would-be alternative conceptions. Jasanoff commented:

> In denying the existence of role ambiguity [between science advisors and policymakers] these discursive repurifications implicitly rely on some objective grounding in nature and scientific roles as their source of authority.
>
> (1987: 226, also see Jasanoff, 1990; van der Sluijs *et al.*, 1998)

The social framings of knowledge and science are therefore enacted politically through the drawing of boundaries around contested and variously experienced interfaces between nature and society. The resulting concepts – identified and produced by the historically powerful sectors of society – reflect, in actuality, a "hybrid" combination of experience, framings, and evaluation of events that give order to complex reality. Such hybrids are then further institutionalized as "fact" by succeeding scientific debate, the regulation of science, and the use of such hybrids to support further political objectives in later debates.

Such processes may be referred to as "coproduction" and "hybridization," and are discussed in relation to environmentalism in the following chapter. Before this, however, it may be useful to summarize different theoretical approaches to understanding the influence of social framings and boundaries on environmental science. These different approaches are used later in the book.

Theorizing the social institutions of environmental science

It is clear, then, that the production of knowledge through mechanisms known as "science" or other means are contingent upon a variety of social processes, involving the framing or purpose of knowledge; the language used to express it; the social groupings and contexts in which it is sought and presented; and the political purposes to which it is put. Before this book analyzes the evolution of debates within the field specifically known as "political ecology" (see Chapter 5), it is important to summarize the implications of this current chapter on theoretical approaches to social framings of environmental knowledge.

There are clearly many alternative conceptualizations of science to the frameworks of "orthodox" science described in Chapter 3. Orthodox science has been described, simply, as the search for universally applicable "laws" of nature based upon practices that guarantee accuracy and lack of political bias. Instead, the alternative models of science are largely based upon the social controls that exist in how science evolves, is regulated, and is applied. Such social controls may be described as "institutions" – or shared norms, language, framings, etc. – that may exist among scientists, their practices, or their objectives (Jasanoff *et al.*, 1995).

The following discussion lists three important approaches to theorizing institutional approaches to the evolution of scientific knowledge about environment. These approaches may be applied to various aspects of environmental science, or a "critical" political ecology. The discussion ends with a summary of four potential models of science that provide a framework for comparing different approaches to scientific explanation.

Pragmatism and institutions of science

The Philosophy of Science known as "pragmatism" is largely attributed to the works of Charles Peirce, William James, John Dewey, and later Richard Rorty (Rorty, 1989a, b). The early applications of pragmatism within philosophical debates referred mainly to its approach to the definition of truth. Yet, increasingly, the term is used to refer to the social institutions that may also uphold supposed truth statements, and hence it is worthwhile noting the concept in this chapter focusing on how social divisions and language support different explanations of reality.

Pragmatism may be seen to refer to three key tenets: the rejection of essentialist concepts of truth; the perception of no epistemological difference between facts, values, morality, and science; and a belief that social networks or solidarities determine scientific inquiry. For pragmatists, "truth" is just the name of a property that all true statements share. The term "pragmatism" refers to the necessary limitations such social solidarities place on the extent to which scientists – or the networks to which they belong – can produce explanations that go further than their own experience and objectives. In this sense, pragmatists seek to understand how social networks (or institutions, or solidarities) may be the determining factor in understanding complex reality, rather than placing innate faith in the predictive power of science itself (see also Light and Katz, 1996; Proctor, 1998; Williams, 2001).

In a well-known quotation, Rorty wrote:

> Those who wish to ground solidarity in objectivity – call them "realists" – have to construe truth as correspondence to reality. So they must construe a metaphysic which has room for a special relation between beliefs and objects which will differentiate true from false beliefs. They must also argue that there are procedures of justification

of belief which are natural and not merely local... By contrast, those who wish to reduce objectivity to solidarity – call them "pragmatists" – do not require either a metaphysic or an epistemology. They view truth as, in William James's phrase, what is good for *us* to believe.

(1989b: 36–37)

The term, "pragmatism," is often used to denote this very focused inspection of social solidarities and truth statements. (This approach is discussed further in Chapter 8.) Yet elements of pragmatism may also be seen in many other approaches to scientific knowledge that emphasize the importance of consensus-building and shared norms and experiences. These other approaches – such as Cultural Theory or narrative and storyline analysis – adopt more sophisticated means of explaining how such social solidarities emerge.

Cultural Theory and the myths of nature

Cultural Theory is a further framework for explaining social solidarities in environmental explanation. It is mainly influenced by the work of the anthropologist Mary Douglas, who argued that the variability of an individual's involvement in social life can be adequately captured by two dimensions of sociality: group and grid. "Group" refers to the extent that an individual feels incorporated into bounded units. "Grid" indicates how far an individual's life may be affected by externally imposed rules of prescriptions (see Thompson *et al.*, 1990: 5). Cultural Theorists usually denote their difference from other forms of cultural theory by using a capitalized C and T.

Cultural Theory is different to many other social and cultural accounts of behavior by proposing that social groupings or individuals may fall into five (and only five) ways of life that may indicate different elements of grid and group. The five groups are hierarchy (or high levels of both grid and group); egalitarianism (high group, but low grid); fatalism (high grid, low group); individualism (both low grid and group); and autonomy (the position where grid and group have least meaning). In many discussions, and in this book, however, only four groups are referred to, as it is commonly believed that the fifth category (of autonomy), by definition, either cannot exist or cannot be discussed alongside the other groups. Consequently, further references to Cultural Theory in this book will refer only to the first four groups.

Some typical examples of the four groups could include a state actor (a facilitating or hierarchical role); a political activist or NGO (egalitarian); powerless and marginalized workers (fatalist); and transnational companies (profiteering or individualist). The existence of these four ways of life indicate that Cultural Theory is, in part, a structuralist form of explanation – although it is worthwhile noting that the structures are not deemed to be permanently rigid, and that different ways of life may be adopted over a period of time by the same group or individual. Yet the structural and definitive claims of Cultural Theorists that different social circumstances may be analyzed according to these ways of life are, at the same time, the main strength, yet also the main criticism of Cultural Theory.

There is much application of Cultural Theory in general debates in society (e.g. Douglas, 1987), but in environmental terms, it has been widely used in explanations of environmental uncertainty, environmental perception, activism, and the varied generation of knowledge (e.g. see Thompson and Rayner, 1998a, b). Indeed, one of the most influential books concerning environmental orthodoxies, *Uncertainty on a Himalayan Scale* (Thompson *et al.,* 1986), was written from the perspective of Cultural Theory, although the book did not make this explicit. Most significantly, Cultural Theory translates the five ways of being to corresponding "myths of nature," which represent visions of environmental stability or fragility according to each way of life. Figure 4.2 shows the myths of nature, and how they relate to the different ways of life.

The different visions of environment are called "myths" because they are both true and false representations of environmental belief and experience, which both form a structure of everyday life but also provide only a partial experience of environmental reality (see Chapter 2).

> The myths of nature, in consequence, are both true and false; that is the secret of their longevity. Each myth is a partial representation of reality. Each captures some essence of experience and wisdom, and each recommends itself as self-evident truth to the particular social being whose way of life is premised on nature conforming to that version of reality.
>
> (Thompson *et al.*, 1990: 26)

The myths may be summarized as follows. *Nature benign* (individualist) presents an image of environmental impacts having little long-term damage, and consequently social and economic policies relating to environment should be as laissez-faire as possible in order to allow the least interference with human actions. *Nature ephemeral* (egalitarian), on the other hand, represents an image of nature as fragile and susceptible to rapid change, and potentially irretrievable degradation. *Nature perverse/ tolerant* (hierarchical) indicates a more managerialist attitude to environment, in which environmental change has some limited impacts, and potentially worse impacts, but both may be managed by careful monitoring and the observance of limits. *Nature capricious* (fatalist) is a picture of a random world; there is no point learning about environmental change or attempting to manage degradation, as change will occur regardless. (The perspective of the hermit, or autonomy, is of *nature resilient*: there is a rejection of dualisms of humanity and nature, environmental change is seen as inevitable, but urgency and need to change policy are eschewed.)

Yet, importantly, the adoption of different myths may change as a result of environmental surprises, or more gradual transitions in social perception and debate (see Price and Thompson, 1997).

The advantages of this framework is that it provides an overview of all possible positions for environmental perception, and links such environmental perceptions to underlying cultural positions, based on the sharing of beliefs. Cultural Theory importantly notes that there are plural environmental rationalities, that may coexist, or replace each other over time, and which account for different forms of environmental understanding. As Cultural Theorists state in relation to the myths, the point is to understand how each myth is based on different cultural standings rather than an absolute and privileged understanding of environmental reality: if you have to ask which myth is right, then you are wrong.

But, in addition, Cultural Theory also highlights the epistemological impacts of dominant political and social institutions in terms of the state (hierarchy); economy (individualism); civil society (egalitarianism); and the disenfranchised (fatalism). Indeed, the group of the fatalists may not usually be identified in social debate because, by definition, fatalists may lack the voice or political power to represent their views in political

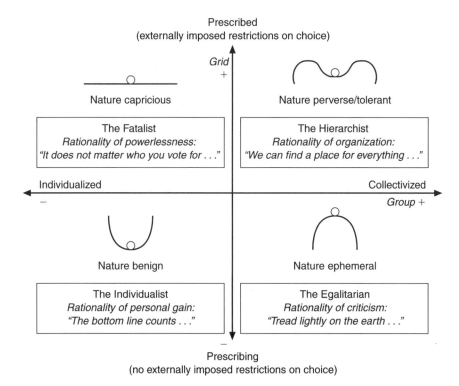

Figure 4.2 Cultural Theory and the "myths of nature"

Source: adapted from Schwarz and Thompson, 1990: 9.

arenas. Environmental research on Cultural Theory and scientific know-ledge has argued that there is a clear link between the framings, laws, and empiricism conducted under different "myths of nature" that can lead to environmental "laws" and explanations that are in different ways "mythi-cal" because they are based on partial experiences and evaluations of bio-physical reality (Schwarz and Thompson, 1990).

Cultural Theory, therefore, offers the potential of a form of analyzing social studies of "nature" and environmental change that highlight the underlying and structural basis of how knowledge is generated, and is then maintained in order to support the social institutions that create it. Critics of Cultural Theory, however, point to two key alleged failings. First, many social scientists working to identify various levels of social differentiation argue that the five ways of life are far too reductionist, and ignore the variety of ways in which social experience may be expressed through differ-ent themes of culture, gender, race, political standpoint, age, wealth, educa-tion, and so on. Furthermore, the linguistic basis of each way of life, or "myth of nature" may also be ignored under Cultural Theory if it is assumed that the definition of each myth is pre-established and universally applicable to different locations worldwide. Linked to this, the historical evolution of alternative conceptualizations of, for example, environmental fragility or environmental resilience, may also be ignored by the (relatively simple) reliance on so few (and pre-defined) ways of life. Cultural Theorists counter such points by arguing that the five divisions are actually more numerous than most other social analyses of change, which commonly place individualists (or market) against either hierarchy (state) or egali-tarian (e.g. social activists) positions. Furthermore, they claim that the myths incorporate historical events and evolution of terms, but that there is a limit to how far each "myth" or way of life is constructed only by history.

The second main criticism often made against Cultural Theory, particu-larly in environmental debates, is that it is relativist, and accordingly gives equal precedence to each "myth" regardless of whether it is possible to establish whether different approaches may be more accurate than others. Cultural Theorists often reject claims of relativism by pointing out that each myth is only a partial vision of reality, and also that all truth claims made under each myth are subject to normal bounds of expectation. As Thompson *et al.* note:

> To say that ideas of nature are socially shaped is not to say that they can be anything at all... "Okay, go and jump in front of the train," say the relativity rejectors ... But of course no one is saying that percep-tion is completely fluid, only that it is not completely solid. Rather, ideas of nature are plastic; they can be squeezed into different configu-rations but, at the same time, there are some limits. The idea of nature that would have us all leaping in front of trains is outside of these limits, that is, it is not a *viable* idea of nature.
>
> (1990: 25, emphasis in original)

Nonetheless, the main contribution of Cultural Theory toward environmental debate and explanation remains at the macro level of overall visions of environmental fragility or resilience. There is little within Cultural Theory to assess the different accuracies or validities of competing land-management schemes that may be questioned *within* one myth or framing, such as by indicating how serious one particular rate of soil erosion may be, or whether to select one species of food tree or another.

Cultural Theory, therefore, provides a framework for fixing multifarious and opposing perceptions and explanations of environment into a classification that reflects some allegedly deep-seated and universal institutions in society. Alternative approaches are less structural and universal, and instead point to the evolution of language and discourse over time and as the result of differential power interests in society.

Narratives, storylines, and Actor Network Theory

The third main approach to environmental constructivism accentuates the historical evolution of environmental discourse and explanation rather than a supposed structure of different shared values associated with different social actors. The third grouping includes more poststructuralist forms of explanation, particularly influenced by the writings of Michel Foucault.

Various concepts may be classified under this third general grouping. At the most general level, "environmental narratives" are commonly heard environmental concepts and explanations that may be described as dominating discourses (e.g. Cronon, 1992; McComas and Shanahan, 1999). The social and linguistic interpretation of environmental narratives is to focus on how such narratives become adopted as "truth" because of social processes, rather than because of a Realist belief that such narratives reflect biophysical reality as uncovered by science. For example, Harré *et al.* wrote:

> We do not tackle things like the "ecological crisis" as if it were a natural phenomenon. The "crisis of our times" is at root a discursive phenomenon. It comes about through a shift in our ways of seeing and assessing what we see, made possible by the taking up into our discursive resources new vocabularies, new judgmental categories, new metaphors and analogies that have promoted awareness of much that was previously overlooked.
>
> (1999: 3–4)

As Harré *et al.* (1999) discuss, the analysis of narratives may also include the individual syntax and structure of statements to gain political power. Included within this analysis is the possibility that explanations and discourse may be structured in order to portray different actors to undertake particular roles of (for example) victim, villain, and savior to enhance the power of the narrative as an explanatory tool, or as a device to enact political implications. Narratives, therefore, are close to stories because they have beginnings, middles, and ends that serve a purpose in ordering social actors (and sometimes physical items) into a causal structure. According to Roe (1994), a narrative policy analysis includes the identification of how different narratives within policy debates may conform to aspects of "storytelling," and then to assess how far such stories may be enforced or resisted by different political actors for the sake of influencing wider political uncertainties. In certain occasions, the purpose of such storytelling may be to reinforce the speaker's own position as powerful in political debate. The phenomena of narratives within political debate have also been called "storylines." For example, Hajer wrote: "Storylines are devices through which actors are positioned, and through which specific ideas of "blame" and "responsibility" and "urgency" and "responsible behavior" are attributed" (1995: 64–65).

A crucial element of a "critical" political ecology, therefore, is to assess how far existing scientific explanations of environmental degradation may – in effect – be storylines that represent alternative political viewpoints and the redefinition of preexisting political debates under environmental guises. For example, it has been argued by some observers that the vilification of some shifting cultivation groups in Asia and Africa has reflected long-term resentment of minorities for a variety of cultural and political reasons, and hence has led to them also being blamed for environmental degradation (e.g. Fairhead and Leach, 1996; Schmidt-Vogt, 1998; Fox *et al.*, 2000). Under such categorizations, cultivators may be identified either as villains or victims according to different evaluations. Similarly, the concern against anthropogenic climate change may often be expressed in terms of unregulated selfish business (villains) and powerless island states vulnerable to rising sea levels (victims). Cultural Theory may approach these different positions from terms of opposing egalitarianism (islands) versus individualism (business). Under a storyline analysis, however, the different evaluations are seen not to reflect universal institutions in society, but instead are considered situation-specific and resulting from different actors' influence over time.

In terms of construction of knowledge, storylines and narratives have importance in ordering society into preexisting structures of "blame" and "reward," that can lead to the framing of further empiricism and explanation. Second, the interaction of different narratives can lead to the formation of further apparent truths and "facts." The concept of "discursive argumentation" (e.g. Davies and Harré, 1990), for example, highlights how social interactions are themselves framed within wider contexts of argumentation, and that each statement made in argumentation is intended to change the other's position. Partly because of such interactive argumentation, the emerging positions of agreement are influenced by the participants in the argument and the nature of the argument, rather than an objective and asocial establishment of "factual" reality. The interactions between different narratives and arguments may, therefore, lead to the enforcement of a perceived reality and framing of the external world that is a product of the argument. Hajer calls these "discourse coalitions": "Discourse coalitions are defined as the ensemble of (i) a set of story lines; (ii) the actors who utter these storylines; and (iii) the practices in which this discursive activity is based" (1995: 65). Similarly, van der Sluijs *et al.* (1998) have identified the term, "anchoring devices" to refer to statistical or "factual" consistencies in political debates about environment or science that reflect the bargaining positions of different political negotiators rather than any certainty about the nature of the statistic or "fact" to begin with. For example, they argue that the climate change negotiations have apparently maintained the predication that global temperatures will rise by between 1.5 °C to 4.5 °C between the next 50 to 150 years, even though a variety of alternative predictions have emerged in the meantime. Yet instead of engaging in debates about the accuracy and meaning of the statistical prediction, negotiators have continued to use these figures in order to maintain constancy from which to negotiate other, more pressing concerns. As Hajer wrote:

> Storylines are essential political devices that allow us to overcome fragmentation and come to discursive closure. ... The point of the storyline approach is that by uttering a specific element one effectively reinvokes the storyline as a whole. It thus essentially acts as a metaphor ... they allow the possibility for problem closure ... a storyline provides the narrative that allows the scientist, environmentalist, politician, etc. to illustrate where his or her work fits into the jigsaw.
>
> (1993: 56)

A further step toward constructivism, and the scientific model of extended translation, is Actor Network Theory (ANT) (Law, 1991; Long and Long, 1992; Law and Hassard, 1999). ANT is an approach to narratives and storylines that highlights the influence of historic actors and networks in establishing both political prioritizations for environmental debate and hybrid boundaries between "nature" and "society." Early work on ANT by Callon (1986) referred to an analysis of the scientific analysis of scallop production and scallop collectors in France. The scientists and scallop collectors could each be seen to be identifying and then recruiting different elements of "society" and "nature" (including the scallops themselves) into their own networks in order to support the particular explanation of production of scallops proposed. In this sense, ANT refers clearly to the extended translation model of scientific knowledge, in which the boundaries between what is considered "real," "natural," and "social," with implications for policy and land management, are a function of the different networks reflecting different actors and purposes of analysis. In this sense, "actor" does not necessarily refer to the overt scientists or individuals involved in each network, but to the French word, *actant*, meaning the active involvement of objects, individuals, or groups on behalf of a purpose. ANT may therefore be seen as a particularly advanced approach to the impacts of historical science and politics on "relational materiality" – or the construction of facts and reality according to different perspectives. It is also increasingly a sociological tool in the identification and reflexive critique of the individual in shaping and reporting reality (see Law and Hassard, 1999).

Different models of science

As a result of these diverse ways of explaining constructivist alternatives to "orthodox" science, Callon (1995) classified science into four main categories. The categories reflect the generative mechanisms for knowledge adopted, and the claimed ability of each model to generalize about the world.

The first, obvious, category is orthodox science, or science as rational knowledge. Under this typology, science adopts the frameworks of orthodox scientific methods and the social regulation of findings by the scientific community. As discussed above, this approach to science implies an overall accuracy, logic, and progression to the generation of knowledge, in which the dynamics of the inner system of conjecture, refutation, and further conjecture is the mechanism by which science advances. This position generally acknowledges the role of social directions of science through the testing of theories (under the critical rationalism of Popper, see Chapter 3). But this model more usually assumes that the workings of the scientific method, and the social regulation of findings, reduce any further social impact on the relevance, social framing, or political manipulation of knowledge. This first category, of course, is the most optimistic and forgiving model of science, although many career scientists do not always accept it uncritically.

The second model refers to the advancement of science according to competition between different scientists, organizations, or research programs. The essence of this second model is that the competition for publication, funding, or public recognition is the underlying dynamic in scientific work and progress. In this sense, the second model is still closely related to the first, rational, model of knowledge generation because it is still assumed that scientists adhere to the scientific method and place faith in the social regulation of findings. Yet unlike the first model, the focus on competition points to the social and political factors in the planning and dissemination of research that are not expected to exist within the Mertonian norms of fair practice and open debate expected of the initial model. The writings most closely associated with the second model of science are Kuhn's (1962) analysis of the social basis of paradigm change, and Lakatos's (1978) discussion of competition and paradigms within research programs.

The third model approaches science as socio-cultural practice. Under this approach, scientific knowledge is seen as reflecting the social, cultural, and political influences that created it. Consequently, there is no possibility of scientific progress in the conventional sense of the gradual uncovering of reality through objective research. While it is possible that science as socio-cultural practice may reflect aspects of biophysical reality, the knowledge about that reality cannot be separated from such contexts, in the framing, the sampling, or the purposes to which such science is put. A further implication is that the nature of scientific knowledge is inherently historical, as it reflects the catalog of social change and influence of historical actors that have influenced the evolution of knowledge over the years (see Pickering, 1995).

Finally, the model of extended translation is the most far-flung criticism of orthodox science. Under extended translation, scientific knowledge is not just a reflection of socio-cultural practices, but also a social shaping of boundaries between "social" and "natural" worlds. Social framings of the "natural" world in essence create hybrid blends of social and physical objects, such as Latour's quasi-objects or Haraway's cyborgs. Science is not seen as a rational and epistemologically unbiased process free from social influence, but instead as part of wide networks linking biophysical entities, technical devices, statements of causality, and humans. "Objects," therefore, cannot exist separately from networks in which they are located, but must be seen as an integral part of the networks in which they are located.

The model of extended translation has important implications. First, if "objects" or explanations exist because of the networks that gave rise to them, then changing these networks might give rise to new objects and explanations. For example, as discussed in Chapter 3, Searle's (1995) concepts of "brute" and "institutional" facts proposed that objects might be relabeled according to the function ascribed to them. This might occur when one object such as a "pen" could equally be used as a "back-scratcher," or when expensive environmental equipment is used by local people as building materials. This multiple use for specific objects may be called symmetrical interdependence of the observed (i.e. objects) and the observers (i.e. science, society).

A second implication is that representations of reality generated through extended translation may only be seen to be accurate when the same conditions within the network are recreated at distant places. Under this assumption, Latour (1983, 1988), for example, argued that Pasteur's scientific experiments on anthrax led to the imposition of his laboratory-based assumptions all over France, on the tacit assumption that the laboratory results only succeeded in "the field" when the same laboratory-type conditions are recreated. Similarly, the Universal Soil Loss Equation (USLE) (see Chapter 2) might also be seen as an exercise in extending the experimental network of conditions used to develop the equation in the Plains of the USA to a variety of locations where either physical conditions of soil formation, rainfall, or social and cultural evaluations of erosion are different. More overtly, transferring such scientific knowledge and networks elsewhere may also reinforce or legitimize those networks. .

These four models are convenient ways to summarize the different positions between the orthodox, or rational model of science, and progressively more constructivist positions. Understanding how different explanations of environmental problems emerge as the interaction between complex biophysical processes and politics is a key objective of a "critical" political ecology. The following chapters now build on these discussions by assessing how science and politics co-evolve dynamically.

Summary

This chapter has built on the discussions in Chapter 3 by describing further ways in which scientific statements are shaped by social processes. The chapter summarized debates in Sociology of Scientific Knowledge that refer to the influences of social framings, problem closure, and language on how aspects of environmental change are perceived and then institutionalized within science. This process is also affected by social divisions such as between class, gender, and race, although sometimes it is difficult to remedy these divisions without also reinforcing these divides.

The chapter discussed the important influence of metaphors and narratives upon how environmental problems are seen. Such narratives, or storylines, of environmental degradation such as "deforestation" reflect a history of different experiences of environmental problems by specific social actors, who have had most impact on defining and giving meaning to different explanations of degradation. The study of "boundaries" in science, in terms of who participates in inquiry, and how divisions are drawn around complex "hybrid" objects and processes, offers a way to understand how environmental science has evolved. Democratizing environmental science, or showing how current explanations reflect such historic actions, therefore depends on showing how such hybrid concepts have evolved, and with (and without) whose input.

The chapter concluded by summarizing some different approaches to theorizing social institutions and science. These approaches included pragmatism; Cultural Theory; and a variety of poststructuralist debates such as environmental narratives, storylines, and Actor Network Theory. Such approaches allow a variety of alternatives to the frameworks of orthodox science, which, as discussed in Chapter 3, has contributed to the creation of environmental orthodoxies, or inaccurate and unrepresentative environmental science. Most crucially, the chapter argued that many alternative approaches to science show that truth statements are dependent on social networks that enable consensus to be reached, or for laboratory conditions to be replicated in different locations. Environmental science may therefore appear to succeed when those networks or conditions are recreated. Yet changing these conditions may also lead to different scientific outcomes, and a rethinking of the purposes of science. The following chapters apply these thoughts to a revitalization of political ecology, in order to make environmental politics more aware of the contingent and constructed nature of many environmental explanations commonly assumed to be "fact."

5 The coproduction of environmental knowledge and political activism

The previous chapters have summarized a variety of debates relating to the social and political influences on science. Chapter 3 described arguments in Philosophy of Science that claim scientific "laws" are not universally applicable, but instead reflect a variety of social and institutional influences on how inference is made. Chapter 4 drew largely from debates in the Sociology of Scientific Knowledge to indicate how supposed scientific "facts" about environment reflect wider social framings and discourses, which have also evolved historically. Now, in Chapter 5, we look at how these themes may be combined to identify how social change and environmental science co-evolve dynamically. The chapter will:

- introduce and define the concepts of coproduction and hybridization that describe how environmental knowledge and politics co-evolve dynamically;
- demonstrate how environmentalism, as a "new" social movement, helped shape many general beliefs and discourses about environment that have since been used to explain the causes of environmental degradation; and
- illustrate how such general beliefs – when used uncritically in new contexts – may fail to acknowledge complex biophysical causes of environmental changes, or alternative framings of environmental change by people not included in the formation of the explanations.

In particular, this chapter focuses upon general beliefs such as linkages between environmental degradation and capitalism; or the association of degradation with political oppression and the "domination of nature." The chapter does not suggest that criticisms of capitalism or social oppression are misplaced, but argues it is necessary to see how political activism linked to the criticism of capitalism or oppression has shaped beliefs about the causes of environmental degradation.

This chapter therefore helps to build a "critical" political ecology by showing how science and politics co-evolve, and by arguing that many common assumptions about environmental degradation need to be reconsidered in order to acknowledge such political influences. Chapters 6 and 7 build on these arguments by exploring in more detail the political agency and globalization of scientific explanations. Chapters 8 and 9 then seek ways to make such political influences clearer through reforms to research practices and ways of debating environmental science.

Coproduction and hybridization

This book has already summarized many ways in which politics and society may influence the production and dissemination of science and scientific "laws" (see Chapters 3 and 4). There is now a need to analyze how science and politics co-evolve dynamically, in order to understand how such political factors lead to the evolution of hegemonic environmental explanations. This understanding is necessary in order to build a "critical" political ecology, which integrates the evolution of science into the analysis of environmental politics.

Perhaps the most useful and all-encompassing conceptual device for explaining the mutual evolution of science and politics is "coproduction." Sheila Jasanoff (1990, 1996b: 393) defines coproduction as "the simultaneous production of knowledge and social order." The term refers to processes by which knowledge, including scientific knowledge, is framed, collected, and disseminated through social interaction and change, and how such knowledge also impacts upon such change. "Social order" does not necessarily refer to a state of apparent political stability, but can also describe the struggle for order, or conditions of enforced order. The important principle of coproduction is that it is a *dynamic* process, in which knowledge and society continually shape each other. The concept reflects earlier insights by Foucault, who wrote, for example: "the exercise of power perpetually creates knowledge and, conversely, knowledge constantly induces effects of power" (1980: 52).

The term "coproduction' is an important backcloth to the influence of political action on the generation and legitimization of scientific knowledge. Some critics of constructivist approaches to science have feared that acknowledging such social influences on science may imply a relativist approach to environment, or the belief that there is no "hard" reality beyond the language and concepts developed by society. These fears are exaggerated. The concept of coproduction does not imply there is no "external reality" or biophysical world that exists beyond human experience. But it does mean that knowledge about such a biophysical world cannot be separated from social influences, and particularly from how society is clustered and organized. Some historic work on coproduction has focused specifically on the relationship of policy processes and scientific knowledge, and especially on how social or political institutions may shape and enforce knowledge because of the interaction of political actors (e.g. Jasanoff, 1990; Guston, 2001).

Another term that may be discussed alongside coproduction is "hybridization." Researchers of the social construction of environmental "facts" have used this term to refer to the (often historical) processes by which social–natural objects become entwined through social discourse, and henceforth become accepted as objects. The concept is based on Bruno Latour's discussion of hybridity, and the emergence of "quasi-objects" through the social bounding of nature–society interfaces. As Swyngedouw illustrates: "Hybridization is a process of production, of becoming, of perpetual transgression..." (1999: 447).

> If I were to capture some water in a cup and excavate the networks that brought it there, "I would pass with continuity from the local to the global, from the human to the nonhuman" (Latour, 1993: 121). These flows would narrate many interrelated tales, or stories, of social groups and classes and the powerful socioecological processes that produce social spaces of marginality; chemical, physical, and biological reactions and transformations, the global hydrological cycle, and global warming; capital, machinations, and the strategies and know-ledges of dam builders, urban land developers, and engineers; the passage from river to urban reservoir; and the geopolitical struggles between regions and nations. In sum, water embodies multiple tales of socio-nature as hybrid.
>
> (ibid.: 445–446)

Hybridization is similar to coproduction because it refers to the current and historical social factors that have contributed to what we refer to as reality. Unlike coproduction, however, many historic discussions of hybridization have focused specifically on the processes of framing and bounding that have led to the evolution of specific objects or terms as socio-natural entities. Discussions of coproduction on the other hand have generally considered the broader framing, legitimization, and dissemination of knowledge within society and policy processes. Both terms, refer to the dynamic co-evolution of knowledge and social change.

The following discussion now applies the terms coproduction and hybridization to the emergence of ecological "facts" and discourses that have emerged simultaneously with political activism related to environment. As discussed throughout this book, much conventional academic analysis of environmentalism has tended to separate politics from science in ways that do not acknowledge their mutual embeddedness. A crucial component of a "critical" political ecology, therefore, is to acknowledge the coproduction of environmental science and political activism in ways that show how assumptions about the nature and causality of environmental problems are interlinked with such activism.

The social framings of environmental concern

Concern about environment has, of course, existed in various forms and locations for centuries. Yet it is widely recognized that the main impetus for the emergence of environmentalism as a topic of national and international significance in recent years has been the political activism known as the "new" social movements in North America and Europe since the 1960s. Many campaigners have claimed that this activism resulted from increasing evidence from science about the threats to the environment (see Chapter 1). But it is perhaps surprising that there has been relatively less attention to how far such science and activism have been inter-connected and influential on each other.

The purpose of this section of the chapter is to assess the interconnection – or coproduction – between environmental activism and environmental science. By doing this, the chapter does not criticize environmentalism, but seeks to demonstrate how many environmental themes are embedded in wider political and social concerns, and how such concerns also shape explanations of environmental degradation. Three key themes are identified: resistance to capitalism and modernity; perceived loss of wilderness; and concerns about the so-called domination of nature. This section summarizes how each of these has been portrayed in debates about environmentalism. The next section discusses how these may have influenced environmental science.

Opposition to capitalism and modernity

The rise of environmentalism as a globally significant political force in the 1960s was, of course, inspired in part by important new findings about human impacts on ecosystems. But it is also important to acknowledge the impacts of contemporaneous social and political changes within the advanced industrial economies of Europe and North America, and the impacts of so-called "new" social movements on political debate.

The term "new social movements" has been used to describe the kind of social activism that emerged in Europe and North America during the 1960s, and which was associated closely with new "identity-based" politics such as women's rights, gay rights, and peace campaigns. They were called "new" because they were seen to differ from "old" social movements based upon material interests often represented by different economic classes (Morris and Mueller, 1992). Many scholars have argued that new social movements are both symptoms of, and solutions to, the contradictions inherent in modern bureaucratic society, and emerge because of tensions between the regulation of society by the state, and the emerging autonomy and diversity of identities experienced within postindustrial societies (e.g. Habermas, 1981; Touraine, 1981; Offe, 1985).

One particular theme associated with new social movements has been the rejection of so-called "instrumental reason," or the domination of social life and politics by functional and managerial principles usually associated with economic exploitation or distant and controlling forms of government (see Calhoun, 1992; Benhabib, 1996). Herbert Marcuse, for example, particularly in his *One Dimensional Man* (1964), wrote of the repression of the erotic and playful aspects of human nature under a society organized toward functional industrial or capitalist objectives. Social activism under such conditions sought to achieve a connection with Eros, or a sense of expressiveness lacking in modern society. Linked to this, many social activists also criticized the dependency of society on science and technology as agents of instrumental reason. Scient*ism* – or the mechanistic, non-emotional evaluation of decisions and rationality – was seen to be a key way in which instrumental reason was enforced on society at large, without relevance or sensitivity to the lifeworlds of individuals (Alford, 1985).

Another theme associated with new social movements is the belief that social activism could benefit all society rather than one specific class, even if most activists came from more educated, middle classes. Giddens (1973), for example, proposed that the "new" middle class of postindustrial societies is "class-aware" but not "class-conscious." Offe similarly wrote: "New middle class politics, in contrast to most working class politics, as well as old middle class politics, is typically a politics *of* a class but not *on behalf of* a class" (1985: 833, emphasis in original).

To this extent, new social movements therefore aimed to speak on behalf of all society, even if the speakers did not come from all society. Berger (1987), somewhat critically, argued that this, in effect, created a so-called "knowledge class" in which a new division of society becomes responsible for the creation and consumption of knowledge about itself. Stehr (1994) similarly argued that postindustrial society could also be characterized as a "knowledge society," or one that increasingly creates knowledge about itself for its own analysis.

There are many links between these characteristics of new social movements and the emergence of environmentalism as a political force. Environmentalism is commonly considered one of the classic new social movements. Indeed, Szerszynski *et al.* wrote: "it is as a challenge to the subjugation of the natural lifeworld by the ravages of state and technology that environmentalism began as a critical discourse, rich in cultural resources and resonances" (1996: 4).

The emergence of environmentalism as a new social movement, therefore, was embedded within a growing field of political activism that was associated with a variety of wider themes concerning the impacts of modern industrial society on the lives of its citizens, and assumptions about the representativeness of activism for all society. As social theorists voiced concern about the suppression of lifeworlds from capitalism and modernity, environmentalists also drew attention to the "natural lifeworld" and the "ravages of state and technology." Such statements have had implications for the evolution of environmental explanation. These implications are discussed later in this chapter.

The loss of wilderness and tradition

It is also well recorded that environmentalism, in its dominant forms as experienced in Europe and North America, has been linked to the growing appreciation of particular landscapes as wilderness, and that such wilderness is both beautiful and fragile. The work of the American historian Roderick Frazier Nash (1973), in particular, argued that the growth of some leading environmental membership organizations in the USA was linked to increasing industrialization and urbanization, especially during the early twentieth century. As has been widely argued elsewhere (e.g. Woodgate and Redclift, 1994), the implication of this historic association between socio-economic trends and environmentalism is that the early framings and objectives of the political force known as environmentalism were set by the concerns of people who were living in increasingly industrial and urban contexts.

Such a transition, however, can also be linked to social changes associated with greater modernization in general, and the emergence of a narrative of "nature" that is influenced by the growth of industry and cities. Giddens, in particular, has pointed to the emergence of environmentalism, or a specifically rural, or "green," conservationist vision of environment, to indicate the perceived loss of heritage and tradition in modern societies. In this sense, environmentalism, or conservationism, offers both a sense of continuity with a lost past, but also a useful antidote to the strictures of modern industrial, urban life: "while environmentalism can largely do without it, 'nature' is as important to ecological thought as 'tradition' is to conservation" (1994: 204).

According to this perspective, the very concept of "nature" as a feature of environmentalism may also be seen as an attempt to protect, or even recreate, essences of wilderness and tradition that are seen to be threatened under modernity or industrialization and urbanization. Similarly, the concept of "ecological crisis" may also be a metaphor for the perceived loss of balance under modernity: "the ecological crisis is a crisis brought about by the dissolution of nature – where 'nature' is defined in its most obvious sense as any object or process given independently of human intervention" (ibid.: 204–206).

Of course, such statements are not meant to deny the existence of the vast changes in landscape or ecosystems that have been caused by industrialization and urbanization, or to suggest that worrying about such impacts is unjustified. Yet, it is worthwhile noting how concepts of wilderness and tradition have influenced environmentalist concerns, and accordingly how far these concerns have affected explanations of how environmental degradation occurs.

Such influences can indeed be identified in much classic writing about environment. Rachel Carson's extremely influential and gripping account of the impacts of pesticides on ecosystems, *Silent Spring* (1962), famously starts with an account of a somewhat traditional and romanticized landscape affected by sudden change. She wrote:

There was once a town in the heart of America where all life seemed to live in harmony with its surroundings. The town lay in the midst of a checkerboard of prosperous farms, with fields of grain and hillsides of orchards where, in spring, white clouds of bloom drifted above the green fields. In autumn, oak and maple and birch set up a blaze of color that flamed and flickered across a backdrop of pines. Then foxes barked in the hills and deer silently crossed the fields, half hidden in the mists of the fall mornings ... Then a strange blight crept over the area and everything began to change. Some evil had settled on the community: mysterious maladies swept the flocks of chickens; the cattle and sheep sickened and died. Everywhere was a shadow of death...

(1962: 1–2)

Similarly, the motives and concerns associated with the new social movements may also be seen in the construction of environmental change with the assault on personal identity and expressiveness resulting from modern instrumental reason. Some writings about trees and forests, as particularly redolent of lost harmony, echo the writings of Marcuse about the loss of Eros under modernity. For example, in a book romantically entitled *The Power of Trees: The Reforesting of the Soul*, Michael Perlman writes:

Our relationships, human and nonhuman, whatever their quality, inevitably involve combinations of similarity, sameness, and intimacy on the one hand; and of difference, distinction and more radical alienation and estrangement on the other. In exploring what trees tell us about these combinations, we explore what they can mean for the nature of ecological relationship in its fullest sense – for Eros in its fullest sense ... a dangerous alienation from our natural, ultimately our *forest* roots, lay at the heart of the global ecological crisis...

(1994: 4, emphasis in original)

Much environmental concern at perceived changes in landscapes may therefore not simply be a reaction just to those changes alone, but to the increasing perception of such landscapes as representative of lost harmony and tradition under the wider process of modernization. Such changing perceptions, of course, do not imply that radical, or degrading, changes in landscape have not occurred. Yet it is important to note the contextualization of such reactions to environmental changes within wider social trends that increasingly perceive lost tradition and harmony as important problems. The implications of these trends for the explanation of environmental issues are discussed later in this chapter.

The domination of nature

One further pervasive influence on environmental concerns has been debates about so-called "domination of nature." Going back to the British philosopher, Francis Bacon (1561–1626), this concept refers to the anthropocentric and exploitative style of government and development that has been associated with the Enlightenment, and dominant streams of thought in Europe since the 1500s. As stated above, the concept of "nature" was originally discussed within the new social movements as an indication of human nature, or the expressiveness and vitality that was lost under oppressive state authority and industrialism. Much environmental debate has built on this theme, and asserted that environmental degradation, or domination of environment and resources, has resulted from the same instrumental causes that dominate human nature (see Agger, 1992).

Marcuse, for example, famously wrote in *One Dimensional Man*:

> Science, by virtue of its own methods and concepts, has projected and promoted a universe in which the domination of nature has remained linked to the domination of man [sic] – a link which tends to be fatal to this universe as a whole.
>
> (1964: 166)

And later: "The domination of nature has remained linked to the domination of man [sic]" (1969: 31; see also Vogel, 1996).

Marcuse was writing about the oppression of humanity under capitalism and modernity. But such comments may also be seen in statements relating specifically to environmental degradation. For example, Alford (1985) spoke of environmental problems as evidence of the "revenge of nature" resulting from social oppression of modern industrial society.

Furthermore, the ecocentric environmental writer Robyn Eckersley wrote:

> The achievement of a rational, democratic consensus by an informed citizenry concerning societal goals is being increasingly subsumed by a technical discussion by a minority of experts concerning means (based on presupposed ends, namely, economic growth, the expansion of the bureaucratic-technical apparatus, and the domination of human and non-human nature).
>
> (1992: 107)

And the former US Vice President, Al Gore, in his book: *Earth in the Balance*, wrote: "we have become so successful at controlling nature that we have lost our connection to it" (1992: 225).

Such statements draw attention to important concerns about democratic governance and the loss of personal meaning under modern economic and social systems. Yet they also include assumptions about environmental degradation that can be questioned for a variety of reasons. First, these statements refer to a sense of "nature" but do not consider what this may mean, or how far this concept varies between different people or cultural contexts. As Soper (1995) notes, the word "nature" denotes essences innate in bodies, something universal and essential, and consequently without diversity. Yet "nature" (and "domination" too) reflects a variety of social factors, and can vary according to individuals, cultures, and societies. Furthermore, as noted above, earlier discussions of "nature" by writers such as Marcuse, emphasized the lost vitality and expressiveness of humanity under oppression, but this usage has diversified over years to also include a reference to environment and biophysical resources (see Castree, 1995, 2001; Braun and Castree, 1998). Many discussions of the word "nature" therefore contain references to both human vitality and environmental resources in ways that do not make this shared heritage of the word clear. (The implications of environmental activism for discourses of nature are considered in more length below, pp. 115–126).

Second, these discussions of the impacts of oppressive economic and political regimes on nature do not refer to how environmental changes experienced as degradation may result from factors outside immediate human experience, or causes unrelated to social systems of oppression. It is somewhat reductionist to explain environmental change and degradation by referring only to social systems of inequality and oppression. It is also simplistic to make such causal links between social oppression and environmental degradation when there is also debate about how we have come to perceive society as oppressed, or wilderness as threatened.

Such concerns have already been discussed by a variety authors concerned that the term "domination of nature" be interpreted too pessimistically and critically in relation to economic development. John Bellamy Foster (2000), for example, has argued that the origins of the debate about domination of nature needs to acknowledge the co-evolution of concepts of ecology and material development that emerged from the seventeenth to nineteenth centuries. Indeed, Foster argues that there is more evidence within the writings of Marx and other theorists of material development to indicate care for environmental protection than commonly thought. Similarly, environmental thought associated with the new social movements is also more influenced by the seventeenth-century scientific revolution. As a consequence, it may be possible to integrate forms of economic development and ecological protection in ways that avoid the simple belief that scientific and technological development necessarily lead to the domination of nature and environmental degradation (see also Leff, 1995). Yet achieving an integration of material development and ecological protection requires reassessing both the social justice of development, and the ways in which ecological degradation is defined. Both concerns are discussed later in this chapter (pp. 116–123).

This section has summarized simply some of the key social framings of environmental concern that may have implications for the coproduction of environmental science. The following sections now expand on these by discussing the implications of such framings for scientific practice, and for overriding discourses of environmental degradation that commonly form justifications for environmental debate and policy.

Implications for scientific practice

The preceding section aimed to show that much environmental concern has attempted to draw attention to human impacts on ecosystems, yet has been framed within wider social apprehensions about the suppression of human vitality and expression under modern industrial societies. The objective of outlining these social frames is not to suggest that humans do not impact on ecosystems, or that impacts are not important, but to indicate that the scientific knowledge coproduced with environmental activism has reflected social concerns. The next sections discuss the implications of these framings for scientific knowledge. The first section discusses implications for science itself. The second section analyzes implications for more general discourses of environment and society that are used to frame environmental debate and lead to the generation of further research and knowledge.

The contradictions of science and environmentalism

As discussed above, much early environmentalism in the 1960s was rooted in wider social concerns about the impacts of modern industrial society upon human nature, and ecosystems. Yet critics have also pointed to how this position presents an apparent paradox of both blaming modern science and technology for environmental degradation, but also relying on natural science for information about environmental change and human impacts (see Alford, 1985; Yearley, 1992; Vogel, 1996; Gandy, 1997).

Marcuse's statement in *One Dimensional Man* that "science ... has projected and promoted a universe in which the domination of nature has remained linked to the domination of man" (1964: 166) is an example of this criticism of science. Alford also summarized this contention from the Frankfurt School that:

> a particular *idea* of science held by Marcuse and Habermas ... suggests that the very nature of modern science, its core concepts, indeed its essence – is one dimensional, necessarily instrumental, indeed domineering, in its orientation toward nature.
>
> (1985: 9)

Environmental writers have reflected this concern about science in various ways. Miller (1978: 98), for example, stated that the subservience of science to commercial technology was proof that much scientific work was instrumental, and linked to the wider systems of capitalism that oppressed society. Yearley (1992: 552) quoted the British environmentalist, Jonathan Porritt, as writing "the scientists are now with us rather than against us" in order to indicate a gradual transition in the relationships of scientists and environmentalists from one of opposition to one of agreement about environmental protection. Indeed, some noted environmentalists have also claimed science to be totally supportive of environmental positions, such as Paul and Anne Ehrlich, in their book, *Betrayal of Science and Reason: How Anti-Environmental Rhetoric Threatens our Future* (1996) (see also Chapters 1 and 2).

Theorists have explained this apparent ambivalence of environmentalists toward science by describing the difference between "science," as a methodology of producing statements about reality, and "scient*ism*," as the use of science and technology for instrumental and exploitative objectives. (For example, the classic debate between Habermas, who criticized scientism, and Marcuse who saw science in general as problematic, see Alford, 1985; Vogel, 1996.) Yet, as noted in Chapter 1, it is still unclear how to reconstruct science in order to reflect less instrumental social concerns. Marcuse hoped to address this problem by developing a so-called "new" science, in which analysis and inquiry could be used to avoid instrumentality, and instead revitalize society toward more humane objectives (this theme is discussed more in Chapter 8).

Similarly, it is also unclear how environmentalists have harnessed scientific findings in order either to legitimize or oppose different positions in environmental campaigns (this theme is illustrated in relation to Genetically Modified Organisms later in this chapter). Any assessment of the coproduction of environmental activism and science therefore requires an acknowledgment of how "science" is identified and portrayed within different political disputes, and under what circumstances it is seen to be beneficial to each argument. These themes are addressed throughout the rest of this book, and specifically in Chapter 8, which considers new approaches to scientific methods; and in Chapter 9, which focuses on the discussion of and participation in science by the public. But first, we can summarize some ways in which coproduction and hybridization have occurred in relation to specific environmental debates.

The coproduction of environmental science and values

Perhaps the strongest links between the emergence of environmentalism as a new social movement and the coproduction of environmental knowledge is in the influence of environmentalism on scientific practice. Some authors have argued that this link is closer than many scientists would like to claim, particularly in relation to the underlying notions of equilibrium ecology, or "nature in balance" characteristic of much historic ecological thinking.

Shrader-Frechette and McCoy (1994), for example, have argued that the non-equilibrium debate in ecology has demonstrated the high dependency of ecology-as-science upon dominant normative judgments about how the world should be, rather than in critically assessing evidence for common framings of environmental concern. In particular, they have pointed to the variety of interpretations and calls for a view of "nature in balance"; the lack of empirical support for this concept; and the various political (and argumentative) reasons conservationists might use to demonstrate a belief in a balance to nature. As a result, Shrader-Frechette and McCoy have argued that much ecological research has been conducted along predefined norms of how nature should be. They wrote:

> Developers could point out that, if there were no conclusive ecological evidence for a "balance of nature" preserved in pristine wilderness, then postulating such a balance would require opponents of development to make contextual and methodological value judgments. Because positing some "balance" is dependent on making such value judgments, then it would be more difficult for environmentalists to argue against such development on the grounds that it might destroy some inherent "balance." Such disputes over development versus preservation illustrate that, because of the magnitude of the empirical underdetermination of ecological theory (regarding balance or stability), scientists have been forced to make methodological value judgments about which, if any, account of balance or stability to pursue. Alternative value judgments about stability, in turn, have different consequences for environmental values and policy.
>
> (1994: 113)

As an example, Schrader-Frechette and McCoy analyzed disputes concerning hunting of wild animals in the USA. Much ecological theorizing on the subject of hunting (in effect, another environmental orthodoxy) has focused on case studies that demonstrate how populations such as deer grow rapidly with the removal of (natural) predators. A well-known case study is Kaibab Plateau, near the Grand Canyon (Leopold, 1933). Using such case studies, theorists have proposed that predators have a structuring influence on the populations of herbivores, and in effect keep the populations in "balance." In political terms, however, this conceptualization has been accepted or opposed in a variety of ways. Supporters of hunting, for example, have argued that the culling of prey populations is necessary because predators are often killed by hunters and livestock owners, and so there is a risk of the prey population rising too high. Some opponents to hunting (such as the National Audubon Society) also accept this model, but claim that "natural" predation is more effective than hunting conducted by humans, because such hunting may have negative impacts on population by selecting only trophy animals. Yet there are also other groups who oppose hunting because they also reject the initial premise that predation is the controlling influence on population (other factors, such as competition, may be more important). The underlying problem is not the need to choose which position is correct, but to identify that the lack of certainty or agreement about the causal link of predation and population has led to the selection of alternative scientific bases for political arguments through the process of methodological value judgments. As the authors write:

Perhaps ecologists and environmentalists have placed too much faith in algorithms, general ecological theories, or hypothesis–deduction methods that would preclude the necessity for tough-minded, situation-specific, methodological analysis in ecology and for sophisticated natural-history knowledge of individual taxa.... The hypothetical–deductive foundation of general scientific theories is undercut by the presence of methodological value judgments.

(Schrader-Frechette and McCoy, 1994: 115–116)

Such research on the coproduction of scientific knowledge and values indicates three important conclusions. First, many supposedly universal and politically neutral statements about ecological reality reflect the social and political circumstances in which knowledge was produced. Second, many scientific statements are used to give added legitimacy and urgency to different political viewpoints when there is actually much debate and uncertainty about such statements. Third, often the interactions of political argumentation provide a direction for knowledge production as different actors seek ways to legitimize and strengthen their political positions through scientific statements.

As discussed in Chapter 4, such conflictual knowledge claims about environmental change may be explained by reference to a variety of social science debates. Cultural Theory, for example, would counterpose hunters and conservationists in terms of different "myths" of nature (see Figure 4.2). More poststructuralist accounts would emphasize the historical evolution of different storylines or narratives of environmental explanation, such as the adoption of equilibrium, or balance-based approaches to environment. Also, the adoption of different scientific research as legitimate also depends upon the upholding of a network of scientists and policymakers that sees such results as authoritative. These themes are illustrated again in the chapter in relation to Genetically Modified Organisms, and later chapters focus on the means to increase public scrutiny of the normative basis upon which scientific statements are established. The next section now examines some of these normative statements and assumptions about environmental change in more detail.

Implications for discourses of nature and society

The preceding section suggested some ways in which environmental activism may impact upon the use of science, and the construction of scientific statements through orthodox scientific practice. But environmental activism has also impacted upon more generalized discourses of environmental degradation that are often discussed as underlying truths in environmental debate.

This section now discusses some of these popular discourses of nature and society in order to show how these two are embedded within wider social and political concerns. This discussion does not suggest that these concerns should be dismissed, but instead seeks to demonstrate that such general discourses can sometimes be simplistic and overlook more complex biophysical causes of environmental degradation, and the differing experiences of environmental change between different social groups, or at different times and places.

In particular, this section questions some common assumptions made in many current studies of Political Ecology, concerning the focus on capitalism as the ultimate cause of environmental degradation, and the role of environmentalism as an emancipatory force in politics. Advancing alternative approaches, and integrating political analysis with an awareness of biophysical factors of environmental change, is the purpose of a "critical" political ecology, and will be expanded in later chapters.

Ecology as a critique of capitalism

As discussed earlier, the new social movements were largely associated with resistance to the oppression felt under modern industrial societies, and the instrumental reason of capitalism. Environmentalism, as another new social movement, reflected these concerns, and many environmental writers made the association between capitalism and environmental degradation, or the so-called "domination" of nature with the domination of people. Such concerns still underlie many general discussions of environmental issues and problems. Immanuel Wallerstein, for example, wrote:

> There has been an unfortunate tendency to make science and technology the enemy, whereas it is in fact capitalism that is the generic root of the problem. To be sure, capitalism has utilized the splendors of unending technological advance as one of its justifications. And it has endorsed a vision of science – Newtonian deterministic science – as a cultural shroud, which permitted the political argument that humans could indeed "conquer" nature, should indeed do so, and that thereupon all negative effects of economic expansion would inevitably be countered by inevitable scientific progress...
>
> (1999: 9)

Similarly, the association of environmental degradation and capitalism has become a defining feature of many recent discussions of Political Ecology. Watts and McCarthy wrote:

> A compelling and liberatory political ecology must begin with an accurate understanding of capitalist dynamics for the simple and profound reason that they lie at the roots of most problems with which political ecology concerns itself.
>
> (1997: 85)

And Bryant:

> Much Third World environmental research is caught up with questions of proximate, as opposed to ultimate, causation ... there is no consideration in such accounts of the possibility that both poverty and environmental degradation may be linked to economic development within a globalized capitalist system.
>
> (1997b: 6–7)

Indeed, O'Connor (1996) and Benton (1996) have also pointed to a proposed "second contradiction of capitalism" resulting from environmental degradation, in addition to original tensions resulting from appropriation and alienation of workers within it (see also Lipietz, 2000). Writing in a similar vein, Luke also links the negative impacts of capitalism with the oppression of nature, and the political instability such oppression may bring:

> The successful establishment of new social relations organized along these ecological lines might radically alter the social construction of nature in relation to society, making nature again into a subject, not an object, an agency, not an instrumentality, and a more than equal partner, not a dominated/subaltern force... At the same time, no rationalizations of nature's continued destruction could be countenanced in exchange for the false promise of more jobs, greater prosperity, added growth, or closer technological control. Guarantees of ecological security should in turn ramify into greater freedom, dignity, and reasonability for the human beings whose own autonomy suffers in nature's abusive indenturing to corporate enterprises' instrumental rationality.
>
> (1999: 23)

Such statements, of course, draw well-justified attention to the marginalizing and oppressive elements of global capitalism, and the implications this brings for poverty, vulnerability to environmental change, or unequal access to resources. Yet the immediate linking of environmental degradation to capitalism *per se* has three major problems that influence how we explain and hence manage environmental degradation.

First, the explanation of environmental degradation through capitalism alone may be unnecessarily reductionist because it does not refer to biophysical factors that exist independently of such conflicts between economy and society. Such simplistic explanations of apparent environmental degradation were pointed out by Blaikie in relation to one well-known environmental problem, desertification:

> The case for the globalization of capital being causal in desertification looks rather amateur, since the scientific evidence of permanent damage to the environment points in other directions ... For want of attention to a large and accessible body of climatological and

ecological information, the case for adding desertification to the long list of other socially induced woes now looks very thin.

(1995: 12)

Second, commonly, many discussions of environmental degradation under capitalism do not refer to how such "degradation" is defined, or the particular storyline or history of the concept as it has been influenced (or hybridized) by previous experience, research, or debates. Indeed, as discussed in relation to desertification (or also soil erosion and deforestation, see Chapter 2), not all stakeholders or affected people may experience such supposed topics of degradation to be actually degrading to land uses. In Luke's statement above, there is also no attempt to define degradation other than an assumed destruction of "nature," resulting from its treatment as a "dominated/subaltern force": words that seem to owe more to the anti-capitalist rhetoric of Marcuse than recent work in cultural ecology or environmental change actually relating to what constitutes degradation.

Third, the adoption of such uniform condemnations of capitalism as a cause of environmental degradation may also work to disempower local forms of industrialization or entrepreneurialism that may provide means for the avoidance of poverty or social vulnerability. Local industrial development may, of course, be considered part of capitalist development in general, but this depends on the distinctions made between industrialism and capitalism, and access to local resources (see Corbridge, 1986).

The implication of these arguments is to suggest that – at times – anti-capitalist rhetoric may also adopt definitions of nature and environmental degradation that may both over-simplify the causes of environmental problems, and ironically lead to the added marginalization of currently disempowered groups. Indeed, as illustrated in Chapter 2, much recent research in marginal environments in developing countries has revealed examples of farmers or cultivators who have been restricted from certain land uses because they are seen to be ecologically damaging. Yet evidence suggests that such groups may be employing methods that actually protect resources, or are not responsible for environmental change as commonly thought. The simple association of capitalism, or economic growth in general, and environmental degradation may therefore add to this problem by failing to pay sufficient attention to the complexity of underlying biophysical factors causing environmental change; and encouraging the rejection of forms of economic activity that may be less environmentally damaging than assumed. In West Africa, for example, Fairhead and Leach (1996) describe numerous penalties available for people who are caught cutting forest or cultivating crops out of place. Yet these penalties do not acknowledge the diversity of practices that can impact on landscapes, or the role that some of these practices can play in protecting forests, or in enhancing the value of landscapes to local users (see Chapter 2, pp. 34, 43).

Generalizations about environmental degradation and economic development are not restricted to land use in developing countries. Box 5.1

Box 5.1 Environmental discourse and ecological modernization

The concept of "ecological modernization" has become widely debated within environmental policy as a potential means to integrate economic growth with environmental protection. Spaagaren and Mol (1992) have defined ecological modernization as the possibility to overcome the environmental crisis without leaving the path of modernization. In its most optimistic form, much attention to ecological modernization has focused on the increased adoption of environmental policy objectives by private business, such as practices of self-regulation, or waste reduction such as recycling; or the adoption of carefully managed environmentally regulatory procedures by the state. More critically, however, some ecocentric authors have questioned whether ecological modernization is a contradiction in terms because it necessarily implies accepting economic development, and hence must continue to enhance environmental degradation at a fundamental level.

For example, Gorz wrote:

> there can be no ecological modernization without restricting the dynamic of capitalist accumulation and reducing consumption by self-imposed restraint. The exigencies of ecological modernization coincide with those of a transformed North–South relationship, and with the original aims of socialism.
>
> (1994: 34)

Harvey wrote:

> As a discourse, ecological modernization internalizes conflict. It has a radical populist edge, paying serious attention to environmental–ecological issues and most particularly to the accumulation of scientific evidence of environmental impacts on human populations, without challenging the capitalist system head on ... it is also a discourse that can rather too easily be corrupted into yet another discursive representation of dominant forms of economic power.
>
> (1996: 382)

Similarly, Ulrich Beck's (1992) account of "Risk Society" describes the process of "reflexive modernization," which refers to economic growth in advanced societies with ultimately self-destructive outcomes. Reflexive modernization, as defined by Beck, stresses the destructive elements of economic growth that are intrinsically linked to the exploitation of resources, and yet which are increasingly difficult to control because of the ability for powerful economic interests to dominate the terms by which environmental policy is defined and measured.

Such critiques suggest that the objectives of ecological modernization are flawed because it may not be possible to address environmental degradation without also empowering the causes of degradation within the capitalist system. Yet not all critics of ecological modernization accept that the problem lies specifically in the reliance on capitalism, *per se*, but also on how development is defined, and who is allowed to define it. Such more discursive approaches to ecological modernization are typified by Hajer, who wrote:

> The challenge [of ecological modernization] does not concern the goal but the process. The challenge seems to be to think of organization of ecological modernization as a process that allows for social change to take place democratically and in a way that stimulates the creation of an – at least partially – shared vision of the future.
>
> (1995: 280)

Under this more discursive approach, it is more important to identify how, and for whom, economic growth can be beneficial or harmful. This approach avoids the uniform assumption that modernization must be ecologically damaging, but instead seeks to redefine more inclusive forms of modernization and environmental protection.

Source: Beck, 1992; Spaagaren and Mol, 1992; Gorz, 1994; Hajer, 1995; Harvey, 1996.

summarizes some statements about the concept of ecological modernization, and its use in advanced industrial economies to overcome environmental threats. On one hand, many authors assume a relationship between industrial development and environmental degradation (e.g. Gorz, 1994; Harvey, 1996). On the other hand, some critics have sought a more narrative-based, or contextual approach to how degradation is defined, and how far the actual discourse of degradation may be used as a way to empower business, or exclude others from the development process (e.g. Hajer, 1995). The implication of this example is that the criticism of capitalist industry as a cause of environmental degradation need not always imply an essential link between industrial development and degradation, but instead, as Hajer (1995) notes, the need to consider how such development may proceed democratically. Such statements bear similarities to the concerns of Foster (2000) and others, discussed above (p. 111), that the assumption that economic growth leads necessarily to the "domination of nature" may exclude some options for combining economic growth with socially just, and environmentally sound practices.

Such statements, of course, are not intended to weaken the criticism of capitalism as an economic and political system, or to suggest that large companies do not cause much overt pollution and destruction of natural resources. Instead, the aim is to suggest that much discourse about environmental degradation and its causes has reflected the historic origins of environmentalism as a new social movement opposed to the oppressive nature of capitalism. This origin has inevitably influenced assumptions about the causes of environmental degradation, and the problems of such generalizations need to be acknowledged.

Ecology as a metaphor for social emancipation

A further discourse underlying much environmental debate is the assertion that environmental degradation is a symptom of social oppression in general, and not specifically related to capitalism. Indeed, environmentalism and resistance against environmental degradation may also be portrayed as ways to overcome such oppression. The "new" social movements of the 1960s and 1970s in Europe and North America were considered forms of social emancipation against overpowering hegemonies of market and state, and environmentalism, therefore, has been portrayed similarly to have this emancipatory impact.

As discussed above, a common feature of much early environmental debate was to closely associate concepts of "human nature," and "environment." Rosemary Ruether, for example, wrote: "Oppression of persons and oppression of environments go together as parts of the same mentality" (1972: 18).

This perspective has also been reflected in later writings specifically concerning domination of specific social groups. Val Plumwood discussed ecological degradation as a metaphor for the suppression of women in her book, *Feminism and the Mastery of Nature*:

> The three stages of justification and preparation; invasion and annexation; instrumentalization and appropriation can be seen as parts of the overall dualizing process in which reason progressively divides, devolves and denies the colonized other which is nature.
>
> (1993: 192)

Similarly, Andrea Conley has used ecology to imply more than biophysical science, but also as a symbol of liberation from oppression: "ecology can and does include the struggle for a (mental or physical) place of seclusion, an *oikos* of thought that is not subject to systematic control by destructive orders or strategic configurations" (1996: 7).

Again, it is important to consider the impacts of such concerns upon what is implied in discussions of environment and ecology. As discussed in relation to critiques of capitalism, such approaches to ecology are based upon discussions of the domination of nature typified by Marcuse and others, rather than more nuanced understandings of the causes of environmental change. Yet, perhaps more importantly, such statements reflect the common assumptions about ecology and domination of nature without also considering the origin and history of such concepts, and who contributed to their meaning. As summarized above in relation to the new social movements, social activism was commonly held to be identity- rather than interest-based, and in this sense characterized a society in which historic class interests were less important. Indeed, as Offe (1985) noted, movements were often of a (middle) class, but on behalf of all classes. This approach has been criticized by many, who have alleged that the initial ethos of environmentalism was biased toward middle-, and not working-class interests, and with an emphasis on preservation of wilderness rather than a more human-based appreciation of environmental risks (Leff, 1995; Foster, 2000). Enzensberger, for example, wrote:

> The social neutrality to which the ecological debate lays claim having recourse as it does so to strategies derived from the evidence of the natural science, is a fiction... In so far as it can be considered a source of ideology, ecology is a matter that concerns the middle class.
>
> (1974: 9, 10)

And similarly, Guha and Martinez-Allier (1997) have called for a new "environmentalism of the poor," as a means to democratize the currently dominant "green" environmental agenda (of, typically, preserving wildlife and habitats) toward more "brown" or "red–green" aspects (referring to industrial or urban environmental problems, or those specifically affecting poor or vulnerable people). Researchers of environmental problems in developing-world cities have also supported these views. Not surprisingly, environmental policy in rapidly growing cities such as Bangkok or Caracas have often followed the interests of the ruling urban elites, for example, by seeking ways to remove waste or polluting industries from city centers to surrounding suburbs or less powerful neighboring cities (Cutter, 1995; Satterthwaite, 1997; Hardoy *et al.*, 2001). Such groups may also resist environmental regulation within factories owned by them. Nationally, they may also form the main civil political pressure for protecting wilderness, and other green-agenda objectives. Under such contexts, emergent environmental agendas may not be inclusive of all society, and indeed may not be emancipatory for all.

The lesson of this discussion is that the word "ecology" is often used in positive terms as a metaphor for the domination of nature under oppressive social and economic regimes. Yet the meaning of nature is commonly unexamined, and the concepts of ecology used often reflect histories and experiences of particular sectors of society rather than uniformly experienced and agreed evaluations of environment. Theoretical approaches to environment dating from the new social movements and the Frankfurt School continue to reiterate essentialist links between capitalism and modernity with environmental degradation. For example, the German environmental campaigner, Wiesenthal wrote: "The nub of the political objectives pursued in green politics ... can be summarized in two postulates: preservation and emancipation" (1993: 56).

And Smith commented:

> Sustainability is the way to overcome the ecological crisis, but as it is presently conceived it will be merely an extension of the general process of rationalization. Green thought has identified rationalization as the expression of an orientation to everyday life which has led to the ecological crisis.
>
> (1996: 25)

These statements indicate the inheritance of the Frankfurt School and early forms of environmentalism by discussing the emancipation from social oppression; the preservation of landscapes and wilderness; and rationalization (or the advancement of instrumental reason and bureaucracy) as an essential cause of environmental degradation. But in seeking to emancipate society from oppression by using notions of ecology that were coproduced alongside the new social movement activism of the 1960s, such enlightened political intervention may only impose other dominating discourses of environment upon people not represented in the evolution of the discourse. A more complete social emancipation requires a greater democratization of the discourse of environmental explanation itself. The challenges of achieving such democratization within scientific practices and political debate are discussed in later chapters.

Narratives of fragility and crisis

The concept of environment in crisis is another theme that underlies much popular environmental debate. As discussed in Chapters 2 and 3, many orthodox approaches to environmental degradation have adopted notions of equilibrium ecology and balance of nature. Under the so-called non-equilibrium ecology, concepts of balance have been questioned for not acknowledging how far complex ecosystems may vary in time or over space; and for how different societies may construct their own notions of normality, balance, and time and space scales (e.g. Botkin, 1990).

Despite such criticisms, however, the notions of equilibrium and balance in ecology are well illustrated in environmental writing associated with the opposition to capitalism and modernity. Enzensberger, who provided an early critique of earlier forms of Political Ecology, gave an example of this theme before much discussion of non-equilibrium ecology took place:

> If ecology's hypotheses are valid, then capitalist societies have probably thrown away the chance of realizing Marx's project for the reconciliation of man [sic] and nature. The productive forces that bourgeois society have unleashed have been caught up with and overtaken by the destructive powers unleashed at the same time.... If the ecological equilibrium is broken, then the rule of freedom will be further off than ever.

(1974: 31)

Enzensberger's statement reveals his underlying reliance on Marcusian concepts of society and nature for ecological explanation, and the presumed role of capitalism and bourgeois society in destroying ecological equilibrium. Such statements are still common. Lipietz wrote:

crimes against nature are on the increase, and every crime against
nature is a crime against humanity ... [capitalism has] saturated our
ecosystem and reduced significantly the time available to adapt to the
disruption which we ourselves cause.

(1992: 51, 55)

And Grimes: "We live today in a time of unprecedented crisis on a global
scale. This is a point of agreement shared by most scientists examining
planetary trends. It is also a point many non-scientists intrinsically feel"
(1999: 13).

Of course, non-equilibrium ecology does not exclude the notion of
environmental crisis, and there is much room for concern about many
human impacts on environment. But the point of this discussion is to highlight that the notion of crisis and lost harmony have influenced environmental writing (see Giddens, 1994), and may therefore affect how
environmental changes are perceived and presented.

Such discourses may also reinforce some of the social framings of
environmental explanation discussed above. The perception of trees and
forests as both wilderness and as fragile, particularly in North America
and Europe, is perhaps one example. In his gripping book, *The Dying of
the Trees*, Charles Little reinforces images of environmental equilibrium,
ecological fragility, and threatened wilderness by describing the impacts
on trees alone: "What has this got to do with trees? The answer is, it has
everything to do with trees ... the more trees die, the more trees will die.
Could, perhaps, the whole of the global ecosystem go spinning out of
control?" (1995: 226–228).

Little also reflects an equilibrium and the balance of nature perspective
when he quotes a meteorologist colleague as saying, "Forests may be
God's strategy in the way they mediate climate change" in performing a
"divine balancing act in nature" (ibid.: 96–97, quoting Douglas G. Fox).
Again, the purpose of questioning these statements is to show how such
guiding narratives have influenced statements about environmental
change (there is no intention to suggest that we should dismiss concern
about forests). Further discussions of forests and climate change in a
global context are presented in Chapters 6 and 7.

A second example of the impact of such discourses on environmental
explanation is in the use of environmental problems by political leaders to
allocate blame and commentary within society. As discussed in Chapter 4,
analysts of environmental discourse have used the concept of storylines to
indicate how constructions of environmental problems have evolved to
indicate blame and responsibility between different social actors. Yet,
environmental crisis and ecological fragility have been used in this way
too. Box 5.2 summarizes one speech by Britain's Prince Charles about the
social blame for unusually high rainfall in Britain in late 2000. At the time,
newspapers questioned whether the rain was proof of global warming, or
just freak weather. According to the Prince, the rainfall proved the exist-

ence of global warming and represented another example of the revenge of nature against society's selfish disregard for nature's balance. In so doing, the Prince reinforced notions of equilibrium ecology and the domination of nature, yet arguably did not contribute to our understanding of how the rainfall occurred. (The relationship of rainfall to deforestation, and other land-use changes is widely debated too; see also Alford, 1992; Calder, 1999; this volume, Chapter 2.)

Box 5.2   Prince Charles and sermons about science

During October–November 2000, the United Kingdom experienced unusually high levels of rainfall for a period of some weeks. The rains came during a time of intense debate about the significance of anthropogenic climate change during the build-up to the Sixth Conference of the Parties to the UN Framework Convention on Climate Change. There was also other widespread discussion of additional environmental questions such as the use of Genetically Modified Organisms in agriculture. Prince Charles, the Queen's eldest son, and future King of England, made a speech to address his concerns on this subject.

The Prince stated that the storms and rains were the result of society's "arrogant disregard" for the delicate balance of nature. He expanded: "We have to find a way of ensuring that our remarkable and seemingly beneficial advances in technology do not just become the agents of our own destruction."

The comments received fast criticism from a variety of scientists who declared it was improper to attribute any known cause of the freak rainfall. The scientist Lewis Wolpert, speaking on national radio in the UK described the Prince as "arrogant and ignorant ... he is anti-science and anti-technology. He abuses his position. He talks about things he knows nothing about ... and he cannot be challenged because he's royal. If he wants to debate science, he should leave the Royal Family or debate more widely."

The journalist A.N. Wilson later commented that the Prince would have been wiser to visit the houses of people affected by floods (perhaps using his personal helicopter), than make blanket statements about morality, or scientific causality that no-one can prove.

The point of this story is to demonstrate three features underlying the coproduction of political activism and environmental knowledge. First, the proposed causality for the perceived environmental change reflected the critique of society first, rather than a detailed knowledge of the origins of the rain. Second, the Prince revealed the ambivalent attitude of many environmentalists to science by first speaking with authority about the causes of change, yet also rejected potentially damaging aspects of science and technology. Third, the political position of the Prince as a speaker of authority on environment was clearly challenged by scientists and media, indicating that the definition of what is scientific or legitimate knowledge is highly debated, and possibly inspired too by the desire to criticize the monarchy as an institution, in addition to concerns about science.

Source: Wilson, A.N. (2000).

Environmentalists, of course, have to use concepts of crisis in order to raise awareness, and gain political will for the precautionary principle of acting before important environmental changes occur. This discussion is not intended to weaken the need to think carefully about how future environmental changes may impact negatively on people and places. Similarly, there is no intention to assert that all environmentalists, or environmental scientists, may support these populist discourses of nature and society, or simplistic generalizations about ecology and environmental degradation. Instead, there is a need to appreciate how far these discourses may contribute to simplistic explanations of environmental change because they do not acknowledge how such discourses are influenced by social factors.

> Much influential ecological theory relies on an inadequate grasp of the past, both theoretically and in the light of many recent findings in the fields of social and environmental history. While many ecological theorists argue that we should return to an earlier and "purer" form of life in nature, historical research is at the same time showing us how romanticized and anachronistic such notions are.
>
> (Mukta and Hardiman, 2000: 133)

Example: public protests at Genetically Modified Organisms (GMOs)

It is worth illustrating some of these concerns about the coproduction of environmental knowledge and political activism by referring to the topical example of Genetically Modified Organisms (GMOs), and the discussion of scientific knowledge associated with it. The GMO debate reflects a variety of social framings including notions of ecological fragility, opposition to capitalism, as well as more instrumental and exploitative approaches to economic development.

Discussion of GMOs developed markedly during the 1990s when it was proposed to use GMOs in various contexts in order to increase agricultural production. The process of genetic modification (GM) refers to the transfer of DNA between species using laboratory techniques, and is sometimes referred to as biotechnology. GM has been used especially with agricultural crops to increase resilience to disease or insect attack. Proponents of GM have claimed that modifying DNA may enhance food productivity, and consequently help overcome serious problems of poverty and nutritional insecurity, especially in rapidly developing countries such as India and China. Proponents have also claimed GMOs will reduce the need for fertilizers and water, and expand the range of lands that can be used for agriculture, particularly in areas affected by salt or poor soil fertility (for example, see the Monsanto website: www.monsanto.com).

Such claims, however, have been challenged by a variety of concerned scientists and activists in terms of potentially very serious impacts on ecosystems, and the implications of GMOs for international trade and development. Environmentalists have feared that GMOs may impact negatively on native flora and fauna via so-called "gene flow," or the transfer of GMO DNA via the spread of pollen. Similarly, GM crops that are herbicide- or disease-resistant may become persistent weeds if their seeds spread to unwanted locations. Local biodiversity may also be affected by the impacts of toxins within GMOs on non-target insect species, or by the need to use extra amounts of herbicide against GMOs when they are seen as weeds. In New Zealand, for example, the Soil and Health Association claimed that GM field trials would require the use of the antiquated toxin, chloropicrin, which was developed in 1917 and has been used as a chemical weapon (Shah and Banerji, 2001). There are also fears that GMOs may enhance food allergies among humans, or allow the transfer of resistance to certain antibiotics from plants to the bacterial flora in the intestines of animals (including humans) eating GMOs, and hence make certain health conditions more difficult to treat (Mayer, 2000).

GMOs also have significant implications for international development and fair trade. Activists have expressed concern that GMOs, by definition, require patents on each species, and therefore restrict ownership of seeds to the companies that have developed them. In India, for example, much concern was expressed that access to well-known species such as Neem, Aromatic, and Basmati rice would be restricted to transnational companies such as Monsanto or DuPont. Campaigners have also opposed so-called "terminator" technology, which results in the seed produced by GMOs being sterile. Terminator technology requires that farmers cannot build up their own supplies of seed stocks, and instead have to buy new seed each season. Despite promises by companies not to use such technology, there is concern that GMOs will raise costs for poor farmers, and simultaneously increase the power for large companies to manipulate profits from food production. Such concerns have led to a variety of protests and destruction of GM crops (Grove-White *et al.*, 1997) (for example, see the critical website: http://www.geneticsforum. org.uk/).

The debate about GMOs, however, has been notable for the different framings used by different political actors, and by the uncertainty and discussion about scientific findings. In particular, much criticism of GMOs has referred to ecological fragility, impacts on an undisturbed nature, and the injustices of patenting food species by transnational companies. Speaking at an anti-World Trade Organization (WTO) meeting in Seattle, the India writer and activist, Vandana Shiva described GMO patenting as "rape": "Biopiracy is the rape of our biodiversity, our intellectual and cultural heritage ... The WTO legitimizes this rape. And the current negotiations, standing where they are, protect that rape. They're protecting the rapist" (Shiva, 1999: 6–7).

Other activists have also emphasized senses of fragility and invasion of nature. One Maori campaigner wrote:

> As a Maori woman I wear the mantle of the *kaitiaki* (guardian).
> Within *tikanga Maori* (Maori culture), Maori women hold unique
> roles in the protection of *mauri* (life force), *tapu* (sacredness) and
> *whakapapa* (genealogy). Our cultural essence and survival demand
> opposition to genetic modification (GM) and biotechnology ... It is
> customarily sinful to allow biotechnology to degrade the state of the
> natural world. Not to object would be against the nature of our *kaiti-
> aki* role.
>
> (Hutchings, 2001)

And similarly, other authors have referred to GMOs within the frame of
opposition to capitalism:

> The world is being spun around by big corporations who have an
> ability to produce more goods than the world can consume. And so,
> they focus on their efforts on consuming each other, along with any
> smaller elements that get in the way ... This is an unsustainable state
> of affairs, and it takes on an even more ominous dimension when you
> look at the world of biotechnology.
>
> (Alan Simpson, MP, 1998: 1)

Against these claims, proponents of GMOs have often described the rigor
and proof of scientific methods and findings to dismiss criticisms of GMOs.
British Prime Minister, Tony Blair stated in 2002 that opponents to GMOs
were "anti-science." Other critics have claimed that companies who have
refused to use GMOs have done so as a knee-jerk reaction to public controversy rather than a full appreciation of scientific results.

But do scientific findings really provide an answer to the uncertainties
of GMOs? A variety of social science debates reveal further information.
Using Cultural Theory, for example (see Chapter 4), the different positions from NGOs, businesses, and governments may be classified according to the "myths" of nature that express varying perspectives on
ecological fragility and social responsibility. Under this system, Vandana
Shiva, clearly, falls into the egalitarian voice; companies such as Monsanto
represent the individualist perspectives of big business; and Tony Blair's
attempt to achieve order is the hierarchical view of the state, which seeks
to establish rules for both industry and its critics. According to Cultural
Theory, proponents of each "myth" of nature will both collect and use
knowledge to support their perspectives, and we need to appreciate that
all such knowledge will be embedded in such ways.

In addition, GMO science may also be assessed through different
approaches and storylines, narratives and networks, and by reference to
the enforcement of political institutions concerning the production of
scientific knowledge (see also Chapter 4). In common with other scientific
bodies, the production of scientific knowledge about GMOs is regulated
by peer review and critical comments by fellow scientists. Yet according to

critics, such regulation has acted in favor of GMOs. Perhaps most fundamentally, some critics have pointed out that the main funding body for Britain's academic biologists, the Biotechnology and Biological Sciences Research Council (BBSRC) – for example – has many senior members linked to large corporations such as DuPont, Rhone Poulenc, and Zeneca (Matthews and Ho, 2001: 1). Less critical observers might say such corporate presence is sensible within such a funding body.

Such fears about the vested interests of scientific bodies were brought out when one doctor was openly criticized for speaking out against GMOs in 1998–1999. Dr. Arpad Pusztai was a scientist conducting research on GMOs at the Rowett Institute in Scotland. In 1998, Dr. Pusztai announced to the press that he had found evidence to question the health impacts of GMOs. Dr. Pusztai had conducted a test to assess the impacts of GM potatoes on mammals by comparing two sets of rats. One set was fed potatoes containing a lectin gene from snowdrop flowers. The other group was fed normal potatoes as a control. After observation, Dr. Pusztai claimed the rats eating the GM potatoes experienced growth and immune system impairment.

Two days after these statements, Dr. Pusztai was fired from his job. He was prevented from speaking to the press, and in time, an independent report was published that dismissed his findings. Yet, support for Dr. Pusztai had grown, and in 1999, some twenty international scientists announced that they supported Dr. Pusztai, and criticized the independent report. The controversy led to a full investigation by the Royal Society of London in late 1999, which then upheld the original criticisms of Dr. Pusztai. The Society claimed he had made important errors in research methodology such as confounding his variables by using potatoes with different protein contents to feed the two groups of rats. In addition, the Society argued that Dr. Pusztai had not acknowledged the inter-dependence of his measurements on separate organs of rats (measurements will not be separate as any underfed rat will experience multiple symptoms); and pointed out that there was low statistical power in his results because they contained high variability. Despite these statements, Dr. Pusztai's work was finally published in the respected medical journal, *The Lancet* at the end of 1999 (Mayer, 2000: 109).

Similar controversy was experienced following the publication in 1999 of research claiming that Monarch butterfly larvae have unusually high death rates fed on pollen from Bt Maize (Losey *et al.*, 1999). Bt Maize contains the Bt toxin, which is one of the more commonly used toxins within GM crops, and it was feared such toxins would impact negatively on the butterflies and the rest of the food chain. In response, the GMO company, Monsanto, claimed the study was invalid because it was conducted in a laboratory and not in the natural habitat of the Monarch butterfly. The company also pointed to other research that suggested there was not much difference between the impact of Bt Maize and non-GM maize on butterflies (both are not ideal food sources). Yet the importance of such high-profile controversies in shaping public opinion was recognized by Monsanto, who stated:

> The data is now starting to pour in, and it is not to the environmentalists' liking ... We now have a string of studies that suggest the hazard is remote. But as the first study was the worst-case scenario, it is the one that everyone remembers.
>
> (The Knowledge Center, 2000)

But some critics have also alleged that Monsanto also plays a part in shaping the storylines of GMOs. Controversially, the British journalist, George Monbiot has alleged that the public debate about GMOs has been actively manipulated by public-relations companies working for Monsanto through practices such as posting pro-GMO messages on Internet discussion boards by fictitious members of the public (Monbiot, 2002). Such messages give the impression of widespread public support for GMOs, and reproduce information that the companies want the public to know. Ironically, Monbiot alleges that Tony Blair's pro-GM speech in 2002 reflected such statements, and hence, in effect, the GM companies crafted Blair's speech for him. Of course, the companies deny such allegations.

Controversies such as these indicate that there are strong institutional influences upon the production and presentation of scientific knowledge concerning GMOs, and that these influences reflect different campaigning positions. Indeed, new networks are now emerging among critics of NGOs. For example, new NGOs such as GeneWatch disseminate critical reports about GMOs, and the Five Year Freeze campaign seeks to unite businesses and other organizations to support a five-year moratorium on growing, importing, or patenting GM crops in the United Kingdom. Some pro-GMO activists have claimed that GMOs are welcome in many developing countries because of the potential to increase food supply where shortages are feared. Some NGOs, such as Action Aid, have vigorously denied this, and have shown public mistrust against GMOs in various developing societies (http://www.actionaid.org/).

In the light of such political forces on the production and presentation of scientific information about GMOs, it is clearly difficult to rely solely on scientific "facts" as they are reported in the press or company reports. Indeed, sociologists of scientific knowledge such as Levidow and Carr (1997) and Wynne (2001) (in a similar vein to Shrader-Frechette and McCoy, 1994), have suggested that many scientific reports have failed to acknowledge the ethical assumptions that have underlain their research, and consequently have contributed to a loss of trust from the public in such scientific knowledge. These sociologists have argued there is a need to discuss general questions of responsibility concerning GMOs more publicly, rather than simply to seek answers from science without context. Such actions need not be necessarily pro- or anti-GMOs, but it may help clarify the institutional and philosophical bases upon which different scientific findings are made. They will also increase public trust in the science conducted on the risks of GMOs, and help reveal what role they can play in enhancing – or restricting – food supply in many countries.

The case of public protests against GMOs indicates one topical example of some ways in which environmental knowledge is coproduced with political activism. Different political parties have sought to influence the debate in various ways with the production or dissemination of scientific information. In keeping with environmental activism associated with new social movements, some well-known concerns about potential oppression by business, or lack of accountability by the state, are clearly apparent in many criticisms of GMOs. Yet public understandings of GMOs has not been affected only by this counter-opposition of political actors, but also by the emergence of a shared storyline about GMOs that has evolved over years as the result of different disputes and the interactions of influential actors. Public understandings of GMOs has both shaped, and been shaped by, controversies associated with GMOs.

This chapter has argued that many popular discussions of environmental problems have been framed because of the influence of the new social movements in Europe and North America since the 1960s, and have particularly reflected wider social concerns about the purpose and impacts of dominant themes of economic development. Not surprisingly, the environmental concerns and values associated with this activism have influenced scientific explanations and assumptions about environmental degradation, and have given especial significance to themes of lost equilibrium, the domination of nature, or of social activism that can represent many diverse groups. As discussed throughout this book, such assumptions and explanations may fail to give sufficiently complex understandings of how environmental changes occur; and may fail to acknowledge the diverse perspectives and needs of less powerful groups in society.

There is a need to acknowledge the social embedding of environmental knowledge and explanations within wider social and political debates, particularly when applying such explanations to a broad range of social or ecological contexts. This task has to be done with some trepidation: questioning the effectiveness of certain environmental narratives such as the role of capitalism in causing degradation should not be taken as a call to dismiss criticisms of capitalism, or to deny the impacts of exploitative development when they exist. Instead, the objective is to draw attention to the failings of such beliefs in contexts, especially in contexts of environmental change where orthodox explanations of environmental degradation are inadequate or may add to social injustices.

The following chapters now discuss the global implications of coproduced explanations of environmental change and problems, and how far these may be challenged or enforced by further social action and scientific practice.

Summary

This chapter has started to apply critiques of orthodox science to the dynamic processes of environmental politics by showing how environmental science co-evolves with politics. In particular, this chapter has achieved this by demonstrating the concepts of "coproduction" and "hybridization" in relation to generalized beliefs about environmental degradation that emerged as a result of environmentalism as a "new" social movement.

Some of the beliefs – or meta-narratives – that have been associated with "new" social movements have included the close association of environmental degradation with capitalism, the "domination of nature," and the loss of wilderness and tradition as the result of modern industrial development. This chapter pointed out some of the problems of using these beliefs as guiding principles for all environmental policy by showing how such statements fail to acknowledge the complex biophysical causes of apparent degradation that may exist beyond social concerns about environment. Moreover, these beliefs can sometimes fail to represent the interests or needs of poorer people who were not included in the social activism that contributed to these scientific assumptions. There is consequently a need to see how apparent scientific statements about the causes of environmental degradation were shaped by environmental values, and to see how far such coproduced science and values are of benefit to all people who are affected by these statements. Commonly, this may mean seeking ways to promote inclusivity in how environmental or developmental objectives are defined, and to question more critically who wins and loses under currently dominant environmental agendas. One possible way to achieve this in the short-term is to provide more attention to the so-called "brown" environmental agenda (relating to industrial and urban risks) rather than the "green" agenda (such as wilderness preservation or the protection of trees) that was most notably empowered by the "new" social movement of environmentalism.

Yet, seeking to understand linkages between political activism and environmental knowledge has to be undertaken with care. Criticizing simplistic associations of capitalism and environmental degradation should not be taken as an opportunity to legitimize exploitative development, but as a chance to seek more diversity within environmental explanations. Many meta-narratives about environmental degradation (and which are often adopted uncritically in much political ecology) do not adequately explain complex biophysical changes, and many critics suggest they do not offer sufficient flexibility to represent problems experienced by all social groups. A more "critical" political ecology seeks to understand the hidden politics within supposedly neutral or scientific statements about environmental degradation, and attempts to reconstruct more transparent and relevant bases for policy. These objectives are pursued in the following chapters. Chapter 6 seeks to analyze in more detail how different actors may work to make some explanations more powerful than others. Chapter 7 then assesses how far such historically situated concepts of ecology and degradation have been applied elsewhere in the world. Chapters 8 and 9 then seek to explore alternative practices that can make social and political influences on environmental knowledge more transparent and accountable.

6 Enforcing and contesting boundaries

Boundary organizations and social movements

This chapter now looks in more detail at the political processes and actors that contribute to the adoption of dominant environmental explanations. The chapter will:

- discuss the dilemmas of analyzing structure and agency in science–policy, or how specific "actors" may replicate, reform, or co-construct the boundaries or networks of environmental science;
- introduce the concept of "boundary organization" as a means of analyzing organizations that shape and enforce linkages between science and policy. The chapter provides examples of boundary organizations from state and non-state sectors, and considers their influence on current topics of environmental debate such as carbon-offset forestry;
- analyze further how social movements, as a potent source of social resistance, may challenge or reinforce dominant forms of environmental explanation. This section also includes a critique of some current approaches to political ecology, including resource mobilization and advocacy coalitions, and the impact of social movements on environmental discourses.

The aim of this chapter is to contribute to a "critical" political ecology by showing how environmental politics and actors may shape, and be shaped by, environmental science. Later chapters discuss how this knowledge may be used to increase the transparency and public participation in the formulation of environmental science.

Structure and agency in science–policy

The discussion in Chapter 5 showed that environmental science does not exist in isolation from social debates. Political activism and social concerns influence how environmental changes are framed, and these then affect the objectives and outcomes of environmental science. When scientific statements are seen to be politically neutral and authoritative, they may reinforce the original concerns and framings, and imply these are universally applicable to all people. Such co-construction of science and politics has been called "coproduction," or the simultaneous production of know-ledge and social order through politics. Scholars of science studies have also used the term "science–policy" to indicate the mutual enforcement, rather than neat separation, of science and policy (Jasanoff *et al.*, 1995; Hess, 1997).

Chapter 5 analyzed how coproduction has occurred in the evolution of some common assumptions in environmental debates such as the criticism of capitalism and social oppression under modern industrial society. But to understand the evolution of environmental science in more detail, re-searchers need to focus on the roles of specific actors within shorter-term debates in environmental politics. Such an inquiry requires looking at how coproduction takes place on a day-to-day basis. It also requires identifying and analyzing both "actors" that influence environmental science, and the "structures" and institutions that constitute environmental science.

This book has already drawn attention to the social institutions that underlie the formation of scientific statements. Chapter 3 highlighted the importance of semantic structures in the construction of truth statements, or the internal regulation of networks of science under the frameworks of orthodox science. Chapter 4 further indicated the social and linguistic norms that lead to the framings of science. All of these structures may be identified as social institutions of science that underlie the logical con-straints on how statements are made, or to whom such statements are meaningful (see discussion in Chapter 4). Such institutions and structures, however, may also be seen as agents of science because they both shape science, and act as enforcing, or exclusive rules of how science or environ-mental causality should be. It is therefore difficult to separate structure and agency clearly in environmental science.

The dilemma of identifying structure and agency in scientific institu-tions may be shown further by exploring the meaning of scientific "net-works" and "boundaries." As discussed in Chapter 4, "boundaries" may be used to denote how complex reality is bounded (or cerned) into hybrid objects; and to describe social divisions between different disciplines or social groups such as experts and non-experts (Gieryn, 1999). "Networks" may refer to the groups of people or scientists who share certain values or practices; or in more complex epistemological senses (such as in Actor Network Theory) relating to how social clusterings may, in effect, draw biophysical objects into existence in order to support social or political objectives (see Murdoch, 1997; Law and Hassard, 1999). Networks and boundaries may be described as institutions when they lead to regularized expectations and behavior in the explanation of environmental change. Indeed, hegemonic forms of environmental explanation, such as para-digms, narratives, storylines, and environmental orthodoxies, may all in effect be forms of networks and institutions.

Box 6.1 summarizes some of the key differences between networks and boundaries with the perspective of identifying elements of structure and agency. On one hand, the metaphor of networks implies unity and

---

Box 6.1   Distinguishing boundaries and networks

BOUNDARIES	NETWORKS
are associated with:	*are associated with:*
Borders and spaces	Nodes and links
Difference	Connectivity
Location	Position
Inclusion/Exclusion	Distance
Center/Periphery	Controlled passage

Source: suggestions by Tom Gieryn (Indiana University), at a session on "Boundary Organizations" at the 2001 meeting of the Society for Social Studies of Science (4S), Cambridge, MA.

---

connectivity between different actors and objects. Boundaries, on the other hand, imply distance and difference for people and objects outside networks. If scientific networks can be identified as forms of agency in the formulation of environmental science, then they can only proceed by the enforcement of boundaries, and consequently by imposing forms of exclusion and structure. Influencing environmental science therefore implies simultaneous elements of both structure and agency. "Actors" within environmental science are therefore conditioned by existing structures such as overriding discourses or accepted rules about what constitutes environmental science. Furthermore, scientists or activists seeking to change existing networks may therefore have to decide between working within such dominant rules, or attempting to establish alternative and competing networks.

This chapter examines the interplay of political agency and structure within environmental science, and especially concerning so-called "boundary organizations" and social movements. Boundary organizations are institutions or organizations that can control the coproduction of environmental science by forming linkages between science and policy networks. As discussed in Chapter 5, social movements are the most effective political means of influencing the social debates that lead to the framing of environmental science.

But first, it may be useful to consider why some current approaches to political agency in environmental science are insufficient. The following discussions summarize approaches from orthodox science and environmental politics concerning the role of science and political actors in shaping science. The need for an alternative, and more integrated focus on science–policy (or science co-evolving with policy) is then described in the section on boundary organizations.

Scientific criticism and the enforcement of paradigms

One of the most common approaches to agency within science-policy relies on the actions of scientists themselves, and in the implications for the growth and criticism of scientific paradigms. A scientific paradigm may be considered a form of network or institution of shared norms and objectives.

Chapter 3 discussed many of the frameworks of orthodox science. According to these frameworks, scientific "laws" may be established through the acts of verification (for logical positivists) or falsification (for critical rationalists). Processes of "conjecture and refutation," or the peer review of scientific results may assist with the regulation of scientific findings. Scientific progress is experienced when verification, or more recently falsification, have contributed to a strengthened belief in an observed trend or theory that may take the position of a presumed "law" of nature (see Chapter 3, also Chapter 8). The orthodox approach to science therefore asserts that scientific debate and criticism can control the adoption of environmental explanations, even though – ironically – scientists expect to test and question such "laws" under normal scientific practice.

Such practices, however, have led to a number of potential contradictions in the linkages between science and policy. Frequently, apparent "laws" of nature are interpreted as being a fair and representative indication of biophysical reality. Yet, as discussed in Chapters 3 and 4, such statements can be easily criticized for overlooking the number of social conventions and influences in the production of such orthodox scientific knowledge. Indeed, alternative models of science (see Callon, 1995), have pointed to the roles of networks within scientific institutions or in society in general as means by which scientific information becomes institutionalized.

Moreover, such social influences have also been acknowledged, to varying degrees, by scholars well known to orthodox scientific inquiry. The work of Thomas Kuhn (1962), for example, argued that scientific paradigms changed through the influence of social – and commonly personal – factors, and created periods of so-called "normal science" in which all research addressed questions consistent with paradigms (see Chapter 3). Even Popper, in his later writings, acknowledged the need for frameworks and networks as a basis for communication and development of science: "A rational and fruitful discussion is impossible unless the participants share a common framework of basic assumptions" (1994: 34).

Furthermore, Imre Lakatos (1978) further developed the links between scientific progress and networks by portraying paradigms as research programs. Lakatos argued that research programs have two parts: the negative heuristic, or hard core, and the positive heuristic, or buffer propositions (soft core). The hard core contains propositions that are considered fundamental to the network, such as (for example) the insistence that population increase leads to environmental degradation. The soft core refers to empirically falsifiable implications of the hard core, such as ways and means that population impacts on resources. Lakatos argued that scientists generally tested the soft-core assertions and took the hard-core assertions as received truths. If enough soft-core propositions are rejected, then there may be grounds for modifying the hard-core assumptions. Under this approach, for example, Tiffen and Mortimore's (1994) study, *More People, Less Erosion?*, may be taken to be one empirical test at the soft core. But opinion is divided whether this study – or ones similar to it (see Chapter 2) – should be seen as weighty enough to dismiss the underlying core belief in the environmental orthodoxy that population growth causes degradation.

The problems with Lakatos's analysis in the context of scientific progress is that his approach does not settle how, or with whose determination, such changes in soft or hard cores of science can take place. As noted in Chapter 2, environmental research on so-called environmental orthodoxies has presented a variety of challenges to the basic tenets of the $I = PAT$ equation, or the framing of environmental change in terms of equilibrium or balance of nature. Furthermore, such research has often been conducted within the same methodology or canons of good practice adopted within physical ecological research. Yet such research that challenges environmental orthodoxies has not replaced the hard core of these conceptualizations, but instead encouraged the development of competing networks that exist side-by-side. For example, as Calder (1999) notes, the orthodox belief that deforestation is primarily responsible for water shortages has been adopted in China, but now no longer adopted in South Africa, Australia, or New Zealand. Alternatively, different scientific networks may exist simultaneously in the same country, but policy be based on only one network.

It is therefore clear that there are many more factors underlying policy changes than scientific research alone. The suggestion that science itself might influence policy in a linear process, as suggested by orthodox scientific models, needs to be replaced with an understanding of how science and policy co-evolve.

Identifying science–policy "actors"

Similarly, the emergence of scientific networks may also be explained through the agency of political actors who have helped shape or enforce common norms of environmental explanation. This approach has been widely adopted, particularly in the context of establishing epistemic communities or new environmental objectives in public policy. Actors have often included NGOs (e.g. Princen *et al.*, 1994), media, or charismatic leaders of environmental organizations.

One feature of much conventional writing on environmental politics has been a further argument that enhancing environmental objectives also implies empowering social actors (such as citizens, grassroots organizations, or non-governmental organizations) against state and/or industry. This approach has often referred to social theory debates about the oppressive nature of instrumental state-based politics or industrial development when applied to society or environment (e.g. Eckersley, 1992; see Chapter 5), or the instrumental use of science as a political force (e.g. Boehmer-Christiansen, 1994). It has also underlain some approaches to political ecology, particularly regarding the role of state–industrial alliances in forging exploitative industrial growth in developing countries. Typifying this approach, Bryant and Bailey wrote:

> States use their legal–political powers to grant businesses privileged access to environmental resources, while businesses use their financial and technical knowledge "efficiently" to extract, produce and market environmental resources and/or consumer goods. Both actors seek to expand commercial activity so as to increase their income and/or power over other actors. In contrast, grassroots actors use their grasp of local political–ecological conditions to resist more powerful actors, whereas ENGOs [environmental NGOs] seek to provide technical and financial support, as well as media coverage, to these location-specific struggles. Both grassroots and ENGO actors here seek to assert the primacy of community environmental management so as to promote social justice and/or environmental conservation.
>
> (1997: 190)

Yet this approach to actors can be questioned for two main reasons. First, the reliance on broad categories of state, society, and industry may encourage the essentialization of actors into pre-identified positions in environmental debate, and may overlook the possibility for certain actors to perform roles in ways such a categorization suggests unusual (although Bryant and Bailey do acknowledge this risk). Second, the focus on actors alone as agents of scientific networks overlooks the ways such actors are identified or constructed as a result of overriding discourses that provide the structure to environmental debate. The construction of "actors" in this way can be explained from a variety of approaches in social theory.

From a Cultural Theory perspective, each actor representing "state," "business," or "NGOs" may be seen to come from hierarchical, individualist, or egalitarian perspectives that provide predictable opinions about environment, and which are not surprisingly incompatible (see Chapter 4). The fourth category, fatalist, may be equated with the voiceless peasant or factory workers who are often represented by grassroots or non-governmental organizations, but who may often be mistakenly incorporated into more powerful perspectives by the projection of the speakers' voices upon them. Thompson noted:

> If you think of the Brent Spar saga, Shell was the actor from the market (or individualist) solidarity, the government experts who

okayed the deep ocean disposal were the hierarchical actors, Greenpeace (whose eleventh hour intervention drastically upset this negotiated outcome) was the egalitarian actor, and those (like myself) who found themselves totally convinced by whoever they last heard arguing the case on television were the fatalists.

(2000: 114)

Other well-known environmental debates can be portrayed in these terms. For example, ex-US Vice President Al Gore's (1992) book, *Earth in the Balance,* may be described as both egalitarian and hierarchical because it sought to demonstrate the fragility of the global environment in conjunction with the ability of a caring state to enact sensible precautions. The resulting counter-publication, *Environmental Gore: A Constructive Response to "Earth in the Balance"* (Baden, 1994) instead contained a variety of chapters that could be called individualist because all argued against the need for central legislation, and the possibility for self-regulation by business, and the innate resilience of ecosystems to reduce impacts of industry. Indeed, the kind of oppositional politics portrayed by these two books indicate a reduction of environmental politics into binary positions that tell us more about the underlying objectives of each side, rather than on the complex reality of environmental change and risk each book is supposed to discuss.

Similar criticisms can be made from other social theory debates. Niklas Luhmann (1989), for example, using an autopoetic approach, also criticized environmentalism for failing to see the structural reasons that cause it to oppose state and industry. Under more poststructuralist approaches to environmental explanation, the role of "actors" such as "state," "business," and "NGOs" needs to be seen alongside how we have come to understand what these groups represent, or what position they hold in narratives of political struggle and justice. For example, as noted in Chapter 5, it has become common to identify "business" actors with negative acts toward environment, and "NGOs" as necessarily beneficial. Taking "sides" in this way may avoid identifying how far such sides reflect wider social concerns that may not be applicable on each occasion, or how we have come to identify such sides (Wynne, 1996b).

Narrative and storyline analysis allow researchers to understand the more historic and culturally situated evolution of different voices (see Chapter 4). Under Actor Network Theory, the very concept of "actor" is questioned by the argument that all "objects" in the network (including biophysical objects such as Michel Callon's scallops) are also actors that have been enlisted in order to enhance the stature of the science or explanation produced by the network (see Mol and Law, 1994; Law and Hassard, 1999). Under these approaches, political "actors" do not shape the discourse; the discourse shapes them. Indeed, this principle is shown more generally when political activists call upon specific events or social groups to illustrate a political position rather than calling on all possible events or people, or questioning how their objectives might influence their selection of evidence.

Analyzing environmental conflicts in terms of battles between these different actors therefore may make the mistake of assuming these voices are somehow independent of the structures of society that lead to these perspectives. Attributing political agency to these actors, without understanding how they replicate discourses, may overstate the power of the actors, and overlook the disciplining power of discourses or structures. Instead, research needs to acknowledge the co-construction of environmental actors and the scientific claims of each.

Boundary organization theory

One approach to science–policy that acknowledges the coproduction of science and politics is the concept of boundary organizations (Shackley and Wynne, 1996; Guston, 2001). The objectives of this concept are to acknowledge the coproduction of science and policy, and the difficulty in identifying political actors outside of discourse. Boundary organizations may be defined as social organizations or collectives that sit in two different worlds such as science and policy, and can be accessed equally by members of each world without losing identity. The term "boundary organization" is based upon the earlier concept of "boundary object" (Star and Griesemer, 1989; Gieryn, 1999). Boundary objects are items that can be used by different networks without losing identity (for example, patents on research can be used by a scientist for commercial gain, or by policymakers for evaluating the productivity of research: see Guston, 2001: 400). The distinguishing feature of both boundary objects and organizations is that they provide sites where different epistemological networks may unite.

The concept of boundary organizations allows researchers to see how, where, and by whom common norms are established in networks of science and policy. Boundary organizations usually have to meet three criteria. First, they provide the opportunity and sometimes incentives for creating and using boundary objects or standardized packages. Second, they involve the participation of actors from both sides of the boundary, and occasionally professionals in a mediating role. Third, they exist on the boundary between science and politics, but have accountability to both. A successful boundary organization may therefore achieve the objectives of two constituencies yet remain organizationally stable despite continual negotiation within the organization regarding the boundary between the constituencies (Guston, 2001: 400–401). The dual agency between scientific research and political action makes boundary objects sites of coproduction of science and policy. As Guston wrote:

> The boundary organization is able to project authority by showing its responsive face to either audience. To the scientific principal, it says,

"I will do your bidding by demonstrating to the politicians that you are contributing to their goals, and I will help facilitate some research goals besides." To the consumer, who is also a principal, it says, "I will do your bidding by assuring that researchers are contributing to the goals you have for the integrity and productivity of research." ... The politicization of science is undoubtedly a slippery slope. But so is the scientization of politics. The boundary organization does not slide down either slope because it is tethered to both, suspended by the coproduction of mutual interests.

(2001: 405)

The attraction of the boundary organization concept is that it acknowledges both structure and agency of political actors working within wider discourses, and the dynamic way in which organizations may establish common norms between scientific and political networks. Furthermore, the concept may be applied to political organizations from state, industry, and society (including NGOs and grassroots organizations), and so avoids any predefined model of "sides" within environmental disputes.

Usually, the concept of boundary organization refers to any agency or institution that lies on the interface between, and influences, science and politics. Yet, the words "boundary organization" may also refer to the role of such organizations in setting the boundaries that define hybrid objects or contested notions of science and policy through practices such as problem closure and framing (see Miller, 2001). Often such boundary work by these organizations takes place where there are many multivalencies, or discourse coalitions, between different political interests and actors (Hajer, 1995, see Chapter 4). An analysis of the epistemological implications of boundary organizations may therefore indicate how science networks or controversial environmental explanations are defined and enforced through the actions of specific organizations or political actors.

Enforcing boundaries: examples of boundary organization analysis

This section provides some examples of boundary organization analysis for a variety of environmental debates. The objective of this analysis is to indicate how science–policy research can provide critical insights into the ways in which science and politics co-evolve, or how specific environmental explanations may emerge as dominant.

Boundary organizations link scientific and political networks by acting as intermediaries on topics of scientific and political controversy. Boundary organizations set the goalposts of environmental political debate by providing definitions or approaches to contested science that are then used as "fact." Seeing how organizations influence scientific knowledge may therefore empower the political analysis of environmental science.

The actions of boundary organizations, however, need not be totally autonomous, and may be in response to other structuring elements in environmental debate such as preexisting narratives or overriding political concerns. Organizations may also act in coordination with each other, and thus reinforce narratives used by a variety of organizations.

There are many potential examples of boundary organization. This section focuses on the actions of specific organizations relevant to the debate concerning climate change, and its implications on related aspects of environmental science such as the role of forestry. These topics are topical yet also indicate the relationship of current environmental debates upon the reinforcement of preexisting, orthodox, environmental explanations discussed elsewhere in this book.

The Intergovernmental Panel on Climate Change (IPCC)

The preceding section defined boundary organizations as organizations that link different constituencies in science and policy, and where this intermediary role may have influences on the framing and legitimization of scientific information. The case of the Intergovernmental Panel on Climate Change (IPCC) provides a variety of contexts in which these influences on scientific information may be shown.

The IPCC was established by the World Meteorological Organization (WMO) and the United Nations Environment Program (UNEP) in 1988. Its main objective was to assess scientific, technical, and socio-economic information relevant to the understanding of human-induced climate change, potential impacts of climate change, and options for mitigation and adaptation. Its main products to date have been assessment reports on the state of scientific understandings of risks posed by climate change, which have been used as the key components guiding political conventions such as the 1992 United Nations Framework Convention on Climate Change (UNFCCC) and the 1997 Kyoto Protocol (e.g. IPCC, 1996, 2001).

The IPCC, of course, has been the target of much overt political criticism by actors who doubt its scientific accuracy, or who wish to avoid the costs of environmental regulation. For example, Senator Chuck Hagel of Nebraska, commented:

> We've had climate change since long before the Industrial Revolution ... the Summary for Policymakers was written by UN environmental activists, not the scientists who wrote the individual chapters... [IPCC summaries are] political documents, drafted by government representatives during intense negotiating sessions.

> (2001)

Such comments have produced counter statements by the IPCC to reaffirm its scientific integrity. For example, Michael Zammit Cutajar, the former Executive Secretary of the UNFCCC, stated: "The science has driven the politics ... if the science is to continue guiding the politics, it is essential *to keep the politics out of the science* (2001, emphasis in the original; also reported in Chapter 1).

Indeed, the IPCC – unlike some other environmental assessments (see Farrell *et al.*, 2001) – requests that scientists involved in researching the rate and impacts of climate change are not included in sessions that write policy recommendations. Moreover, Senator Hagel's comments that the IPCC Summary for Policymakers is written by "UN environmental activists" is technically inaccurate because such reports are written by political representatives of parties to the UNFCCC who have to agree on every word.

Yet this overt separation of science and policy overlooks a variety of ways in which science and politics are mutually constructed through the actions of the IPCC. Such mutual construction does not imply that all findings of the IPCC should be dismissed, or that action against climate change should not be taken. Yet it is important to see such tacit influences in order to see how the science–policy process represented by the IPCC affects the production of supposedly politically neutral assessments of ecological change.

A variety of themes may be discussed as examples. First, some authors have suggested that the IPCC has influenced the coproduction of science and politics through its overt focus on the General Circulation Model (GCM) as a new methodology for assessing atmospheric changes (Shackley and Wynne, 1995; Demeritt, 2001). Indeed, a preceding research program, the International Geosphere–Biosphere Program (IGBP), established in 1986, was claimed to be reliant on new technologies of remote sensing and vegetation indices in ways that allegedly transformed ecological science (Kwa, 1987, 2001). A variety of authors have questioned how the use of GCMs might make scientific and political uncertainties less apparent to users of information provided by GCMs. There have also been assumptions that GCMs provide the only effective means for estimating the risks posed by climate change. For example, in a testimony to the US Congress, the Director of the US Department of Energy's Office of Science stated:

> It is only through such general circulation models that it is possible to understand current climate and climate variability and to predict future climate and climate variability, including prediction of the possible effects of human activities on the global system.
>
> (Martha Krebs, 1999, in Demeritt, 2001: 319)

Such statements suggest that scientific simulation and prediction may be used interchangeably, and overlook other means of assessing the risks or vulnerability to climate change (these are discussed more in Chapter 7). Furthermore, other "black-boxing" of uncertainties of GCM findings may occur with respect to the practices of correcting fluxes in models (the extremities of predicted outcomes) to levels seen to be more credible to users. As Demeritt noted: "In this way, modelers' tacit beliefs about downstream needs and identities legitimate their assumption about the need for intensified physical reductionism and continued GCM development" (2001: 319).

A further element in which the IPCC and associated bodies may act as boundary organizations lies in their role as mediators and facilitators of political agreement. For example, the UNFCCC's Subsidiary Body for Scientific and Technological Advice (SBSTA) was established in order to achieve political consensus on aspects of scientific and technological dispute as a prelude for political negotiation (Miller, 2001). Yet, although the language of SBSTA rules request that SBSTA members will be government experts, many countries send the same delegates to SBSTA as to UNFCCC negotiations. SBSTA has also negotiated various elements of scientific and political uncertainty such as the measurement of greenhouse gas emissions, and the definition of carbon-offset forestry (see below, pp. 147–154) that are fundamental to the ecological causes and impacts of climate change and associated policy. Such questions are inherently epistemologically hybrid because they reflect both natural and social elements of framing and experience. The advice coming from SBSTA, however, avoids making such framings clear. Miller (ibid.: 493) cites as an example the decision to include methane emissions from cattle as anthropogenic, and to state that emissions from deer are not, even though both cattle and deer populations are ultimately decided by human policies.

It is important to note that many climate modelers and SBSTA negotiators acknowledge the problems associated with GCMs or other aspects of climate change debate. Models and assessment methods are also advancing in complexity. Yet the overt focus on GCMs as the key component of predicting the risks of climate change have been shaped in part by expectations of how that risk is framed by policymakers, and this influences the evolution of how climate change is understood. In turn, such co-constructed science and policy reinforces the overriding perception that the "solution" to anthropogenic climate change is the mitigation of atmospheric greenhouse gases. As discussed later in this chapter, and in Chapter 7, such approaches to environmental risk may overlook a variety of social and political factors that underlie how biophysical changes are experienced as threats, and may even legitimize environmental policies that can interfere with strategies to lessen such vulnerability.

National and international forestry organizations

Boundary organizations may exist in various arenas of environmental debate. The case of forestry has attracted much attention for the diversity and influence of state or international expert bodies, and for the impacts of these bodies on diverse ecosystems known as forests, and for the multiple groups of people who use them.

Many authors, for example, have examined the influence of state forestry agencies established during colonial times in both defining the objectives of forest management and the people included in this process (e.g. Jewitt, 1995; Bryant, 1997a; Fairhead and Leach, 1998; Robbins, 1998; Leach *et al.*, 2002). The historic co-evolution of organization structure and scientific forms of forestry has important influence on modern-day environmental struggles over forest resources by state and social actors.

Sivaramakrishnan (2000), for example, assessed the history of state forestry in Bengal, India, in order to indicate the tacit politics contained within scientific practices considered to be beyond political debate. The very concept of "scientific" forestry is intended to denote a sense of order and optimal use of resources as according an efficient colonial administration. Such approaches, however, make assumptions about the types of forest and forest uses that impact on local shifting cultivators whose uses of forests may be seen to be inimical to commercial timber extraction, or the maintenance of reserve land. Indeed, the quantification of wood mass was one important aspect of scientific forestry, and the cultivation of specific timber species such as teak and sal. Hill forest areas were also identified as less valuable for timber production and therefore were burnt to encourage the cultivation of less valuable products such as sabai grass. Such practices both adopted and reinforced notions of wilderness management and orthodox explanations of the impacts of forest disturbance on environmental degradation and watershed management.

The institutionalization of scientific practices favoring specific viewpoints has formed the basis of many current environmental disputes between state agencies and social groups who seek to establish alternative forms of forest management such as Joint Forestry Management or community forestry (e.g. see Cline-Cole and Madge, 2000; Fox *et al.*, 2000). Such conflicts may also be seen in the activities of some international advisory organizations (Dover, 1992; Brechin, 1997). For example, the International Center for Research in Agroforestry (ICRAF) is a autonomous research organization supported by the Consultative Group on International Agricultural Research (CGIAR), focusing on ways to enhance the adoption of agroforestry. The Consultative Group also oversees the "Alternatives to Slash and Burn" (or shifting cultivation) initiative, which is carried out through various agencies of the group. According to ICRAF's publicity material, "the consequences of [slash and burn] are devastating, in terms of climate change, soil erosion and degradation, watershed degradation, and loss of biodiversity" (ICRAF, 1999). As discussed in Chapter 2, such statements represent classic environmental orthodoxies about the impacts of shifting cultivation, and are challenged by a variety of research (e.g. Alford, 1992; Fox *et al.*, 2000; see Box 2.2). They also fail to include more localized framings of environmental problems such as declining soil fertility.

The mutual enforcement of such scientific framings and policy were illustrated at a workshop organized by ICRAF in northern Thailand in 1999. The workshop was entitled *Environmental services and land use*

change: bridging the gap between policy and research in Southeast Asia.
The workshop was opened, however, by a keynote speech by the Director
General of the Thai Royal Forestry Department, who urged that farmers
in the highlands caused environmental problems. One month before the
workshop, the Forestry Department had forcibly broken up a demonstration of some 5,000 such farmers who had been protesting about land-use
policies in northern Thailand. Protestors were particularly resisting the
planting of tree plantations on agricultural land, and the refusal of the
Thai government to allocate citizenship to many farmers, thus preventing
them from achieving more diversified livelihoods (Johnson and Forsyth,
2002).

Many members of ICRAF, of course, are aware of the problems of
many orthodox scientific approaches to shifting cultivation, and are
seeking to reform the organization or orthodox views from the inside. But
the role of such organizations as supposedly neutral arbiters of scientific
knowledge and expertise needs to be drawn into question. Similar concerns have also been stated about the Forest Stewardship Council (FSC)
or International Tropical Timber Organization (ITTO), for example, as
other international bodies seeking to advise on various uses and classifications of forests (e.g. Naka *et al.*, 2000). This chapter has insufficient space
to consider them all. The next section discusses how boundary organizations within climate change and forestry management may complement
each other in the case of carbon-offset forestry.

Interlocking boundary organizations: the case of carbon-offset forestry

The preceding examples might be seen to be isolated cases of boundary
organizations that reflect the specific interests of each organization in satis-
fying different constituencies in science and policy networks. Yet, in addition, different organizations can also complement each other and enforce
scientific networks that affect more than single organizations. Under such
circumstances, the actions of individual boundary organizations may reflect and contribute to broader discursive structures in environmental politics such as discourse coalitions and storylines about the objectives and
responsibility for environmental policy (Hajer, 1995; see Chapter 4).

One important example of interlocking boundary organizations and discourse coalitions is the mounting use of reforestation (and/or afforestation) as a means of mitigating climate change, usually as a means of
offsetting industrial emissions elsewhere on the globe. Such carbon-offset
forestry is also known popularly as "carbon sinks" or "carbon sequestration," although both "sinks" and sequestration can be achieved through
means other than forestry (for example through soils, see Olsson and
Ardö, 2002). The topic is also a source of deep divisions about environmental policy and justice between many developed and developing countries (Gupta, 1997; Kelly and Adger, 1999; Lohmann, 1999, 2001). It also

relates to other debates about the politics of reforestation (e.g. Dove, 1992; Brechin, 1997; McManus, 1999), and the replication of supposed "myths" about environmental degradation concerning tropical forests discussed in Chapter 2 (e.g. Fairhead and Leach, 1998).

In essence, the purpose of carbon-offset forestry is very logical: if anthropogenic climate change is a global problem caused by the increasing concentration of atmospheric greenhouse gases, then any measure to reduce these concentrations is worthy of consideration. Furthermore, carbon-offset forestry offers, in principle, cheaper means of reducing concentrations than many alternatives such as replacing fossil fuels with renewable sources of energy, or retrofitting coal- and gas-fired power stations with emissions reduction technology.

Since the mid-1980s, advisers on international environment have suggested controlling deforestation as a means to mitigate climate change. At the First Conference of the Parties to the United Nations Framework Convention on Climate Change (UNFCCC) in Berlin, in 1995, it was agreed in principle to investigate the possibility for achieving future national greenhouse gas emission reduction targets through investing in projects in other countries. This tentative agreement led to the establishment of so-called Activities Implemented Jointly (AIJ) as a pilot phase for international cooperation on reducing greenhouse gas concentrations, and which saw a number of carbon-offset forestry projects established in a number of countries by large emitters such as Japan and the USA. Such projects were also supported by additional organizations such as the United States Initiative on Joint Implementation, and by sections of the World Bank (now named the Prototype Carbon Fund), and environmental think tanks such as the World Resources Institute. The 1997 Kyoto Protocol finally gave approval for such international offset projects to go ahead. It was agreed that Annex I countries (countries with greenhouse gas abatement targets) could achieve targets partly through national forestry budgets; or internationally through so-called flexible mechanisms such as Joint Implementation (JI) (referring to investment within Annex I countries), and the Clean Development Mechanism (CDM) (referring to investment in non-Annex I countries, usually the developing world). These mechanisms are not restricted to forestry projects, but much discussion has focused on carbon-offset forestry because these projects – in certain forms – offer lower marginal costs than alternative projects such as those involving renewable energy or energy efficiency (see Grubb *et al.*, 1999).

The arguments in favor of carbon-offset forestry, however, were supported by a variety of additional environmental claims and interests that constitute a number of discourse coalitions to support this policy option. These too were met by a variety of counter claims. Indeed, it is possible that carbon-offset forestry may now be a new epiphenomenon of environmental debate – a new arena within which older political debates are replayed, yet from which apparently neutral environmental explanations emerge (Hajer, 1995; see Chapter 4). Furthermore, the emergence of carbon-offset forestry as a policy option, with its associated focus on forestry as the chief means of carbon sequestration, as opposed to other forms of sequestration, may also be analyzed in terms of network theory (including Actor Network Theory, or extended translation). This approach emphasizes the importance of political and social networks in both identifying and explaining biophysical objects or processes (Callon, 1995; see Chapter 4).

Some of the most potent claims to support carbon-offset forestry came from campaigners for forest conservation who saw convenient synergies between climate change mitigation and managing other problems such as biodiversity loss. For example, the rainforest biologist Norman Myers wrote:

> One of the most cost-effective and technically feasible ways to counter the greenhouse effect lies with grand-scale reforestation in the tropics as a means to sequester carbon dioxide from the global atmosphere – provided, of course, that the strategy is accompanied by greatly increased efforts to slow deforestation.
>
> (1990: 399)

Furthermore, carbon-offset forestry has also been welcomed by other environmentalists who have considered reforestation to be something of a "magic bullet" to address a range of environmental problems comprising climate change, declining biodiversity, controlling erosion and water shortages, and the aesthetics of lost wilderness. For example, the British explorer and popular writer, Robin Hanbury-Tenison, wrote, somewhat romantically and controversially: "Carbon sinks ... these are exactly the elements of the Kyoto protocol that offer our last hope of saving the rain forests" (2001).

And Lester Brown of the Worldwatch Institute wrote:

> Restoring forests ... means reversing decades of tree cutting and land clearing with forest restoration, an activity that will require millions of people planting billions of trees.... A small area devoted to plantations may be essential to protecting forests at the global level.... At present tree plantations cover some 113 million hectares. An expansion of these by at least half, along with a continuing rise in productivity, is likely to be needed both to satisfy future demand and to eliminate one of the pressures that are shrinking forests. This, too, presents a huge opportunity for investment.
>
> (2001: 82, 85, 95)

Carbon-offset forestry, and some of these additional claims, however, have been met with great opposition. Clearly, there are many controversies associated with some of the benefits claimed from carbon-offset forestry.

First, as argued throughout this book, many alleged additional benefits of reforestation have been thoroughly questioned by a wide range of research about the nature of deforestation, biodiversity loss, and watershed degradation (see Chapter 2 for full references and discussion). Much initial discussion of carbon-offset forestry has referred to reforestation or afforestation in the form of plantations (or monoculture forests). Yet, plantation forestry has not been closely linked to the restoration of biodiversity; it has been claimed to increase levels of erosion (see Calder, 1999); and it restricts local livelihoods by excluding many forest users from entering and using forest areas. Indeed, in Thailand in the early 1990s, for example, the then-military government sought to control large areas of the northeast of the country by establishing plantations, leading to both political and environmental turmoil as angry farmers burnt trees and marched on Bangkok (Pasuk and Baker, 1995: 390). Some proponents of carbon-offset forestry have urged that plantations (or monoculture forests) should not be used. But if such diversification occurs, the more complex forms of reforestation often take longer to grow, and may have fewer opportunities to lead to forms of sustainable harvesting, and as a result, such forestry will be more costly than plantations, and may even be of a comparable cost to some energy-related or industrial strategies for climate change mitigation.

Second, some developing-world environmentalists have argued that carbon-offset forestry in non-Annex I countries unfairly places the emphasis for abating greenhouse gas emissions on activities in the developing world, rather than at source (Agarwal and Narain, 1991; Rocheleau and Ross, 1995). For example, one official at the US Department of Energy was quoted as saying in 1994 that "tree planting will allow US energy policy to go on with business as usual out to 2015" (in Lohmann, 1999: 2). Some developing countries, notably Costa Rica, have openly welcomed carbon-offset forestry, but other countries, such as India, China, and Brazil, have been more critical. Such concerns were shown at one climate change meeting in London before the Kyoto Summit, in which one angry African negotiator told the audience, "Our countries are not toilets for your emissions!" (Forsyth, 1999b: 255). Similarly, other critics, such as the Uruguay-based NGO, the World Rainforest Movement, have claimed that carbon-offset forestry is equivalent to "CO_2lonialism."

A third theme of criticism has come from environmental scientists and environmentalists who have questioned the ability for forestry projects to influence climate change (although these claims are often rejected by scientists in favor of carbon-offset forestry). One common concern has been the difficulty of establishing clear baselines for measuring the impacts of forestry projects on emissions. (Problems include measuring how far projects have sequestered carbon, and deciding whether there is a counterfactual problem of calculating what would have happened to reforested land if there had been no carbon-offset forestry.) Some critics have overtly accused forestry projects of overstating mitigation impacts of reforestation

(Cullet and Kameri-Mbote, 1998). Other concerns have included the need to avoid "leakage" in projects, or the tendency for reforestation (or forest protection) in one zone to encourage enhanced deforestation elsewhere.

Such concerns, of course, have had immense impacts on how scientific explanations of reforestation and sequestration have been portrayed, and similarly, on the presentation of other scientific debates that may or may not support forestry projects. For example, the discussion of carbon-offset forestry has led to simplified accounts of complex forest systems. Between 1997 and 2000, for example, Parties to the Kyoto Protocol and the IPCC sought to reach a uniform definition of forest in order to help define how carbon-offset forestry could proceed. Some 130 definitions of "forest" were mentioned in debates among states alone, before a universal definition of forest was defined by the Conference of the Parties to the UNFCCC at The Hague in November 2000:

> "Forest" is an area of land of 0.3–1.0 hectares (ha) with tree crown cover (or equivalent stocking level) of more than 10–30 percent with trees with the potential to reach a minimum height of 2–5 meters (m) at maturity *in situ*. A forest may consist *either* of closed forest formation where trees of various storeys and undergrowth cover a high proportion of the ground; *or* open forest formations over an area of 0.3–1.0 ha with a continuous vegetation cover in which tree crown cover exceeds 10–20 percent. Young natural stands and all plantations which have yet to reach a crown density of 10–30 percent or tree height of 2–5 m are included under forest.
>
> (UNFCCC/SBSTA/2000/CRP.11, p. 7, November 2000;
> also see Fogel, 2002)

Such a definition is a further example of hybridization as it shows the reduction of complex biophysical systems to a definition driven by social and political needs. Some forms of vegetation previously considered to be "forest" may be excluded from this definition. This definition may also overlook occasions when fire may form clearings within forest areas at a variety of scales (see Chapters 2 and 3), or when such clearings are formed by agriculturalists. The definition of "forest" is also related to the debate concerning official certification of forests through other means, such as by the Forest Stewardship Council, which has provided official certification to several monoculture plantations, to the concern of NGOs such as the World Rainforest Movement.

Similarly, portrayals of carbon-offset forestry projects in marketing campaigns have also simplified biophysical processes. The Japanese company, Mazda, for example, launched a car in Great Britain in the late 1990s that came complete with a year's "carbon-neutral driving" (Lohmann, 1999: 6). This strategy allowed Mazda – a company relying on the consumption of fossil fuels by itself and its customers – to appear climate friendly without changing that reliance. As one critic commented:

"Pretty soon, it may be expected, every time you turn an ignition key, flip a switch, take a holiday, or cook some food, you will not only be using up fossil fuels but also planting trees on someone else's land" (Lohmann, 1999: 6).

The implication of the evolution of carbon-offset forestry is that one type of climate-related activity (reforestation) is being increasingly equated with the mitigation of others (emissions). Yet the underlying uncertainties and political controversies of this link are subsumed under the enforcement of the scientific boundary, or black box (see Chapter 4), that links the two. This boundary, however, does not refer to the experiences of a variety of local forest users or people affected by reforestation, and whose livelihoods may be negatively affected by the ostensibly positive acts of carbon-offset forestry. Furthermore, it does not refer to the numerous uncertainties and contested truth claims about the biophysical benefits of reforestation. It is important to note that the UNFCCC itself has acknowledged some of these uncertainties, and urged that carbon offset forestry be adopted only when negative social impacts can be avoided. But, as Fogel noted:

> While such visions [of carbon-offset forestry] were captivating, and to some extent accurate, they were also misleading, as they were based on partial knowledge and a sanitized, "outsider's" understanding of rural dynamics in most of the world.
>
> (2001: 4)

Indeed, in 2000, a number of NGO activists, including representatives from Greenpeace and the Rainforest Action Network, signed the "Mount Tamalpais Declaration" (after the site in California) to oppose the use of the CDM for supporting plantations, and to urge greater consultation of local users of forests in decisions about climate change policy.

Yet perhaps the greatest controversy came when the IPCC published its "Special Report on Land Use, Land Use Change and Forestry" (LULUCF) in 2000 (IPCC, 2000). The report was intended, in part, to settle some of the outstanding scientific disputes, and form a basis for carbon-offset forestry to continue unchallenged. Some commentators have stated that the report represents some progress in sophistication over the IPCC Second Assessment Report (e.g. Fogel, 2002). But the World Rainforest Movement published a damning rebuke of the report, alleging a conflict of interest because lead authors of the report also had stakes in private-sector consultancies seeking to promote carbon-offset forestry (WRM, 2000). The criticism was summarized as follows:

> The World Rainforest Movement (WRM) says that many of the authors and editors of the Intergovernmental Panel on Climate Change's "Special Report on Land Use, Land Use Change and Forestry" (LULUCF) – unveiled in June before international climate

negotiators – were businesspeople in a position to profit financially from the tree-planting schemes likely to follow in the report's wake. . . . The WRM's charge of conflict of interest is particularly stinging since the Intergovernmental Panel on Climate Change responsible for the report prides itself on being an independent body providing "neutral" scientific, technical and economic information to the parties to the Framework Convention on Climate Change and the Kyoto Protocol.

(*Multinational Monitor*, June 2000)

The statement about the IPCC Special Report suggests that the supposedly neutral scientific advice from the IPCC may be flawed by personal interest. These criticisms are, of course, denied by the IPCC. Yet, as discussed in Chapter 3, such combinations of influences on supposedly neutral political advice make it very difficult to see how the environmental science underlying carbon-offset forestry does not reflect political or social factors. As listed in Box 3.1, one ardent defender of orthodox science listed ten canons of good scientific practice, including the statement: "['Factual science' includes] the ethos of the free search for truth, depth and system (rather than, say, the ethos of faith or that of the bound quest for utility, profit, power or consensus" (Bunge, 1991: 246).

It is important not to let disputes like this lead to a simplistic acceptance or rejection of carbon-offset forestry, but to appreciate the political pressures upon the presentation of scientific information that is portrayed as being neutral and accurate. There are possible forms of carbon-offset forestry that may be beneficial for climate change mitigation as well as local development. Yet it is clear that the emergence of carbon-offset forestry has been based on a powerful network of national and intergovernmental organizations, investors, climate change negotiators, national governments, scientists, and some conservationists who, together, have presented considerably more power than the coalition of critics who have tried to oppose it. One implication of the network seeking to establish carbon-offset forestry as an acceptable option has been the over-simplification of much underlying biophysical uncertainty, and the reinforcement of many environmental orthodoxies associated with deforestation (see Chapter 2). In this way, the new debate of carbon-offset forestry, and its associated boundary organizations (in both public and private sectors), have acted both as science-policy agents – by shaping debate about climate change mitigation – yet also as structures by reinforcing environmental discourses about the role of forests in environmental degradation.

Critics of carbon-offset forestry are faced with various options against this relatively strong network. Some activists may continue trying to oppose the use of forestry in this way. Less confrontationally, activists may seek to reform how sequestration is achieved. One option is to restrict the amount of land offered to monoculture plantations, especially in locations where land may be used for agriculture. Another option is to integrate carbon sequestration with local livelihood strategies. Forms of agroforestry, for example, may allow local food and tree-crop production as well as carbon offsets. A further means is to enhance soil fertility through soil-carbon sequestration (Olsson and Ardö, 2002). Such strategies, however, may not be as cheap as monoculture plantations, and would not appeal most to activists who see reforestation as a way to restore lost wilderness, or state and commercial interests that wish to maximize income from plantations. Consistent with Actor Network Theory or the extended translation model of science (Callon, 1995; see Chapter 4), either of these approaches might constitute attempts to work within existing networks in favor of carbon sequestration, but to refocus their main attention from trees alone to other potential biophysical objects.

Chapter 7 further discusses questions of carbon sequestration, the CDM, and vulnerability to climate change. This section has sought to identify some ways in which current boundary organizations have impacted on environmental science and policy. The next section now assesses potential challenges to networks and boundaries from social movements.

Challenging boundaries: social movements and reframing science

Boundary organizations are one approach to analyzing the coproduction of environmental science and policy. Yet the power to shape environmental science in this way is not restricted to formal organizations or expert agencies, and can include less formal forms of political activism such as social movements. Indeed, much discussion in environmental politics or political ecology has pointed to the positive role played by social movements – or a vibrant civil society in general – in forming a more ecologically aware, or more socially just form of development (e.g. Princen *et al.*, 1994; Taylor, 1995; Peet and Watts, 1996; Bryant and Bailey, 1997).

Chapter 5 has already discussed the importance of environmentalism as a "new" social movement, and its impact upon environmental discourse and some generalized beliefs about the causes of environmental degradation. In Chapter 6, however, the discussion looks more critically at the claims of some debates in political ecology about the positive impacts of social movements in achieving social justice, especially in the developing world. The objective of this discussion is to examine how far social movements may act as agents to reframe environmental science, or whether they are constrained within, and may even replicate, existing structures and narratives of science.

The discussion is divided into two main sections. The first section summarizes some of the optimistic approaches to social movements within political ecology. The second section then discusses social movements in the context of the coproduction of political activism and environmental knowledge. In particular, this section considers how far the participation in social movements by different groups may influence environmental epistemology – or the knowledge generated about environmental risks and problems.

Ecological enlightenment and rationality

A variety of authors have pointed to the potential benefits of environmental social movements on entrenched and ecologically degrading forms of development and politics. Such authors have generally been influenced by the debates in Critical Theory referring to the instrumental rationality and social oppression of undemocratic government and impersonal economic growth described by Habermas and Marcuse.

As discussed in Chapter 5, such approaches have adopted an approach to ecology that has seen instrumental rationality as the overriding cause of environmental degradation. Furthermore, social activism through social movements has been identified as a way of redressing this oppression. For example, the German environmentalist, Helmut Wiesenthal (1993: 17) urged that the purposes of green politics should be seen as "preservation and emancipation." Such views reflect an approach to ecology that is influenced by concepts of equilibrium and threatened wilderness, and the symbolic use of ecology as a reversal of instrumental rationality on the lives of citizens.

Similarly, other authors have suggested that environmental social movements offer means to revitalize environmental discourse that is increasingly dominated by state and industrial interests (e.g. Blowers, 1997; Mol, 1996; see also the discussion of ecological modernization in Chapter 5). Such activism may be considered "green" not just because it reflects the findings of environmental science but also because it offers a reaction from society to the domination of policy debates by the interests of economic growth (Eder, 1996). For example, Blowers wrote:

> Environmental movements are one set of interests in civil society which have begun to thrive... The rise of such movements, their influence on policy and their impact at the level of consciousness-raising and value-shift has been a process of social and institutional learning ... [it] is arguably a major factor in the incorporation of the environmental dimensions into all levels of policy making.
>
> (1997: 865)

Consequently, many writers see environmental social movements as key ways to extend environmental networks, and to revise policies in general. These beliefs are rooted in an approach to ecological rationality that frequently highlights the fragility of environment in the face of industrial growth and modern society (e.g. Dryzek, 1990; Eckersley, 1992; Murphy, 1994). The implications of such approaches for the contestation and enforcement of environmental science networks are discussed more in the following section.

Advocacy coalitions and resource mobilization

A further approach to the reforming power of social movements has focused on the advocacy coalitions and political impacts of environmental social movements worldwide. Advocacy coalitions have assisted in communicating and empowering environmental policy in a variety of national and international debates (Sabatier, 1987; Sabatier and Jenkins-Smith, 1993). Coalitions have also been influential in resource mobilization, or the combination of different political forces to win environmental struggles, particularly concerning resources and livelihood struggles in developing countries. Concerning advocacy coalitions, Keck and Sikkink write:

> Like activists in social movement organizations, activists in transnational advocacy networks seek to make the demands, claims, or rights of the less powerful win over the purported interests of the more powerful.
>
> (1998: 217)

And concerning local livelihood struggles, Bryant and Bailey write:

> Grassroots actors derive their power primarily from the combination of a detailed local social and environmental knowledge, and a willingness and determination to use such knowledge through covert and public means to promote their interests.
>
> (1997: 189)

Sometimes the advocacy coalition and resource mobilization approaches to social movements also adopt the counter-opposition of social groups against state and/or industry. For example, Bryant and Bailey (1997: 190) further argue that grassroots and non-governmental organizations make a "natural alliance" in environmental politics against the dominating (and commonly allied) interests of the state and industry. (Indeed, such an alliance represents the model of political roles of state, business, and society described in this chapter, pp. 138–141.)

The coordination of political activism by grassroots actors, NGOs, and associated allies, such as academics, media, or experts, may therefore form the best chance to subvert the interests of state and industry in exploitative or undemocratic development. Yet, combining different actors in complex coalitions may often imply compromise and power struggles between partners. A key concern in political analysis of such coordinated social movements is therefore to examine how far alliances and activism on behalf of marginalized groups may enforce their concerns, or those of more powerful partners. These concerns are discussed further later in the chapter.

Liberation Ecologies

Similarly, other theorists have argued that, in addition to opposing the interests of state and industry, social movements may also democratize underlying environmental discourses. Of the three approaches to social movements discussed in this section, this approach comes closest to the challenging of the boundaries of environmental science.

This approach to environmental social movements has been most associated with Arturo Escobar (1996, 1998). For example, in discussing the role of social movements in the conservation of biodiversity, Escobar (1998) asked, "Whose knowledge? Whose nature?" Similarly, clearly echoing Foucault's (1980) call for an "insurrection of subjugated knowledge," Escobar wrote:

> We need new narratives of life and culture. These narratives will likely be hybrids of sorts; they will arise from the mediations that local cultures are able to effect on the discourse and practices of nature, capital, and modernity. This is a collective task that perhaps only social movements are in a position to advance.
>
> (1996: 65)

And Peet and Watts, describing an approach they call "Liberation Ecologies," suggested:

> movements are collectivities organized around common concerns and oppressions. But as well as being practical struggles over livelihood and survival, they contest the "truths," imaginations, and discourses through which people think, speak about, and experience systems of livelihood ... Rather than "speaking for" subaltern peoples, the idea is to help uncover discourses of resistance, put them into wider circulation, create networks of ideas. Rather than saying what peasant consciousness should be, were it to be "correct," the idea is to allow discourses to speak for themselves.
>
> (1996: 37, 34)

These arguments acknowledge, more than the previous approaches, that environmental discourses underlying social movements may not always support the interests of marginalized actors. Moreover, this problem may not simply be addressed by paying attention to power relations within an advocacy coalition, but by reframing environmental discourse in general. Attempting to "speak on behalf of others" may result in projecting values onto such groups.

In response to these problems, "Liberation Ecologies" seeks to allow marginalized social groups to speak for themselves. The approach offers a way to engage political activism with social constructions of science, and therefore is potentially an important means of achieving locally relevant environmental science-based policy. Yet such positive engagement also means reassessing the objectives, arenas, and languages of environmental activism in order to avoid imposing external framings. Peet and Watts suggest that one useful concept in enhancing local framings of environmental problems is the so-called "environmental imaginary." They wrote:

> The environment is an active construction of imagination, and the discourses themselves assume regional forms that are, as it were, theoretically organized by natural contexts. In other words, there is not an imaginary made in some "separate" social realm, but an environmental imaginary, or rather whole complexes of imaginaries with which people think, discuss, and contend threats to their livelihoods.
>
> (1996: 37)

Social movements may therefore be inspired by, and give institutional strength to, environmental imaginaries. Yet such imaginaries are changeable, contested, and can be shaped by other powerful interests. The following section discusses this approach in more detail. (Chapter 7 also considers further problems of "speaking on behalf of others.")

Rethinking social movements and environmental epistemology

The preceding section listed approaches to environmental politics that have generally identified social movements as powerful and beneficial agents in shaping environmental discourse and policy. These approaches have included discussions of social movements as a source of state reform within industrialized economies (e.g. Mol, 1996; Blowers, 1997); and as livelihood struggles within developing countries (e.g. Peet and Watts, 1996; Bryant and Bailey, 1997). While there is no denying that social movements have impacted greatly on the adoption of environmental values, and influenced the evolution of environmental discourse, this chapter argues that these approaches have tended to overrate the role of social movements as autonomous, powerful agents. Instead, it is necessary to see how far social movements have operated within preexisting structures imposed by social institutions of environmental science. Indeed, on some occasions movements may reinforce these structures. As a result, social movements may not be as liberatory as some scholars have argued.

This section now examines these concerns about environmental social movements. In particular, the section focuses on the social inclusivity of social movements, and hidden implications of environmental activism for the construction of risk. Together, these themes indicate how social movements contest or enforce environmental science.

Social inclusivity of environmental alliances and discourses

One common theme adopted by optimistic approaches to environmental social movements is that their actions may lead to positive outcomes for people affected by environmental degradation or oppressive actions by state and industry. Yet each of the approaches listed above vary in the degree to which social movements may include all sectors of society. In general terms, discussions of ecological rationality assume that ecologism offers a uniformly beneficial outcome for society if it opposes instrumental rationality and exploitation by the state and industrialism. The advocacy coalition and resource mobilization approaches also assume that movements can unite people against oppressive regimes – although writers acknowledge that differences and rivalries may occur within partnerships. The discursive or "Liberation Ecologies" approach suggests that when social movements are allowed to emerge from subaltern voices they can successfully reframe environmental discourse.

A number of questions can be asked concerning these assumptions about social inclusivity. In turn, these questions impact on the ability for social movements to be seen as effective and autonomous agents in re-shaping the boundaries of environmental science. First, as initially discussed in Chapter 5 and above (pp. 138–141), there is a tendency for much discussion of environmental social movements to be based on assumptions about general differences between "state," "industry," and "society." These assumptions are most obvious in relation to the "new" social movements and the ecological rationality approaches to social movements that have generally assumed a direct and essentialist causality between capitalist development and environmental degradation. As mentioned in Chapter 5, such an approach to causality may reflect social concerns about life under oppressive political and industrial regimes rather than a more long-term and complex understanding of ecological change. Furthermore, such criticism may be described as coming largely from middle-class activists, even though they are occasionally portrayed as avoiding class differences, and speaking on behalf of all society (Offe, 1985).

These beliefs about the presumed causes of degradation, and the inclusivity of this kind of activism, however, overlook two crucial concerns. First, such activism may give especial attention to particular aspects of environmental degradation that are of concern to the more powerful voices in the movement, rather than the types of risk experienced by less powerful voices. For example, Satterthwaite (1997) has argued that "new" social movement activism has favored interests of urban middle classes in, say, protecting wilderness (the "green" environmental agenda), rather than industrial and urban risks affecting poor city dwellers (the "brown" environmental agenda). Second, the belief that capitalist development *per se* causes environmental degradation may also contribute to the opposition of some forms of industrial development that may not have such degrading impacts. Indeed, some scholars in development studies have shared this concern and suggested that "industrialism" and "capitalism" need to be separated in order to indicate how far each may lead to social exploitation. Such a suggestion should not be interpreted as a way to legitimize exploitative development by multinational corporations, but a desire to see that criticisms of capitalism do not result in disallowing less exploitative forms of industrial development, or in damaging the livelihood prospects of poor entrepreneurs in developing countries (Corbridge, 1986; Schuurman, 1993).

A similar question about inclusivity of social movements may be directed at the advocacy coalition and resource mobilization approaches. Keck and Sikkink (1998), for example, urge that advocacy coalitions may support the interests of the marginalized. Yet if these interests are defined in general terms as "opposing environmental degradation," such a definition does not allow for the different priorities afforded to varying kinds of risk from different social groups. Differences in the experience of risk may exist along lines of class, race, and gender, or in the resistance to the activities of one social group that may carry impacts for other social groups. Covey (1995), for example, noted that environmental alliances between middle-class NGOs and grassroots organizations in the Philippines tended to emphasize political objectives of the middle-class groups rather than grassroots actors. For instance, the generation of municipal waste in city centers such as Manila may lead to environmental concern within cities that leads to the establishment of waste dumps, and their associated environmental risks, on city outskirts. Such results may leave grassroots organizations asking whether the extra political power coming from alliances are worth the consequent loss of focus. As Lohmann wrote: "powerful when its voice (i.e. a group) is joined to a variegated chorus of others, it loses its power, and risks being targeted by forces of repression, if it demands that everyone sings in unison with itself" (1995: 226).

Of course, such problems of achieving unity in political alliances are well acknowledged by writers on environmental politics in developing countries (e.g. Bryant and Bailey, 1997). Yet there is a need to consider how far such differences between coalition partners may influence how the social movement or political activism may result in reframing the boundaries of existing environmental science in favor of different groups. Social movements or advocacy coalitions may successfully challenge exploitative development from state and industry, and may result in a greater adoption of environmental policy. But if the new policies and values resulting from this activism replicate environmental discourses and explanations based upon the experiences and agendas of only selected members of the alliance, then this may not necessarily work in favor of groups who are less well represented in these discourses and explanations.

Hajer (1995) outlined three main problems with the advocacy coalition approach to environmental politics. First, advocacy coalitions assume an individualist ontology, or that social movements and coalitions may be autonomous and active agents of political change. Instead, coalitions should be seen to be reflective of preexisting institutional bases of knowledge and belief that condition the purpose and actions of advocacy and activism. Second, advocacy coalitions are often based upon a priori definitions of beliefs and social norms. There is insufficient awareness of how such beliefs may be time and context specific, or how they may be interpreted more locally in social contexts different to those where the values were initially communicated. Third, the advocacy coalition approach assumes a rational model of cognitive change, as though values can be communicated separately of the circumstances in which communication takes place. Instead, there is a need to appreciate that agreements and communication will reflect local circumstances of language, shared interest, and perceived purpose between different parties, rather than be the absolute transfer of clearly defined concepts from one group to another.

As a result of these problems, Hajer (1995: 65) argued that the alternative concept of "discourse coalitions" is a more useful way of understanding how political activism may reinforce existing boundaries and networks of science (this concept was introduced in Chapter 4). Discourse coalitions are the interactions between different narratives and environmental storylines that allow different political parties to reach agreement about the subject matters of debate while continuing their political agreement or disagreement, and hence partly define that subject matter. (Indeed, such areas of agreement may be called "multivalencies," see Chapter 4.) As a result, consensual positions on topics of environmental concern may be more influenced by the interactions between different actors than by a realist understanding of the "factual" reality of these topics. Advocacy coalitions may therefore create a variety of opportunities for such interaction and alliance that shape, or adopt, existing structures of environmental science, rather than act externally to such structures.

The implication of these criticisms is that social movements may not be as autonomous agents of environmental reform as is often discussed. Moreover, social movements may even replicate existing environmental discourses, or reshape these according to the objectives of the more powerful voices in alliances. Such reshaping may also occur in the case of "Liberation Ecologies," or the approach to environmental social movements that seeks to reframe environmental discourse along more locally determined lines. Indeed, Peet and Watts's concept of "environmental imaginaries" (1996) suggests that local environmental concerns and perceptions may be the impetus for effecting change in environmental political objectives. Yet concepts such as discourse coalitions and the contextual and dynamic nature of communication and argument suggest that such imaginaries may be reshaped or co-opted by powerful narratives within stronger social networks. We now look at how such interactions may shape (and be shaped by) environmental explanations.

Implications for the construction of risk

The previous section argued that optimistic approaches to analyzing social movements in environmental politics have often failed to appreciate how social activism and environmental discourses are interlinked. If social movements are to challenge the boundaries of dominant science networks, there must be greater attention to how far movements revise, reject, or replicate the assumptions and norms of these networks. Failing to appreciate how social movements may engage with and reframe the institutions of environmental science may mean that some environmental struggles may end up reinforcing rather than democratizing hegemonic forms of environmental explanation (Jamison, 1996). Indeed, Eyerman and Jamison (1991) have described this process as the "cognitive praxis" of social movements – a term similar in many respects to coproduction.

Two examples from Thailand illustrate how environmental activism from social movements influences the production of environmental explanation (see Forsyth, 1999c, 2001a, 2002). These examples are then compared with the case of activism by people with AIDS in the USA, and the relatively more beneficial influence such activism had on reframing science (Epstein, 1996).

The first example in Thailand concerns an environmental dispute in an industrial estate in the province of Lamphun in the north of the country. During the years 1993–1994, between ten and twenty workers at the industrial estate, and some family members, died suddenly from unexplained causes. The majority of deaths reported were suggestive of industrial poisoning, as they all involved rapid deaths of generally young workers who had been exposed to the materials used in soldering and cleaning activities conducted during the manufacturing of electronic circuitry. The factories were generally owned by Japanese investors, and included the well-known fax manufacturer, Murata (Forsyth, 1994, 1998b).

The sudden deaths of the workers immediately induced claims and counter claims by different actors. Local workers believed the deaths were caused by lead poisoning or inhalation of solvents. The local government authorities and factories blamed the deaths on AIDS, or on other causes unrelated to the factories. One doctor experienced in industrial poisoning claimed that the majority of deaths were more likely to have been caused by solvent poisoning, or the inhalation of the powerful chemicals used to clean electronic circuits before soldering and packaging. Testing for the causes of death was made difficult by the local practices of cremating bodies quickly, the lack of local hospital testing equipment, and the difficulty of detecting solvents after inhalation.

In the following months, however, most public debate surprisingly focused on lead poisoning as the cause of death. This cause seems unlikely as most victims died suddenly and without previous symptoms. Yet local NGOs preferred to discuss lead poisoning because it was already a well-known cause of concern, and because it also supported other worries about lead mining in supposedly protected forest areas. On the other hand, government officials seemed happy to discuss lead poisoning because it was easier to deny than solvent poisoning. The debate about lead poisoning therefore allowed NGOs and state to oppose each other along well-known lines of opposition, even if lead poisoning did not seem the most likely cause of death. The topic even became an inspiration for pop songs about the perils of industrialization.

A further example concerns public opposition to the filming of the Hollywood movie, *The Beach* in Thailand in 1998–1999 (Forsyth, 2002). The movie, which starred Leonardo di Caprio, was based on the sensationalistic novel about backpackers and drugs, and required a setting on a remote beach environment. The government allowed the filming to proceed at Maya Bay in the Phi Phi Islands national park in the south of Thailand, but overlooked the law in Thailand that prevents any economic activities or disturbance inside national parks. In response, a variety of campaigners sought to prevent the filming, using street demonstrations, Internet and media campaigns, and lawsuits. In particular, the protestors sought to demonstrate the devastating effect of the filming on the local fragile environment. The film's producers, however, took pains to demonstrate that plants, sand, and coral were restored as fully as possible, and claimed that they had to remove some two tons of garbage left by tourists on the beach.

The dispute concerning *The Beach* was portrayed in media as a further case of environmental degradation resulting from a corrupt alliance between the government and a foreign investor. Yet a more critical analysis reveals deeper themes. First, the campaigners used arguments about ecological fragility in order to add greater urgency and concern to other arguments about the apparent abuse of Thai laws by the government. Indeed, the statements of the campaigners, and the legislation concerning national parks, reflect an equilibrium, or "balance of nature" approach to ecology that has been criticized by many ecologists (see Chapter 3).

Second, the arguments against *The Beach* reinforced an existing narrative and network of environmental science in Thailand that have other political implications. As discussed in Chapter 3, discourses of ecological fragility and "balance-of-nature" have frequently been used to support politically repressive policies (Zimmerer, 2000). In other parts of Thailand, this discourse has been used to support claims to relocate hill farmers from a variety of forest or watershed areas, even though such claims have been criticized by a variety of research (also see the discussion of ICRAF above, pp. 146–147). This scientific network was further enforced in practical terms by the fact that one of the main protestors against *The Beach* was also a family member of a well-known activist in northern Thailand aiming to resettle farmers from upland areas. The activism associated with *The Beach* therefore sought to enhance democratization by criticizing a government that did not uphold its own laws. Yet it was based on an environmental discourse that was constructed in less democratic ways, and which may even restrict livelihoods for some less powerful people.

Social movements and political transparency

Environmental social movements may therefore not necessarily lead to a radical reframing of environmental discourses, but instead may co-opt and replicate existing narratives in order to increase their political power. What can be done to avoid such pitfalls?

This question is considered throughout the rest of this book. Indeed, under a "critical" political ecology, there is a need to assess how far different knowledge claims about the environment reflect hidden assumptions, and on the means to expose and reframe these assumptions. Such an approach, however, does not imply rejecting a role for social movements, but in developing new ways of understanding their impacts on environmental discourses and the boundaries of environmental science.

This final section suggests some positive ways of engaging with environmental social movements that can focus more clearly on the ways they engage with environmental science. These suggestions are then discussed in more detail, especially in Chapter 9. In particular, they consider how far social movements may themselves form boundary organizations in their engagement with the producers of scientific knowledge.

One optimistic example of politics of social movements comes from the USA. Steve Epstein's (1996) study of AIDS activism and the politics of knowledge is one case where activism was able to reframe the scientific basis of explanations and achieve a greater representation of the needs of affected parties. Epstein's study assessed the impacts of social movements on the study and dissemination of medical research in the USA. He wrote: "The case of AIDS activism suggests that social movements can pursue distinctive forms of participation in science and, conversely, that the engagement with science can shape movements in powerful ways" (1996: 332).

Epstein argued that the activism concerning AIDS focused on public discontent with the lack of consultation from medical practitioners on both the formulation of scientific statements (e.g. that HIV causes AIDS); or in how people with AIDS are treated as a result. Epstein (ibid.: 28) drew upon Bruno Latour's (1987) discussion of "black boxes" in science, in which "observations" are presented as "discoveries," which then become "facts" and finally "common knowledge" (see Chapter 4). AIDS activists were able to challenge these black boxes by undertaking strategies such as portraying themselves as the potential population of research subjects; adopting the language and communication styles of the biomedical sciences; and using allies within the establishment. Such tactics led to the establishment of credibility within the scientific network. Furthermore, the use of activists themselves as the potential population of research subjects presented them as obligatory passage points for scientific research (Epstein, 1996: 335–336).

The result of this activism was to gradually shift medical science concerning AIDS from a focus on stopping HIV (as the key "cause" of AIDS), toward also considering medical techniques that address the symptoms of AIDS. This approach did not challenge the role of HIV in causing AIDS, but provided alternative means for addressing the "problem" caused by AIDS. Epstein wrote:

> In their critiques of "pure" or "clean" or "elegant" science, and in their invocation of the "real world" and "pragmatic" decision making, AIDS activists have emphasized the *local* and *contextual* character of usable scientific knowledge ... In the alternative conception that develops out of activist critiques, reliable knowledge is produced through close attention to the concrete social, moral, and political context: better science comes about *because of* the focus on individual patients and their needs, desires, and expectations. This alternative conception of science is willing to surrender claims to universal validity in exchange for knowledge that nears some local and circumscribed utility.
>
> (1996: 342, emphasis in original)

The implication of Epstein's study is that social movements can be successful in reshaping the institutions of science. But this example also indicates that success came when activists were able to negotiate – and to some extent join – dominant networks, rather than challenge the network outright. Furthermore, negotiating with the network may mean adopting the same language and negotiating style in order to gain credibility and legitimacy. Indeed, the activism concerning AIDS benefited from having a relatively focused purpose, and by comprising activists from educated professional classes, with a variety of personal and professional linkages to the targeted science network. Such advantages, clearly, do not exist for uneducated poor farmers in rural zones, or factory workers in rapidly industrializing countries, where the arenas for intellectual and political debate are poorly developed.

A further example of a more considered analysis of the use of scientific assumptions by social movements concerns protests against Genetically Modified Organisms (GMOs). Protests have often reflected elements of political opposition to the unregulated activities of industry, or the unethical tampering with nature (see Chapter 5). Popular debate about GMOs has often focused on the scientific uncertainties of whether or not they pose a risk, and to whom. Frequently, activists have refused to believe the scientific findings of industrial scientists whom they believe are unlikely to reveal misgivings about GMOs.

Instead of focusing on the public outrage against GMOs, some observers have instead considered how the presentation of science has caused this reaction. Levidow and Carr (1997), for example, have argued that the regulation of GMOs has made hidden distinctions between "risk" and "ethics" that many people consider problematic. Similarly, Wynne (2001) has suggested that science has become the culture of GMO policy, rather than its key resource, and that this has implied that urgent questions of responsibility about GMOs are contained within the scientific debate rather than in public arenas. Such factors have added to, rather than mitigated, public alienation from discussions about GMOs. Instead of seeking to oppose GMOs in their entirety – as some environmental NGOs do – this more considered approach to scientific uncertainty suggests that there is a need to increase public participation in how decisions about GMOs or environment are made. In essence, this means increasing the links between science and political debate rather than seeking to justify "science" by its separation from politics. It also means seeking to construct political fora that allow the public consideration of topics of environmental concern in ways that allow greater transparency of scientific arguments used by each side. These concerns are discussed in the rest of the book.

Summary

This chapter has advanced the discussion of a "critical" political ecology by examining detailed ways to analyze the political influences on the enforcement or contestation of environmental science. In particular, the chapter looked at the ways in which the boundaries of environmental science networks and institutions can be affected by the activities of so-called boundary organizations and social movements.

Boundary organizations are those agencies or organizations that connect different networks in science and policy, and consequently have the ability to influence shared behavior in both parties. Boundary organization analysis offers a more epistemologically sensitive approach to the coproduction of science and policy than the focus on political "actors" adopted by much conventional political ecology. The chapter described some examples of boundary organization analysis, particularly concerning the case of carbon-offset forestry, which has relied upon the interactions of a variety of boundary organizations and orthodox environmental science assumptions.

As a potential challenge to powerful networks, the chapter then assessed debates concerning environmental social movements. Some conventional approaches in political ecology are optimistic about the ability for social movements to reframe environmental policy or discourses in favor of marginalized social groups. Yet this chapter argued that these approaches – such as those involving notions of ecological rationality, advocacy coalitions, or "Liberation Ecologies" – do not pay sufficient attention to the coproduction of environmental activism and the scientific knowledge that is used to add legitimacy to campaigns. As a consequence, social movements may replicate and reinforce preexisting structures of environmental debate, rather than reframe these in favor of marginalized groups. Instead, it may be more productive to understand how different actors use science, or are involved in the scientific process. The chapter presented examples of social movements from Thailand, and concerning AIDS in the USA, to illustrate different cases of the relationships between social movements and science.

The chapter's most important conclusion is that political activists and scientific networks should be seen in terms of complex and interrelated structure and agency, rather than as the simple opposition of clearly identifiable "actors" using predefined explanations of environmental degradation from orthodox science. A "critical" political ecology stance seeks to understand how such interaction leads to different constructions of environmental reality. Not seeing how political activism coproduces environmental scientific discourses may potentially result in the replication of oppressive environmental discourses rather than the successful democratization of discourses. It may therefore be more important to increase public participation in scientific inquiry than to attempt to prove the authority of science by keeping it apart from public involvement.

The following chapters now discuss these themes further. Chapter 7 examines the role of scientific methods and organizations in extending environmental assumptions to different locations. Chapters 8 and 9 consider potential solutions to unproblematized science by seeking ways to increase public participation.

7 The globalization of environmental risk

Chapters 5 and 6 discussed how environmental science and politics have co-evolved over time in a dynamic fashion. Now, Chapter 7 examines how environmental explanations have become extended across space through such means as global models or the projection of understandings of risk onto different societies. The chapter will:

- discuss the evolution of debate about "global" environmental problems as a paradigm in environmental science. Some observers have claimed that global environmental problems are major new risks that are increasingly prevalent worldwide. Against this, critics have suggested that our understandings of "global" problems still reflect the values and practices of scientific networks, and fail to acknowledge how global "changes" present "risks" at local levels.
- examine how research about environmental problems in remote societies has often replicated predefined assumptions about environmental risk and degradation from outside. Ironically, such replication has often occurred with debates that seek to highlight "local," or "indigenous," knowledge.
- analyze environmental vulnerability – or exposure to risk – by comparing approaches that seek to mitigate biophysical changes as the presumed cause of risk, and those that incorporate understandings of local livelihoods and ability to adapt to risk. Worryingly, some approaches that attempt to mitigate biophysical change alone might actually increase local vulnerability to environmental changes.

This chapter therefore adds to the discussion of "critical" political ecology by examining the political factors through which explanations and representations of ecology are assumed to apply across different spatial scales. Making such factors more transparent, or developing alternative forms of explanation, are discussed in later chapters.

"Global" science and risk

Much of this book has examined the role of historic actions by researchers and specific societies in forming the tacit politics within many environmental scientific statements. But what factors influence the spread of environmental science across space? This chapter considers these points, with particular reference to the emergence of "global" environmental problems as a site of research and political concern; and the implications of global generalizations and "laws" for people not represented in the formulation of such explanations.

"Global" environmental problems such as anthropogenic climate change or atmospheric ozone depletion are now commonly discussed as dangerous challenges for the world at large. Indeed, the Global Environment Facility, for example, was established in 1990 as the first international initiative to address the transboundary environmental problems of anthropogenic climate change, oceanic pollution, and depletion of biodiversity and atmospheric ozone. But there is still controversy about the meaning of the word "global," and the extent to which it can be used to describe these and similar problems.

The most common usages of the word "global" in relation to environmental problems are to refer to biophysical changes that threaten the stability or status of the planet as a single unit; or problems that result from changes occurring "globally." The process of economic "globalization," or the increasing impact of global investment and industrialization is an example of the second kind of changes that are often considered to have such globally prevalent environmental impacts. Against these approaches, however, sociologists of scientific knowledge have questioned how far these ways of seeing "global" problems also imply a growing sense of the world as one unit (see Yearley, 1996). Such questions do not deny the possible existence of environmental problems at the global scale, or of globally prevalent causes of environmental degradation. Instead, they point out that the production of knowledge about so-called "global" problems will itself reflect politics and culture and hence reflect how we see such problems. In this sense, the process of "globalization" can be seen both as increasing global investment and industrialization, and as the increasing discussion and conceptualization of the world as a single unit: "globalization as a concept refers both to the compression of the world and the intensification of consciousness of the world as a whole" (Robertson, 1992: 8).

In environmental terms, there is much evidence to indicate that the increased perception of the world as one unit has evolved at the same time as concerns about environmental degradation. The world's first photographs of the Earth from space, for example, were televised in 1969 during the emergence of environmentalism as a new social movement, and a new force in international politics. Indeed, Jasanoff (1999: 146) considers this to have been a paradigm shift in environmental debate and research. The

British environmentalist Jonathan Porritt neatly summarized this new perception:

> Those shots of the Earth taken from the Apollo spacecraft didn't exactly change my life then and there in August 1969, but undoubtedly helped shape my early environmental interests. Not so much because I started thinking of the Earth as "fragile," as so many contemporary commentators seemed to do, but because of the wholeness of what I saw. The Earth as one system, all of a piece, not broken into continents, countries, poles, weather zones, ecosystems, and so on...
>
> (2000: 133)

Furthermore, there have also been many accounts of threats to environment resulting from globally prevalent causes such as population growth or economic "globalization." Martin Khor, director of the Third World Network (an organization based in Malaysia campaigning for greater recognition of developing countries in international politics) wrote: "the ecological crises that threaten the survival of Earth continue to unfold at breakneck speed under the influence of commercial interests, driven even further by the competitive pressures of globalization" (1997: 3).

Third, there is also much discussion of "global" elements of environmental risk in social theory debates about Ulrich Beck's (1992) concept of "Risk Society." In *Risk Society*, Beck argues that advanced societies have progressed from an industrial stage of production and organization to a new stage in which everyday life for individuals concerns the allocation or avoidance of risks from a variety of personal, financial, or environmental sources. Yet, under Beck's initial description of "Risk Society," environmental risks are seen to be global, unchallenged, and resulting again from the production of wealth:

> In advanced modernity, the social production of wealth is systematically accompanied by the social production of risks ... With the globalization of risks a social dynamic is set in motion, which can no longer be composed of and understood in class categories ... The tangibility of need suppresses the perception of risks, but only the perception, not their reality or their effects.
>
> (Beck, 1992: 19, 39, 45)

The objective of this chapter is not to suggest that there is no need to consider "global" environmental problems; or that economic globalization is environmentally unproblematic; or that science cannot assist in understanding such problems. Instead, the aim is to suggest that the three assumptions contained in the approaches to "global" environmental problems described above (of seeing the world as a single unit; of seeing universal causes of globally prevalent degradation; and that global risks are universally acknowledged and unchallenged) may be questioned on a variety of grounds. The aims of this questioning are to highlight how the scientific or causal statements underlying assessments of "global" problems or proposed solutions may – knowingly or otherwise – reflect the political and social values of the societies or networks that created them. In keeping with a "critical" political ecology, the objective of such discussion is to make these scientific approaches more transparent and accessible in order to increase science's relevance to the needs of all social groups, and to increase our understanding of complex biophysical changes.

Challenging the global emphasis

The most common approaches to "global" environmental problems, as discussed above, have assumed that the Earth may be understood as a single entity; or that global changes may be occurring universally as the result of similar causes; or that the understanding and knowledge of risks are increasing globally without challenge. The aim of this next section is to challenge these approaches to so-called "global" problems by highlighting how they also infer a system of seeing the world that has co-evolved at the same time as our understanding of these problems. It is important to note that this challenge does not suggest there are no such things as "global" environmental problems, but that many common scientific approaches to them reflect "local" rather than "global" perspectives.

This section lists different ways in which we should be concerned at uncritically accepting definitions of "global" problems. First, there is an analysis of how many statements about "global" change reflect local framings and practices contained within the societies and scientific networks making the statements. Second, this section looks at how such "global" statements may suppress a number of important differences and insights at the local level that can either contribute to understanding the nature of risks, or indicate the local meaning attached to environmental changes often referred to as problems. This section outlines how it is problematic to refer to "global" environmental problems. The following section now looks more closely at how such "global" statements have been made in such uncritical fashion.

Seeing the local in the global: situated practices in global environmental models

The language associated with "global" environmental problems tempts us all to see the problems and environmental change discussed as unquestionably global. Yet as this book has argued throughout, all acts of inference and explanation cannot help to reflect social values and framings by which observed changes are made meaningful. A more critical, and politically transparent, approach to such scientific statements comes from understanding first what assumptions are built into global explanations of change, and how far these coincide (or not) with alternative framings for changes at both local and global levels.

As noted in Chapter 3, orthodox scientific inquiry has tended to encourage the generation of universalistic statements of causality or explanation. Yearley wrote:

> Given the centrality of science to the diagnosis and analysis of these global environmental issues, it is understandable that the discourse of science will affect the way that environmental problems are conceptualized. Typically, science aspires to universal generalizations. Unless there are powerful reasons to the contrary, scientists assume that natural processes are consistent throughout the natural world.
>
> (1996: 85)

Yet such statements contain assumptions and framings that may reflect "local" practices and beliefs contained within the societies or scientific networks that create scientific statements rather than be a universally accurate representation of reality for the entire globe.

These points were argued in an influential paper by Peter Taylor and Frederick Buttel (1992). As noted in Chapter 1, Peter Taylor was one of the original members of the Oxford-based Political Ecology Research Group during the 1970s. In the paper, Taylor and Buttel pointed out that in global environmental discourse, moral and technocratic views of politics have been privileged as urgent and potent solutions to perceived problems. Yet such discourse has also assumed common interests of society without also looking at how such communality may be experienced. They wrote: "We know we have global environmental problems, in part, because we act as if we are a unitary and not a differentiated 'we'" (Taylor and Buttel, 1992: 406).

Taylor and Buttel illustrated this argument in relation to the evolution of global modeling from the Limits to Growth (LTG) study published in 1972 (Meadows *et al.*, 1972), and the emergence of climate change modeling in the 1980s. The LTG model, developed by system dynamics modelers at the Massachusetts Institute of Technology, typified and shaped ecocatastrophist concern in the early 1970s by drawing attention to the finite nature of world resources resulting from unchecked economic growth. (The model was criticized first by economists for overlooking how scarcity might stimulate technological innovation and hence effectively allow continued growth; and second by political scientists for not acknowledging the unequal distribution of consumption, growth, resources, and technological adaptations between different countries and classes.)

The essence of the LTG system dynamics model, according to Taylor and Buttel, is the moral basis of decision-making at the level of the individual or firm. In keeping with rational choice or economistic modeling in general, modelers do not rely on recorded data as a guide for how decisions have been made, but instead rely on what is assumed to be common sense knowledge of how individuals react when faced with information and choices. In this sense, the LTG approach was similar to the political method advocated by the initial group of "political" ecologists such as Paul B. Sears (1964) who considered ecology a "subversive subject" because it sought a means of political analysis at a level higher than the individual, and which was seen to be within biophysical ecological limits. As Taylor and Buttel reported, this approach therefore assumed that "catastrophe is thus inevitable unless 'everyone' – all people, all decision makers, all nations – can be convinced to act in concert to change the basic structure of population and production growth" (1992: 408).

Yet, as Taylor and Buttel argued, such an approach assumes both similar ecological limits for all society, and that all society action can be modeled from the hypothetical moral choices of an individual. This blending of moral and methodological principles in the estimation of environmental problems therefore represents another example of the co-production of ecological values and environmental science. As Taylor and Buttel warned:

> The science of global environmental change continues to reflect, and in turn reinforce, the moral-technocratic formulation of global environmental problems … Inattention to the national and localized political and economic dynamics of socio-environmental change will ensure that scientists, both natural and social, and the environmentalists who invoke their findings will be continually surprised by the unpredicted conflicts and unlikely coalitions [in response to these supposedly politically neutral models].
>
> (ibid.: 409, 406)

Since the LTG modeling of the early 1970s, modeling of global environmental change, of course, has changed greatly and has largely abandoned system dynamics in favor of more empirically grounded projections based on a variety of more accurate and diverse knowledge sources such as General Circulation Models (GCMs). The history and politics of some GCMs were discussed in Chapter 6 as examples of different organizational approaches to knowledge production in environmental policy. Yet despite the greater complexity of GCMs, they too can be described as reductionist and as overlooking both the diversity of how risk is experienced around the globe, and of how policies developed in conjunction with models may have differential political impacts.

Circulation models may be considered reductionist for two main reasons (see Demeritt, 2001: 316). First, the analytical findings of climate models have generally been based only on physical properties of greenhouse gases, such as atmospheric residence time, radiative signature, and photochemical reactivity. Second, such biophysical measurements give little insight to the social context, meaning, or ability to adapt by different peoples to such predicted changes. By privileging the physical over the social, the global models portray risk only in terms of projected biophysical change to ecological parameters, rather than how such changes may create problems for people. Yet much research in cultural ecology, non-equilibrium ecology, and hazards theory, for example, has indicated that ecological changes (or biophysical events) by themselves may not represent hazards to all people in uniform ways. Instead, a fuller understanding of risk also needs to incorporate the vulnerability to, and perception of, changes taking place rather than to assume a priori that such changes are of necessity problematic (see Chapter 2, and the discussion of vulnerability later in this chapter, pp. 191–200).

The emphasis upon biophysical changes as a key guide to environmental problems, rather than social contextualization and vulnerability to changes, has two further implications. First, if the objective of models has been to quantify and predict changes to biophysical parameters, some models have overlooked the politics of equifinality – or how the same changes may be produced by different causes. The second problem is that assuming biophysical changes indicate risk overlooks how such changes may actually present problems locally, to different people, in different social and economic circumstances. These two implications are discussed throughout this chapter.

The first problem, of overlooking the politics of equifinality, is perhaps best indicated by the well-known example involving an early estimate of national responsibilities for greenhouse gas reduction calculated by Washington, DC-based environmental think tank, the World Resources Institute (WRI) during the early 1990s (see Box 7.1). In 1990, the WRI created a model to predict rates of global climate change that gave high weighting to current rates of tropical deforestation and methane fluxes from developing countries (WRI, 1990). This model was later criticized by the Delhi-based think tank, the Center for Science and Environment (CSE) (Agarwal and Narain, 1991) because the underlying assumptions placed no political evaluation of its assumptions – or, in essence, for producing an apparently neutral "black box" explanation of climate change that contained many controversial, yet tacit implications. As shown in Box 7.1, the model was developed at a time when concern about tropical deforestation was high in North America and Europe. Yet the assumptions in the model were alleged to overlook, for example, the politics of how far current deforestation should be taken on equal terms as historic industrial emissions from industrialized countries, or how far the implication of restricting deforestation might be fair when such views reflect "northern" visions of nature, and do not acknowledge questions of rural poverty.

Indeed, a further example of spatial generalizations may be seen in a second WRI publication, although not at a global scale. The report, *Watersheds of the World: Ecological Value and Vulnerability* (Revenga *et al.*, 1998), presented one-page summaries of different watershed basins with simple statistics of how far each basin was "degraded" based upon the analysis of aerial photographs and satellite imagery at a macro scale. Such

Box 7.1   The WRI–CSE controversy over modeling tropical deforestation and greenhouse gas emissions

This book has already pointed to the importance of deforestation as an apparently urgent and important environmental problem, and as a focal point in the rise of conservation-based social movements. Often these views hide a variety of controversies. Chapter 2 highlighted how "deforestation" may be experienced in many forms, sometimes unproblematically. Chapter 3 discussed the challenge of non-equilibrium ecology to visions of "lost wilderness." Chapter 4 summarized how "tropical rainforests" have been discussed in linguistic terms to denote a variety of values not necessarily related to the forests themselves. Chapter 5 questioned the class basis in social movements seeking to protect wilderness. And Chapter 6 highlighted how interactions between different actors and scientific assessments have reinforced orthodox visions of deforestation.

The perceived importance of tropical deforestation has also affected, and has been strengthened by, the discussion of "global" environmental problems. As Taylor and Buttel (1992: 411) wrote:

> [the] rainforest connection has ... been central in the scientific and popular construction of global change knowledge. At the level of environmental science, it has led to greater stress on the conservation biology of rainforest biodiversity, not only as a subordinate theme within the global environmental change framework, but also as a glamour topic in its own right.

Perhaps the most well-known example of controversies concerning the hidden politics of scientific assessments involving tropical deforestation was the debate between the World Resources Institute (WRI) and the Indian NGO, the Center for Science and Environment (CSE) in the early 1990s. During the 1980s, agencies such as the WRI were noteworthy for their ambitious approaches to influence government environmental policy, especially concerning topics that were of high perceived relevance to the general public (Thompson, 1985). In 1990, the WRI published one of the first reports that allocated potential national responsibilities for greenhouse gas emissions, in the build-up to the Rio Earth Summit (WRI, 1990). WRI used an index later published in 1991 (Hammond *et al.*, 1991). The index gave substantial weight to current deforestation rates and to the predicted release of methane from wet rice and livestock, and put three developing countries, Brazil, India, and China, among the top six emitting countries.

The publication of this report created much resentment among developing countries that had not expected to bear any major share in the reduction of greenhouse gas emissions. In particular, the CSE, and especially Agarwal and Narain (1991), contested the report on various grounds. First, the national allocation bore no reference to per capita emissions, which, of course, were much smaller in developing countries than in developed countries. Second, the index used estimates for both deforestation and methane emission that contained great simplifications. For example, wet-rice methane estimates were extrapolated globally from studies in Italian rice fields. There were no acknowledgments of the diverse ways in which deforestation may occur, or accurate guides to the potential growth of secondary forest or replacement ecosystems. Third, and perhaps most importantly, the index was criticized for its weighting system that placed current tropical deforestation so high. Critics suggested that historic deforestation in developed countries should also be included (as greenhouse gases have lives of many years). Furthermore, there was no reference to questions of poverty or access to land necessary for agriculture that deforestation may bring.

The WRI report of 1990 has since been widely recognized as simplistic and flawed. But the underlying questions of responsibility for climate change mitigation, and the relative importance of industrialization and deforestation, or of current and past deforestation, are still controversial. As noted in Chapter 6, much of the concern about deforestation – or the potential benefits of carbon-offset forestry – have been stimulated by environmental orthodoxies relating to deforestation, such as that simple reforestation may promote biodiversity or act as the "lungs of the earth," despite much discussion of how these concepts are simplistic (see Chapters 2 and 4).

Source: Brookfield *et al.*, 1995: 144–146.

descriptions, however, avoided discussion of how such conceptualizations of degradation were defined, or how they may be contested. The apparent implications of this publication were to suggest that degradation should be defined as loss of forest cover, when a variety of research has questioned orthodox links between forest cover and orthodox functions of watersheds such as water supply (e.g. Hamilton, 1988; Alford, 1992; see Chapter 2). Furthermore, equating land cover changes with "degradation" also overlooks how different land-cover changes might present threatened livelihoods, for whom, and under which circumstances.

The second implication concerned making predictions of risk based upon projections of biophysical change alone (such as changing greenhouse gas concentrations; or more frequent storm events) rather than by acknowledging the local circumstances that influence how such changes are experienced as problematic. Indeed, this implication is not surprising in the context of orthodox scientific frameworks, which – as noted by Yearley (1996) – tends to explain change in terms of universalistic statements operable in all locations, rather than referring to the distinctiveness or complexity of all locations. The purpose of this criticism is not to deny that, for example, rising concentrations of greenhouse gases may have impacts throughout the global atmosphere, but instead to state that the risk deriving from those changes are mediated locally through social, economic, and political factors influencing the exposure of societies to such changes, and their ability to adapt.

One example of the attention given to ecological change above local adaptive measures is the debate concerning the role of anthropogenic climate change on vector-borne diseases (e.g. Tol and Dowlatabadi, 2001; Casman and Dowlatabadi, 2002). Vectors are insects, viruses, or animals that can transmit a disease such as malaria or dengue fever to new locations. Climate change, clearly, may affect the underlying edaphic controls of temperature and moisture necessary for such vectors to survive, and therefore may increase the areas where such vectors can live. Indeed, Chapter 18 of the IPCC Second Assessment Report stated that vector areas are likely to increase, and some commentaries have suggested such changes represent some of the greatest threats to humanity resulting from climate change (Ross, 1996; Gelbspan, 1997).

In contrast to these views, however, other analysts have proposed that any changes to the areas accessible to vectors needs to be seen alongside local institutional controls on disease, such as availability of medical staff and sanitation or the influence of forced migrations in locations with poor sanitation. For example, one study of the state of Texas and three biophysically similar neighboring states in Mexico by the US Center for Disease Control (Dye and Reiter, 2000; Patz *et al.*, 2000), indicated that the Mexican states had more than 62,000 cases of dengue fever between 1980 and 1999, yet Texas had only 64 cases. Studies such as these indicate that vectors do not stop at national borders, and that local public health policies control disease.

Furthermore, other critics have claimed that the best way to control malaria is to raise per capita income, and accordingly increase funds for healthcare and protective measures. Placing the mitigation of climate change alone as a priority, rather than understanding how societies may adapt to climate change, may therefore not reduce vulnerability to disease. Indeed, if climate change policies increase costs, or restrict development, as some claim the 1997 Kyoto Protocol will do, then, ironically, such policy may even increase vulnerability to vector-borne disease (Dowlatabadi, 1997; Reiter, 1998; Tol and Dowlatabadi, 2001; also see the debate between Martens, 2000, and Reiter, 2000).

In theoretical terms, the reliance on biophysical changes as agents of risk, rather than the local contextualization of such changes, may be represented conceptually by Bruno Latour's (1987) term of "immutable mobiles." Immutable mobiles may be defined as socially identified objects, representations, or processes that are considered the same in different locations of cultural settings (see Box 7.2). Concepts of environmental change that insist that changes such as rising sea levels, erosion, or increasing concentrations of greenhouse gases are either "global" in extent, or universally "problematic," might then convert these changes to immutable mobiles by asserting these factors have equal meaning in all locations. Yet it is not always clear if the "mobiles" (i.e. the neat descriptions of biophysical change) are "immutable" because they are indeed considered problematic by every social group that experiences them, or because they are increasingly adopted by both experts and the public as unquestioned representations of reality, regardless of local experience. The identification of "risk" as a biophysical change regardless of social contexts in different places may therefore be considered a mobile immutable. But the adoption of this definition of risk may be more the result of the social and political

Box 7.2   Immutable mobiles

The concept of "immutable mobiles" has been adopted by sociologists of scientific knowledge to describe objects, representations, or processes that remain unchallenged when moved between different cultural settings, usages, or locations. The concept of immutable mobiles assumes that objects are hybrids: neither "natural" nor "social," but a combination of social experiences and framings of biophysical objects that evolve over time as the result of different political and social factors. "Immutable mobiles," by their very definition, however, have not changed in this manner when faced with different social and political factors, and are therefore apparently fixed. (The word "mobile" refers to the ability of any object to move in any direction, between different cultural groups or applications. "Immutable" suggests that mobility does not affect the structure or meaning of the object.)

The concept of immutable mobiles raises a number of questions for Sociology and Philosophy of Science. Realist epistemologists might interpret the apparent existence of an immutable mobile as an indication of some unerring "real" aspect of the natural world that cannot be affected by social perceptions. For example, the existence of "clouds" as a concept that is universally recognized and adopted by all societies may indicate that "clouds" really exist as a globally prevalent object. Observers who are more skeptical would argue that immutable mobiles do not indicate underlying biophysical "reality" but, instead, social structures and networks that give rise to the continued belief in the object despite diversity of experience about the underlying object. For example, so-called environmental orthodoxies (see Chapter 2) such as concepts of degradation, including desertification, erosion, or deforestation, might be considered immutable mobiles if they are universally considered problematic – even though much experience in many locations would suggest the opposite. In such cases, the perception of such concepts as a product of supposedly objective and politically neutral science would increase their status as immutable mobiles because there would seem to be little reason to question their accuracy. The use of environmental models or maps further as unquestioned representations of reality might also be considered immutable mobiles when their status as legitimate and accurate is unchallenged.

Source: Latour, 1987; Jasanoff *et al.*, 1995.

networks that wish to adopt this definition than any realist understanding of how risk is actually experienced by people in different locations.

The social structures that may lead to the adoption of particular scientific models or approaches above others are discussed in the next section. Before, it is important to note how such universalistic approaches to risk and "global" environmental problems may have negative impacts on both scientific explanation and people living in zones where such universalistic approaches are applied.

Imposing the local on the global: are risks universal?

The preceding section considered how so-called "global" environmental problems might actually reflect framings and approaches to environmental science that may be better described as "local." Now, we turn our attention to the experience of risks at the local level, and how far these can be described as "global." Much discussion of "global" environmental risks implies a uniform or shared threat or experience. Yet some analysts have also argued that such risks cannot be called "global" because they are experienced in diverse ways by individuals and different social groups.

First, there is the problem of globalized, universalist statements about environmental problems and risk. These statements can be challenged at the local level by alternative experiences that suggest different causal links, often at smaller time and space scales. In Chapter 3, it was mentioned how some common statements such as "water always flows downhill" can be challenged under specific circumstances (such as, in this case, in the event of the partial-area runoff approach to streamflow). Yet even though such universalistic generalizations can be challenged in this way, the act of making such statements can have the effect of making such generalizations appear an accurate representation of nature when they are in practice simply conventions among the scientists or scientific authorities that create them. In these circumstances, scientific generalizations may have the political effect of reification (see Box 1.1), or of enforcing a rigid definition of what is risk, or the impacts of projected biophysical change. While this effect may act against finding a more accurate and flexible form of explanation at a variety of time and space scales, it also may disempower alternative conceptions of risk, often coming from more local sources.

One classic example of this phenomenon was recorded by Brian Wynne (1994, 1996a) in reference to the interaction of "expert" versus "lay" accounts of the impacts of radioactive fallout on sheep farming in the Lake District of northwestern England. Wynne was another early member of the Oxford-based Political Ecology Research Group of the 1970s (see Chapter 1). After the Chernobyl disaster of 1986, the isotope Cesium 134 was deposited from rain on land used by sheep farmers. Government scientists from outside the region came to assess the risk posed to food production. Yet farmers often rejected the advice given by these experts because it showed little understanding of how sheep lived on hillsides. Some official advice, such as suggesting to farmers that they feed their sheep on hay instead of grass, was rejected out of hand by farmers who knew that sheep rarely – if ever – ate hay. Wynne reported on the uniform method of explanation as an imposition of predefined method and categories on circumstances that deserved more complex treatment. Furthermore, the policies of restricted land use and sheep management resulting from these generalizations were also considered unfair and potentially damaging to the farmers' livelihoods. He wrote:

one farmer caught by the Chernobyl restrictions lamented in this respect: "this is what they can't understand: they think a farm is a farm and a ewe is a ewe. They think we just stamp them off a production line or something."

(1994: 176)

The importance of this case study is that it recorded how scientific advice, formulated by outsiders, without reference to local experience within the region – or more importantly, local framings and priorities – led to scientific advice that was seen as useless by people in that region. Crucially, the scientists had not realized that their rigid methodology had effectively privileged their classifications and explanations of risk as more accurate representations of reality than those of the local farmers. Wynne commented:

This brief glimpse indicates that the scientific knowledge is not naturally determined; it could have been organized differently and still have respected the evidence from nature. Yet social commitments to such organizing epistemic principles as the levels of aggregation of entities into uniform conceptual classes and categories are so deeply enculturated into the scientific canons of given specialities or fields that they are mistaken as being completely determined by nature.

(ibid.: 176)

A second challenge to global notions of risk may come from indicating the individuality and diversity of risk perception and experience among different people in different social groups and places. In this sense, risk perception may vary from the globalized certainty of risk described, for example, by global models based on projected biophysical changes, or Ulrich Beck's account of "Risk Society." As discussed above (p. 170), Beck (1992) has argued that under Risk Society, environmental risks may become more global and prevalent as a result of reflexive modernization – or the kind of economic growth that is ultimately self-destructive because it is intrinsically linked to the exploitation of resources and a lack of ability to control this exploitation (see also Box 5.1). As quoted at the start of this chapter, Beck has argued that such risks can be seen to be global and real, yet increasingly experienced in terms of individual choices and dilemmas, rather than in historically significant terms of classes or other social groupings.

Against these generalizing views of risk, a variety of sociologists of science have instead claimed that responses to risks may be influenced more by local culture and historical experiences than by predefined notions of what constitutes risk or how risk is experienced. Irwin *et al.*, for example, wrote: "we view local understandings of risk as dynamic and discursively negotiated rather than as free floating attitudes or as a producer of reflexive modernization" (1999: 1312).

In a study of potential chemical pollution in northeastern England, Irwin *et al.* noted that citizens living in close proximity to factories or sites of industrial waste acted upon conceptions of risk that are integrated with a variety of *ad hoc* local concerns such as declining economic prosperity, crime, and absence of local leisure facilities. These concerns then draw upon (and contribute to) wider discourses of environment and change in the locality. Such public understandings of risk should not be seen as offering an alternative or supplemental technical appraisal of risk to those of scientists or experts, but instead are premised on different epistemological assumptions concerning what counts as legitimate knowledge in this context. As Irwin *et al.* wrote:

> From the expert perspective, lay reasoning about environmental and risk issues may appear to be ill informed or fallacious, and to include little distinction between what is relevant and what is not. From the lay perspective, meanwhile, the view of experts may appear to be unduly narrow and to ignore what, to the citizen, are crucial aspects of their everyday experience of environmental problems.
>
> (1999: 1324)

In this respect, such locally embedded understandings of perceptions and experiences of risk allow a more locally relevant account of risk than the abstract social processes discussed by Beck and related theorists.

A third criticism of Beck's universalistic approach to risk is the potential overlooking of important political factors in the unequal distribution of risk. As discussed in Chapter 5, some critics have argued that mainstream environmentalism has reflected middle-class interests and failed to acknowledge the widespread risks affecting poorer people or people undergoing industrialization (e.g. Enzensberger, 1974; Satterthwaite, 1997). Beck (1992), as quoted above, has argued that the realities of "risk society" have implied a blending of new global discourses of risk with a new individualistic manner of rationalizing and adapting to them. Under these conditions, Beck (ibid.: 39) argued that risk "can no longer be composed of and understood in class categories."

Such statements, of course, may bear little relation to the experience of a variety of different risks among different forms of affected peoples worldwide. As one critic notes: "Beck's focus on the equalizing effect of global risks, while provocative and supportive of a classless risk society, misdirects our attention from the more acutely damaging impacts of technological hazards" (Marshall, 1999: 270).

It may be premature to dismiss class – or political empowerment more generally – as a determining factor in who experiences different risks. While global discourses of risk are increasingly important in determining policy and debate about environmental problems worldwide, there are still significant political and socio-economic factors behind risks experienced, for example, by factory workers in poorly regulated industries; poor farmers without secured access to agricultural land; or street children in rapidly growing cities. The suggestion that risk is unrelated to class categories may even encourage policies that overlook how such poorer people or classes experience greater environmental hazards, including risks to health, than relatively richer or more secure people. Class analysis – in either a Marxist or post-Marxist context – is still relevant to the understanding of risk.

This section of the chapter has argued that it is important not to accept uncritically environmental problems or risks as "global" for two key reasons. First, the projection of such problems as "global" may be based in framings, problem closures, and practices that might more accurately be described as "local." Second, the experience and meaning attached to risks may be better explained and approached by acknowledging the local contextualization and knowledge of them by diverse societies and individuals. In both situations, the insistence on a uniform, universalistic approach to risk may both result in policy recommendations that lack meaning and practicality for local users; and in the missing of important information concerning the biophysical elements of change that are usually seen as the origin of the risk.

Such criticisms suggest that a key element of overcoming such problems of accepting "global" conceptions of risk uncritically is to understand (as Taylor and Buttel, 1992, note) how we know we have "global" problems, and who helps identify them. Indeed, as discussed above (pp. 179–181), the very definition of who is allowed to be "expert" in framing, measuring, and addressing risks is crucial in determining which knowledge or alternative conceptualizations of problems are accessed. In this sense, knowing how such notions of "global" or "local" environmental problem have been established is a vital part of making the political implications of different scientific explanations more transparent and open to negotiation. The next section assesses the various means by which scientific research may seek to construct notions of "local" and "global" environmental problems and knowledge, with implications for how far alternative environmental explanations or definitions of risk may be extended across different space scales.

Speaking on behalf of others

The point of the discussion so far in this chapter is not to suggest that risks do not exist, or that they exist only in localities or in the minds and discursive practices of people who experience them. Instead, the aim is to indicate that universalistic generalizations about risks may often be taken as accurate representations of global environmental problems, but they may explain change in simplistic or inaccurate terms, and appear irrelevant and intrusive to people who are told to accept them. Such resentment may occur in part because the explanations reflect the priorities and framings of outsiders rather than be compiled in participation with people living more locally. The assertion of "global" environmental problems without such local representation effectively forms an attempt to speak on behalf of others. Yet as Tariq Banuri noted in regard to reactions against imposed forms of development: "what we have learnt from the persistence of unrest, of the unimaginable fury, against this endeavor is the indignity of speaking on someone else's behalf" (1990: 67).

An important step, therefore, to making science more transparent and relevant to different localities is to see how problems are considered "global," how environmental change is framed, and how information is gathered. Such actions are key steps in the globalization of discourses of environmental risk, and the establishment of scientific networks that may promote the notion of global risk and responses.

This section now analyzes these steps as a way to see how knowledge about environment has been extended across space, and the different nationalities or social groups who live there. The section looks at attempts to both expand the adoption of environmental perceptions and understandings (through epistemic communities, and the pre-definition of specific environmental processes as degrading); and at attempts to democratize these explanations through accessing overtly "local" knowledge (but which may also reflect outside framings). The aim of this section is to look at some unintended problems of establishing global projections of risk. More positive suggestions to address these problems are made later in this chapter in relation to redefining vulnerability; and in Chapters 8 and 9 in relation to alternative scientific and political practices in general.

Epistemic communities and their critics

Epistemic communities are perhaps the most well-known concept that describes structures underlying the spread of ideas of science across space or by different nations, societies, or expert bodies. Haas defined an epistemic community as follows:

> An epistemic community is a network of professionals with recognized experience and competence in a particular domain and an authoritative claim to policy-relevant knowledge within that domain or issue area ... What bonds members of an epistemic community is their shared belief or faith in the verity and the applicability of particular forms of knowledge or specific truths.
>
> (1992: 3)

Epistemic communities are often described as positive developments, possibly referring to the increased adoption of progressive values such as human rights, gender, or racial equality, or environmental principles. Indeed, the emergence of environmental concerns as a topic of international political debate during the 1960s and 1970s was seen to be progress in formulating an alternative political voice against global capital or national states, both with myopic views about environment. The importance of environmental organizations or campaigners in creating environmental awareness has been, in effect, the creation of an epistemic community about environmental politics at the global level. In this context, making the community focused and unified as a voice is clearly valuable in resisting the more established forces of states or industry. For example: "NGOs are increasingly prominent forces in framing environmental issues. They help establish a common language and, sometimes, common world views" (Princen *et al.*, 1994: 226).

Similarly, advocacy coalitions and expertise provided by such NGOs for local citizens in specific zones or countries is another means by which an epistemic community may be established. As Keck and Sikkink (1998, 215) note, international advocacy coalitions between environmental NGOs and campaigners in different countries allow "ecological values to be placed above narrow definitions of national interest." Lester Brown similarly urged media organizations to reorganize in order to accelerate the adoption of sources of environmental concern. He wrote:

> In the past, when virtually all news was local, when there were no perceptible climate changes, ozone layer depletion, or collapsing oceanic fisheries, there was no need for global coverage. Today the key stories are global in scope, but there is no global desk to deal with them systematically.
>
> (2001: 260)

Such comments, of course, reflect controversial assumptions about what has been "global" news in the past, and the rates of climate change experienced in different years (such as the so-called "Little Ice Age" during the seventeenth century). But it is clear that, without some form of clear and unified communication, it is arguable that political concerns such as environmentalism would not be as widely recognized today as potent forces.

Such political pragmatism behind epistemic communities, however, hides a variety of epistemological implications and problems (see Jasanoff, 1996a). First, the establishment of a "common" or "world" view through political activism or epistemic communities of course begs the question as to whether such a view accurately or fairly represents all people such views are claimed to stand for. Proponents of epistemic communities like to portray the spread of the community as the increasing conversion of more and more people to a scientific or normative judgment in a progressive manner, and where the greater size of the community indicates the higher possibility of the judgment being universally accurate or politically acceptable. Critics of the concept, however, suggest that more attention should be given to how the community is created, and how the sense of unity is created, rather than assuming acceptance or accuracy have been achieved in general.

A second common question concerning epistemic communities is how far they may refer only to so-called "experts" or whether "lay" people may also participate. As noted above, Haas defined the community as composing only of "a network of professionals with recognized experience and competence in a particular domain and an authoritative claim to policy-relevant knowledge." Yet as noted from the case study of sheep farming in northwestern England earlier in this chapter (Wynne, 1994, 1996a), "experts" may be defined in various ways. There may be differences between the expertise and knowledge of officially sanctioned "experts," whom may often represent the state, and so-called "lay" people whose own knowledge may be more relevant but who lack that official status as "experts." If epistemic communities can only be composed of officially sanctioned experts, then there is a need to understand who these experts are, and how they may reflect other forms of political power or people with similar backgrounds. Indeed, many chief scientists and academics in rapidly developing countries may have studied overseas at the same universities as other international experts, and hence the networks they create may be more spatially situated than they seem from looking at nationalities alone.

One powerful criticism of the epistemic communities concept was provided by Karen Litfin (1994) in her analysis of international policies to combat ozone depletion in the Montreal Protocol of 1987. Litfin argued that much environmental discussion has identified the adoption of environmental policies to limit ozone-depleting substances as an example of positive political action following the evidence of scientific research (e.g. Benedick, 1991). This interpretation, however, suggests an orthodox approach to science preceding environmental policymaking, rather than also acknowledging the role of politics in the formulation and adoption of science. Litfin did not deny the important role of scientists in facilitating political agreement. But the role of science was mediated by two key factors. First, negotiations were facilitated by a group of ecologically minded knowledge brokers associated with the US Environmental Protection Agency (EPA), NASA, and UNEP. Second, the successful negotiations to adopt protective measures were shaped largely by the controversy and publicity surrounding the discovery of the Antarctic ozone hole. No body of consensual knowledge from existing computer models or other environmental assessments supported the need to reduce emissions of ozone-depleting substances in 1987. The epistemic communities approach underplays the multidimensional relationships that exist between science and different scientists, or the interface of specific scientists and policy negotiations. It also fails to acknowledge the structuring role of preexisting political debates within the framing of both scientific research environmental negotiations. As Litfin wrote:

> Without taking into account the political implications of scientific discourse, an interest group approach alone does not contribute much to an understanding of the evolution of the ozone regime ... epistemic

community approaches underestimate the extent to which scientific information simply rationalizes or reinforces existing political conflicts.

(1994: 184, 186)

Because of such criticisms, many sociologists of scientific knowledge have argued that the concept of epistemic communities needs to be refined in order to acknowledge the contextual circumstances that facilitate the communication of particular scientific findings to policymakers. Epistemic communities may also hide political conflicts within their membership. Political analysis should therefore assess who creates epistemic communities, for which purposes, and at the costs of excluding which other actors, rather than simply seeing communities as automatically progressive advancement of universally agreed "environmental" concerns.

"Local" and "indigenous" knowledge

If epistemic communities and discussions of "global" environmental problems have tended to overlook more local environmental perceptions and experiences, then the concept of "local knowledge" has often been sought as an alternative and more sensitive concept. Like pouring water on flames, some reformers have hoped that so-called "local" knowledge may be able both to reverse the pernicious effects of globalizing discourses and increase the development impacts of policy, by providing insights into environment and local needs that can only come from local people. So-called "local" knowledge may refer to knowledge that is clearly spatially bounded within an identified locality; or it may be grounded in a culture or ethnicity (often associated with specific places). The very essence of "local" knowledge is that it is something that is specifically bound to a people or places that are not immediately accessible to outsiders. Yet, for many observers, the term is also symbolic in representing a resistance to oppressive "global" notions of environment, or in the suggestion that "local" (or "indigenous") knowledge is somehow more nuanced and accurate than the widespread beliefs of outsiders.

In a volume entitled *Global Ecology*, Wolfgang Sachs (1993) and other authors spell out their objections to globalized notions of risk. For example, the well-known Indian environmental writer and activist, Vandana Shiva commented:

The "local" has disappeared from environmental concern. Suddenly, it seems only "global" environmental problems exist, and it is taken for granted that their solution can only be "global"... The "global" in the dominant discourse is the political space in which a particular local seeks global control, and frees itself of local, national and international restraints. The global does not represent the universal human interest, it represents a particular local and parochial interest which has been globalized through the scope of its reach.

(1993: 149)

Similarly, Larry Lohmann wrote:

> Green globalism appeals strongly to many Northern environmentalists
> ... because it tries to translate all important "environmental" practices
> and insights into a common, comfortably modern vocabulary. This
> globalism, being both geographical and intellectual, satisfies a deep-
> felt Western (and probably largely male-associated) need for contain-
> ment and control.
>
> (1993: 159)

Such comments, of course, are in broad agreement with points already
made in this chapter. Yet terms like "local" or "indigenous" knowledge as
a response to the "global" risk and environmental problems described
above raise a new quandary. How far can defining "local" knowledge or
people also reflect social and political concerns in the same way as the
word "global" has been shown? As noted by Arun Agrawal (1995), the
definition of any locality or social group as "local" or "indigenous" often
implies that such groups are less powerful, or subaltern, voices. Framing
such groups, or their knowledge, in this way may therefore already repeat
this presentation of groups as less powerful, and thus may help to reiterate
these power relations.

There is, of course, a great desire from many researchers and development workers to represent disadvantaged groups in order to democratize
or reframe environmental policy toward more locally relevant needs (e.g.
see Hecht and Cockburn, 1989; Scoones and Thompson, 1994; Chambers
1997). Subaltern, or less represented groups in this respect may include
poor people, women, or ethnic minorities. But critics have suggested that
seeking to represent these groups using existing communication structures
or political arenas may also reinforce disbalanced power relations. The
postcolonial theorist, Gayatri Chakravorty Spivak, adopting a Marxist and
feminist position, famously wrote that it was foolish to attempt to redress
political imbalances simply by identifying who is not represented and inviting them to take part:

> The Subaltern cannot speak. There is no virtue in global laundry lists
> with "woman" as a pious item. Representation has not withered away.
> The female intellectual as intellectual has a circumscribed task which
> she must not disown with a flourish.
>
> (Spivak, 1988: 308)

Ironically, this problem has also led to controversies among intellectuals
who have sought to represent more local and less represented peoples in
environmental discourse. For example, as noted above, Vandana Shiva has
become well known as a spokesperson for the disadvantaged South, and
for the need to look beyond what she calls the "universal bullying" of
global ecology (Shiva, 1993: 154). Yet Shiva herself has been criticized for

not effectively representing these subaltern groups, and for using allegedly romanticized notions of poverty and oppression, or of notions of ecofeminism that reflect outsiders' views of nature and women (Jackson, 1995; Jewitt, 1995, 2000). In one outspoken criticism of how Vandana Shiva has been adopted as a spokesperson for the South, Cecile Jackson wrote:

> But since the postmodern acid bath has also dissolved criteria such as truth or objectivity as means for choosing between competing versions of reality, these now seem to be valued entirely according to the perceived status of the observer. Why else is Vandana Shiva accepted so uncritically by Western feminists? . . . Authors with Indian or African names are taken uncritically to represent "Third World Women" or "feminists of the South."
>
> (1995: 139)

Similarly, the identification of colorful "tribal" people as representatives of "local" or indigenous knowledge may also be criticized for reflecting outsiders' perceptions of what may be called "local." Under such romanticizing notions, "locality" may also mean "exotic and remote," or "rare and endangered" (Cohen, 1989). While such groups may indeed appear exotic or endangered, a more critical approach would question why these groups are considered to contain specialized or localized knowledge when others do not. Indeed, in a report published in 2001, the United Nations listed examples of "local" knowledge as comprising the actions of the Turkana tribe in Kenya, who plant crops according to observed behavior of so-called "prophets of rain" such as frogs and birds (including the exotically named ground hornbill and spotted eagle owl). Other examples cited included the Aka pygmies of the Central African Republic (UNEP, 2001). Undoubtedly, such groups do have environmental knowledge and adaptions that help protect resources and local livelihoods. But the attempts to define them should beware imposing romantic images of underdevelopment on these people. Under such circumstances, the "locality" perceived in such groups may be imported or be another form of mobile immutable that is transferred from one location to another.

These criticisms, of course, do not suggest that there is no such thing as "local" knowledge, or that it is pointless trying to record it or the views of subaltern peoples. Instead, these controversies are warnings to pay attention to how the term "local" is defined, by whom, and for which political objectives (see also Box 7.3).

Translating environmental "change" into environmental "crisis"

Finally, concepts of "global" environmental risk and degradation can also be globalized by the linguistic and cultural framings within scientific discourse itself. As noted in Chapter 2, there have been many examples of apparent environmental degradation in developing countries associated

Box 7.3   Romanticizing and de-romanticizing women's environmental knowledge in India

Much popular debate within environmentalism has suggested "local" or "indigenous" knowledge has been more effective in protecting environmental resources than much orthodox science or development strategies. While most observers have criticized exploitative development, some critics have also suggested that the desire to exemplify "local" environmental knowledge in this way may be counterproductive to development, and may reflect the agendas of political actors who are not local. Some critics have applied these comments to environmental activism by women in India.

Some environmental disputes in India have become very well known. For example, the actions of the "tree huggers" in the Chipko movement in Uttarakhand; women's actions in community forestlands of Jharkhand; or the activism to prevent the construction of dams on the Narmada river in central India. Commonly these disputes have been used to indicate a sense of common unity between environmentalists in developed and developing countries, or of an unspoiled bucolic image of rural life. It is often reported how women villagers would chant, "What do forests bear? Soil, water, and pure air," as they formed human chains to prevent loggers cutting down trees.

Such romantic images have been questioned by a number of scholars. Rangan (2000: 181), for example, commented:

> Chipko became a symbol of popular environmentalism by invoking discourses of "global environmental protection" and "national defense," which successfully gained the active sympathy and support of national political parties, urban-based environmentalists, academics, and international environmental lobbies.

Rangan further argued that the Chipko movement also became framed within the wider political debate for statehood within Uttarakhand.

Other observers have also questioned the use of disputes to illustrate wider themes of women, environment, and development (WED). Jackson (1995), for example, has argued that the ecofeminist representation of "women's knowledge" as necessarily "closer to nature" essentializes women, is inimical to gender analysis, and leads to conservative and regressive conclusions. Similarly, Jewitt (1995, 2000) has suggested that many interventionist approaches to environmental management in Jharkhand have tended to adopt predefined notions of women–environment relations that overlook how such discourses have been formed and the complexity of gender variations in agro-ecological knowledge. Jewitt (2000: 980) wrote: "there is an urgent need to reassess the contribution made to development rhetoric and policy-making by undifferentiated ecofeminist/WED discourses and simplistic participatory approaches."

Potential solutions to these problems include adopting site-specific approaches to environmental management, and to increase the employment of locally recruited female agricultural extension staff. More generally, there is a need to appreciate that well known examples of environmental conflict and resistance in developing countries may become sites for the imposition and appropriation of environmental discourses by more powerful global debates and political actors.

Sources: Jackson, 1995; Jewitt, 1995, 2000; Rangan, 2000;
also see Rocheleau *et al.*, 1996.

with processes such as erosion or deforestation, but which either present no problem to people living there, and/or may be related to long-term bio-physical processes unrelated to human activities. Much non-equilibrium ecological thinking has also highlighted how landscape flux has been inaccurately regarded as environmental degradation under older, equilibrium-based approaches to environmental explanation.

In recent years, some researchers have begun to indicate how these early experiences of apparent environmental problems have also led to the establishment of spatial boundaries for the existence of environmental or health risks around the world. Richard Grove (1995) and David Arnold (1996), for example, noted how particular forms of western medicine co-evolved with overseas exploration and colonization, and that the new study of so-called "warm climate" diseases led to maps of the world that demarcated where such risks were prevalent. Indeed, these kinds of studies contributed to the form of environmental determinism adopted during colonial times such as the belief that geo-medical boundaries restricted races to what were considered "ancestral environments" (Harrison, 1996; Bankoff, 2001). Simultaneously, the "tropics" became defined as zones unsuitable for Europeans because of the strong ultraviolet rays. According to Arnold, the concept of "tropicality" was used superficially to refer to the spatial zones within the Tropics of Cancer and Capricorn, yet had deeper implications for risk and suitability for inhabitation. Tropicality, he wrote, was: "a Western way of defining something culturally and politically alien, as well as environmentally distinctive, from Europe and other parts of the temperate zone" (Arnold, 1996: 6; Bankoff, 2001).

Similar impressions can also be seen in the interpretations of environmental problems in many developing countries by western experts during the twentieth century. Chapter 2 summarized how many colonial scientists (e.g. Stebbing, 1937) identified aspects of environmental change that they attributed to land mismanagement by local inhabitants. Newer, alternative approaches to assessing land degradation in developing countries might attempt to consider how far local people evaluate such apparent degradation, how far they were responsible, and whether such apparent degradation may result from long-term biophysical processes that may exist regardless of human activities. Furthermore, it should also be asked how far the supposed symptom of land degradation might be a regular feature of landscape change, and therefore be considered "normal" experience by local inhabitants. For example, in the Himalayas in the 1970s, Eric

Eckholm (1976) observed high rates of soil erosion, and considered this evidence of rapid land degradation (see Chapter 2). Later researchers, on the other hand, have noted how far such high rates of erosion are – without overlooking the important problem of erosion in many localities – consistent for a region experiencing rapid tectonic uplift and monsoonal rainfall (e.g. Ives and Messerli, 1989). As a consequence, the projection of crisis onto the Himalayan region by Eckholm reflected his own cultural background of experiencing soil erosion as a major problem in the Dust Bowl, rather than necessarily posing the same threat to hill farmers in Nepal.

In response to these kinds of alternative readings of risk, some researchers have referred to the cultural embedding of different kinds of biophysical change. Jon Anderson (1968: 302–303), for example refers to the "normalization of threat." The more the threat is perceived as chronic, the greater this threat is interpreted as normal experience. Each experience of specific biophysical changes or processes such as fire, erosion, or periodic drought have their "own context of geographic, topographic and cultural variety" (Lewis, 1990: 247). Yet where scientific networks fail to acknowledge how far local inhabitants and external experts have different interpretations of such changes, then there is a chance such changes may be reported as problematic by outsiders in ways that may not find agreement with local perceptions. Such events do not, of course, suggest that local interpretations of biophysical change should always take precedence over more globalized explanations of risk (this question is discussed in more detail in Chapter 8). But the automatic classification of biophysical events as problematic by outside experts is clearly one way in which landscape flux considered "normal" by local inhabitants can be branded as "crisis" by more global networks.

Rethinking environmental vulnerability

So far, this chapter has discussed a variety of dilemmas in establishing environmental explanations at long distance. These problems have generally been in the unwitting replication of local values and practices into supposedly "global" scientific statements, or in the lack of attention to how societies or social groups elsewhere may experience and value projected biophysical changes. But what are the implications of these dilemmas for environmental policy?

In practice, predictions of risk in different locations give an indication of environmental vulnerability. Vulnerability is increasingly discussed as a guide to how far localities or peoples are exposed to negative impacts following biophysical changes or events of high magnitude. Yet there is much controversy concerning how best to address environmental vulnerability. Under orthodox approaches to environmental science, vulnerability may be best addressed by mitigating biophysical changes considered the main causes of risk. Alternative approaches emphasize reducing vulnerability by also increasing the ability of societies to adapt to such changes through reducing the exposure of specific social groups to biophysical changes.

This section is divided into two discussions. The first examines arguments that define vulnerability in terms of impacts on physical landscapes compared with more constructivist accounts of local impacts. The second discussion presents a more detailed analysis of social vulnerability, and the implications of increasing adaptive capacity as a means of reducing vulnerability. These comments build on the examination of carbon-offset forestry in Chapter 6 as one means of addressing "global" environmental risks, yet which may, in certain circumstances, actually increase local vulnerability for some social groups.

Regions or people at risk?

As discussed throughout this chapter, much orthodox environmental science has sought to explain the existence of risk in different locations as the result of projected biophysical changes or high-magnitude events. Indeed, much historic research in ecology has described vulnerability also in physical terms, as one of the key properties of ecosystems such as inertia, resilience, and elasticity (e.g. Goudie, 1990).

Commonly, ecological research has focused on the concept of ecological vulnerability as the opposite of ecological resilience. Resilience has been expressed in terms of the relationships between ecosystem structure, diversity, and disturbance, and refers specifically to the functioning of the ecosystem as a system, rather than the stability of component populations or the ability to maintain a steady state (e.g. Holling *et al.*, 1995; see Adger, 2000: 349). This approach has supported a variety of discussions of ecosystem vulnerability. For example, some tropical terrestrial ecosystems such as tropical lowland evergreen rainforest have stable and diverse populations, but have low resilience because of their relative inability to function following disturbance (Whitmore, 1984). Some coastal and estuarine ecosystems, on the other hand, have experienced high levels of physical disturbance, but have been claimed to maintain high levels of functional diversity despite low species diversity (Costanza *et al.*, 1995).

Under such approaches to ecosystem resilience, vulnerability may refer to the ability for ecosystems to recover standards of structure and diversity after human or non-human disturbances. Some appreciation of risk in ecosystem disturbance was proposed by Holling's "theory of surprise," which is based on the concept of discontinuities in non-equilibrium ecology. Such random disturbances can also surprise human users of ecosystems, even if ecological models can predict such flux (Kates and Clark, 1996). This approach, however, has generally been used in the context of biophysical properties of ecosystems rather than social experiences of vulnerability (although Price and Thompson, 1997, adapted "surprise theory" from a Cultural Theory perspective in order to indicate how such random physical disturbances may also trigger changes in perceptions of environment as stable or fragile).

In an important volume, *Regions at Risk*, Jeanne and Roger Kasperson and Billie Lee Turner presented a summary of more human-oriented debates concerning environmental vulnerability. They argued that the spatial distribution of vulnerability relied upon the concept of "criticality":

> [criticality] refers to situations in which the extent or rate of environ-
> mental degradation precludes the continuation of current use systems
> or levels of human well being, given feasible adaptations and societal
> capabilities to respond.
>
> (Kasperson *et al.*, 1995: 25)

Criticality can itself be divided into so-called geocentric criticality, or an emphasis upon the biophysical impacts of human-induced environmental change; and anthropocentric criticality, or a focus on social or political impacts on people (ibid.: 6). According to this classification, an extreme geocentric interpretation might focus on the impacts of human populations on purely physical factors such as species loss, changes in soil and water composition, or land-cover vegetation. An equally extreme anthropocentric reading might alternatively claim that environmental losses or catastrophes are caused principally by social or political structures that could have prevented them. Orthodox ecologists (e.g. Myers, 1984) have generally adopted the geocentric model. Political ecologists and social theorists of hazards have generally adopted the anthropocentric approach (e.g. Hewitt, 1983).

Yet despite explaining the difference between geocentric and anthropocentric criticality, Kasperson *et al.* still adopt a mainly geocentric approach in defining environmental vulnerability. First, much analysis in the book broadly reflected a linear model of causality for environmental hazards, in which human impacts on biophysical properties were seen to create further implications for aspects of environment valued by humans. This model is described briefly in Figure 7.1 (Hohenemser *et al.*, 1985; Norberg-Bohm *et al.*, 2001). Second, Kasperson *et al.* (1995) defined risk in clear spatial terms, and with a list of nine specific regions where criticality is seen to be high. The nine regions comprise: Amazonia; the Aral Sea basin; the Middle Hills of Nepal; the Ukambani region of Kenya; the Llano Estacadao of the American Southern High Plains; the Basin of Mexico; the North Sea; the Ordos Plateau of China; and the eastern Sundaland region of Indonesia.

There are a number of dilemmas posed by the geocentric, and spatially defined, approach to the evaluation of environmental risk. First, as discussed in Chapter 2, there is much environmental research at a variety of scales to indicate that inhabitants of these regions have developed means to reduce impacts of economic activities on the degradation of resources (these practices are referred to as "environmental adaptations," see

Netting, 1993; Batterbury and Forsyth, 1999). Second, the labeling of specific regions as vulnerable may draw attention away from the underlying political factors that can expose people to environmental risks. Third, using a region-wide approach to risk may also repeat the errors of many "global" projections of risk, by making generalizations based on estimates of biophysical rates of change, without understanding how such changes are interpreted locally, or how far the region-wide projections of risk may reflect the assumptions and framings of outsiders.

There is no suggestion, of course, that these important regions of the world do not experience environmental problems, or that rates of biophysical change are not important. (Indeed, problems such as erosion in Nepal and salinization around the Aral Sea present a variety of dilemmas.) Yet focusing on biophysical changes, and at a regional scale, may avoid different experiences of risk and degradation at smaller scales. Such smaller-scale definitions of risk may refer both to different experiences of degradation within different spatial units such as varying watersheds or forest regions, and to alternative conceptualizations of space or environmental problems resulting from more phenomenological or constructivist approaches to environment change.

Figure 7.2 shows an alternative approach to defining environmental impacts using a non-linear constructivist approach. The aim of this approach is to acknowledge the political factors underlying the rigid definitions of "cause" and "effect" displayed in the linear model of Hohenemser *et al.* (1985) in Figure 7.1. For example, the non-linear model acknowledges that many "human" impacts on environment cannot be easily distinguished from long-term biophysical processes that preexisted human settlement. Second, definitions of landscape and resources valued by humans are not clearly defined and universally agreed, but instead hybrid blends of human and biophysical experience defined by historic research agendas and social concerns (see Chapter 4). Finally, the non-linear model aims to indicate that causal statements associated with this model (indicated by arrows) are freely floating on top of hybrid objects and processes. These statements are not clearly defined and uni-directional (as indicated

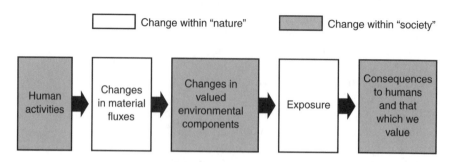

Figure 7.1 A linear model of causality for environmental hazards

Source: adapted from Hohenemser *et al.*, 1985; Norberg-Bohm *et al.*, 2001.

on the linear model), but reflect historic experiences of biophysical events by specific social groups, and the consequent "black-boxing" of apparently proven linkages (see Chapter 4).

In some respects, the differences between Figures 7.1 and 7.2 are similar to the two models of environmental vulnerability to environmental hazards discussed by Blaikie *et al.* (1994). The linear model may be described as similar to the model of "pressure and response," which emphasizes linear causality between changes in physical systems and trends in social systems. The more complex, "access" model assesses the mutual construction of hazards through the existence of poverty, and lack of access to resources that reduce vulnerability such as education, employment, or land tenure (this approach is discussed in more detail below, pp. 196–200).

As discussed throughout this book, a failure to appreciate how dominant environmental explanations reflect partial experiences of physical change and historic social agendas will generally mean two important problems. First, the policies based on these explanations may not address the underlying causes of environmental change because they do not adequately address the complexity of biophysical change beyond human experience. Second, policies may also impose unnecessary restrictions on livelihoods of people living in affected areas. The following discussion now questions alternative means of addressing vulnerability, and with implications for local livelihoods, using geocentric and anthropogenic explanations of environmental criticality.

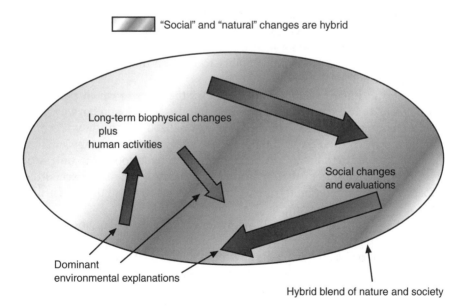

Figure 7.2 A non-linear, constructivist model for environmental hazards

Source: the author.

Vulnerability or adaptation?

So, how can environmental vulnerability be assessed more locally, and with reference to non-linear, constructivist arguments? One way is to look at the local experiences of risk, and to assess how people reduce exposure to biophysical changes. In essence, this approach is to assess adaptations to biophysical changes, rather than to mitigate the biophysical changes themselves (although both can be attempted at the same time).

Adaptation may be defined as the means by which individuals, social groupings, or indeed countries may be able to live alongside, and therefore lessen the impacts of, an environmental change or events that might threaten livelihoods. The concept was introduced in Chapter 2, concerning so-called "environmental adaptations," which have generally referred to small-scale land management practices and strategies that allow increased agricultural production or other economic activities despite rising population or environmental change (Netting, 1993; Batterbury and Forsyth, 1999). Yet, environmental adaptation in general need not be at this small scale, and can refer to a variety of local, national, or international strategies and capacities to withstand ecological change. Such larger-scale adaptation might range from engineering or hydrological infrastructure projects to reduce biophysical impacts of flooding or drought, to political and economic strategies to reduce dependency on resources potentially affected by environmental changes, or to provide safety nets and support for people most affected by change.

Adaptation, nonetheless, is controversial because it is seen by some critics to be a way to avoid restricting the activities that lead to environmental degradation. Ex-Vice-President Al Gore once commented: "believing that we can adapt to just about anything is ultimately a kind of laziness, an arrogant faith in our ability to react in time to save our skin" (Gore, 1992, in Pielke, 1999: 162).

But this attitude to adaptation overlooks the ways in which perceived "problems" only exist when adaptations fail us; or, in a similar way, how environmental processes can become normalized when they become embedded in cultural expectations (see Anderson, 1968). Many adaptations make environmental "processes" invisible as "problems," unless of course the magnitude of the environmental events or changes exceed the capacity of the adaptation to reduce impacts. (This was also the conclusion of Holling's "theory of surprise.")

As discussed in Chapter 2, many environmental adaptations have allowed people to live in marginal lands without damaging resources. Yet an approach to risk that explains environmental vulnerability through biophysical changes alone, or sees risk to be prevalent throughout a specified "region," may overlook the value of these adaptations in reducing the actual experience of risk to people living there. Indeed, some land-use policies resulting from such explanatory approaches might actually increase the risks experienced by people living in regions if these policies aim to restrict livelihood strategies that are seen to increase biophysical changes. As summarized in Chapter 2, policies that seek to restrict herds, forest uses, or cultivation of mountainous land may unwittingly increase the vulnerability of people in the mistaken belief that their experience of risk comes from the existence of biophysical events or changes alone (see Leach and Mearns, 1996; Batterbury *et al.*, 1997; Box 2.2).

The addressing of risk, as experienced by poor inhabitants of developing countries, may therefore require reframing definitions of environmental risk from predefined notions of environmental fragility to an understanding of what provides livelihoods for poor people, and maintaining these as the best defenses against environmental change. The reframing may also require looking at poor *people* as a social group, rather than simple *regions,* or countries, where poor people live. As Kates notes: "If the global poor are to adapt to global change, it will be critical to focus on poor people, and not on poor countries as does the prevailing North–South dialog" (2000b: 16).

In this sense, vulnerability may also be defined in similar terms to the entitlements approach to food security introduced by Nobel Laureate Amartya Sen (Sen, 1981; Drèze and Sen, 1990). Sen argued that development economists were adopting the wrong approach to explaining the existence of famine by looking only at macro-economic questions of food production for countries and regions as a whole, rather than the factors that govern the access to food for different individuals within those regions. Sen's proposed solution to this problem can be summarized simply as assessing each individual's endowments and entitlements to food. Endowments include direct access to food, such as ownership of food, or the land needed to produce it. Entitlements comprise indirect access to food, such as money to buy it, or the ability to work for it. Clearly, people without access to either endowments or entitlements are those most vulnerable to famine (for example, refugees with no residency or work permits could be considered more vulnerable; the aged or very young with no families are other potential examples). The concept of entitlements has also been used in an environmental context as a way to ensure local access to, and protection of, resources in poor communities (Leach *et al.,* 1999) (this concept is discussed in more detail in Chapter 9).

A more anthropocentric account of environmental vulnerability, therefore, draws more attention to the social, economic, and political factors that may reduce people's access to resources to withstand biophysical events or changes, rather than the biophysical changes themselves (see Ribot *et al.,* 1996; Liverman, 1999). Indeed, approaches to risk that seek only to reduce the biophysical changes may actually increase social vulnerability. Blaikie *et al.* wrote:

> By "vulnerability" we mean the characteristics of a person or group in terms of their capacity to anticipate, cope with, resist, and recover from the impact of a natural hazard. It involves a combination of

factors that determine the degree to which someone's life and livelihood is put at risk by a disaster and identifiable event in nature or in
society.

(1994: 9)

In particular, and influenced by Sen, Blaikie *et al.* argue that access to
resources is crucial in determining who is vulnerable:

Access involved the ability of an individual, family, group, class or
community to use resources which are directly required to secure a
livelihood. Access to those resources is always based on social and
economic relations, usually including the social relations of production, gender, ethnicity, status and age.

(ibid.: 48)

This "access" model of vulnerability to environmental hazards stands in
contrast to the "pressure and release" model that focuses mainly on linear
causes between "natural" and "social" systems. The access model seeks to
illustrate the underlying causes of vulnerability within social, economic,
and political factors rather than only within the magnitude of biophysical
events. According to this approach, environmental changes and shocks
(such as floods, storm events, drought) may contribute to vulnerability, but
the ability to withstand such shocks may be better controlled by focusing
on social development and poverty alleviation than mitigation of biophysical changes alone.

The access model of vulnerability has generally not been adopted in
many international discussions of environmental risk. Blaikie *et al.* (ibid.)
have argued that such analysis has been missing in the United Nations
International Decade of Natural Disaster Reduction (IDNDR). It has also
been missing in many early approaches to risks of climate change. For
example, an early Dictionary of Global Climate Change (Maunder, 1992)
did not refer to vulnerability at all, nor did the summary of climate change
issues in *The Kyoto Protocol: A Guide and Assessment* (Grubb *et al.*,
1999). Some early approaches to addressing climate change vulnerability
also adopted "regional," or ecosystem-based frameworks for projecting
risk. The United States Country Studies Program (USCSP), for example,
is one of the largest and most comprehensive current assessments of
climate change impacts and recommendations concerning developing
countries. The Program has provided technical and financial support to 56
developing (or transitional economy) countries to assist them in conducting studies of potential vulnerability and adaptive capacity to climate
change, with particular reference to eight important sectors of coastal
resources; agriculture; grasslands/livestock; water resources; forests; fish-
eries; wildlife; human health (USCSP, 1999: 73). Furthermore, there is a
focus on macro-level changes in (projected) biophysical climatic properties, population, and economic projections, which are assessed at the

national level. Such studies have tended to indicate social vulnerability to climate change in terms of calculating effects of projected changes on current economic activities, rather than on the social structures that may lead to vulnerability, or on the adaptive capacities to potential changes. As Kates noted: "To date, almost all efforts to address global climate change focus on preventive action to limit greenhouse gases rather than adaptation" (2000b: 5).

Some progress in changing these assumptions was made in the IPCC Third Assessment Report, which stated: "The ability of human systems to adapt to and cope with climate change depends on such factors as wealth, technology, education, information, skills, infrastructure, access to resources, and management capabilities" (IPCC, 2001: 8).

Furthermore, in 2001, other actions were taken to enhance the influence of climate change policy on environmental vulnerability. At the second part of the Sixth Conference of the Parties to the UNFCCC (COP 6 bis), negotiators agreed that host governments should approve carbon-offset forestry projects, and that projects should contribute to development (although these terms can be interpreted in various ways). Furthermore, a new "Adaptation Fund" was created from the allocation of 2 percent of Certified Emissions Reduction Units from the Clean Development Mechanism (CDM), plus an element of additional funding from the Protocol's Annex I (developed) countries. This fund is aimed to assist with objectives such as technological upgrading, training, or the improvement of local skills to address environmental changes.

Critics, however, have suggested that this Fund will act as a deterrent on CDM projects because the allocation of funds from Emissions Reduction Units will act as a tax on investment. Moreover, the fund may enact an artificial separation of projects that seek to *mitigate* climate change, and those that enhance *adaptation* that may give the impression that mitigation projects such as carbon-offset forestry may not have potentially damaging effects on local vulnerability. Such a separation reflects a linear model of environmental causality, and fails to appreciate how different individuals or social groups experience various environmental "changes" as "problems" – or how policies reflecting these definitions may create further problems for certain people.

Environmental vulnerability is a suitable question to end this chapter's discussion of the tacit politics underlying projections of environmental degradation across space. The chapter has shown that many attempts at global modeling or projections of risk have often reflected values and assumptions that are not necessarily shared by people in remote locations. Furthermore, this tendency has been shown in attempts to speak on behalf of people in different locations. The problems of these oversights are shown most in approaches to explaining environmental vulnerability. If assessments overlook the local economic and political conditions creating exposure to environmental changes, they may be ineffective in addressing environmental problems as they are experienced. Indeed, in worst cases, simply mitigating biophysical changes (such as, for example, using carbon-offset forestry simply for sequestering greenhouse gases), without also seeking adaptation may in some circumstances actually increase local socio-economic vulnerability to the changes. Under a "critical" political ecology, research might seek to highlight how different accounts of environmental risk and vulnerability may reflect the interests of different political actors and social groups. Yet such an approach would also require the critical assessment of claims about the adaptive or mitigating influences of different proposed strategies.

This chapter concludes the book's analysis of problems underlying current explanations of environmental change. The following chapters now consider potential solutions to these problems through means of reformed scientific practices and political debate.

Summary

This chapter has looked critically at the means by which environmental explanations or conceptions of degradation have been transferred across space to new locations and people. In particular, the chapter has questioned the growing debate about so-called "global" environmental problems. The chapter argued that many assessments of "global" environmental problems rely too much on projections of biophysical changes across the globe, rather than through understanding the ways in which these changes may be experienced, or present problems for different people.

The chapter also considered how much research in developing countries, and ironically on so-called "local" (or indigenous) knowledge, might replicate the assumptions and priorities of researchers rather than those of local people. Indeed, this problem is similar to some of the discussions of social movements as means of environmental democratization in Chapters 5 and 6. It is difficult to represent the knowledge and needs of powerless people when inquiry is both framed by research, and reshaped though communication with other groups. Optimistic concepts such as "epistemic communities" may actually refer to growing networks around predefined ideas, rather than growing consensus in each country or location. The more critical concept of "immutable mobile" instead draws attention to the scientific concepts and policies that can be transferred between places without being questioned.

The implications of these problems were discussed in relation to different approaches to environmental vulnerability. Orthodox scientific approaches to environmental vulnerability have emphasized exposure to biophysical changes. Such approaches have also been associated with assessments of risk at a regional scale, or through using linear models of environmental causality, which show clear linkages between human activities, biophysical impacts, and human loss. The problems of these approaches are that they overlook how risks may be interpreted more locally than at the regional level; how some people may lessen the impacts of environmental changes through the adoption of strategies such as environmental adaptations; or how social vulnerability to environmental change may be created through economic and political processes. Moreover, these approaches also fail to acknowledge how environmental policies based on linear models or at the regional level may impose restrictions on livelihoods and hence increase social vulnerability to environmental change.

As an alternative to these approaches, social and political theorists of environmental change have emphasized the importance of adaptive capacity to biophysical changes as a controlling factor in environmental vulnerability. Using non-linear models of environmental causality, they have argued linear models of environmental impacts have often used causal links that may reflect only limited framings and experiences of environmental change. Under this alternative approach, environmental vulnerability may be identified as exposure of individuals or social groups to biophysical changes (rather than of regions). The chapter illustrated this principle in relation to much current climate change policy that still seeks to mitigate biophysical changes at the global scale (by sequestering atmospheric greenhouse gas concentrations), yet often by practices that might increase vulnerability at local levels (such as through unregulated carbon-offset forestry).

The implications of these criticisms are not to dismiss the idea of global environmental problems, or to suggest that environmental policy should be based on adaptation alone. The aim is to warn that scientific assessments of environmental problems over space should not be taken as accurate and final, and that they will always reflect the actions of scientific networks and specific framings according to who participated in their creation. Instead of researching the existence of global problems, "critical" political ecologists may therefore also look at the social institutions and networks that allow dominant explanations of "global" environmental problems to be accepted and seen as authoritative.

The following chapters now explore potential solutions to these problems by reexamining scientific practices and political debates in order to increase the transparency and relevance of environmental science.

8 Democratizing environmental explanations

This chapter now starts seeking solutions to some of the problems described in this book. So far, the book has outlined the problems of environmental science, and the difficulties of explaining environmental degradation without also reflecting social and political framings. But is it possible to explain environmental problems in ways that do not reflect politics? How can scientific practices be reformed to overcome the problems of environmental "myths" (or environmental orthodoxies) that have been shown to be inaccurate and unhelpful in so many contexts?

The chapter will:

- summarize new approaches to environmental explanation that acknowledge the institutional basis on which causal or truth statements are built. These "institutional" approaches to explanation may be considered more democratic than the generalizations of orthodox environmental science because they reveal the tacit politics contained within causal statements, and allow greater possibility for reframing explanations in favor of localities or under-represented social groups.
- present case studies of institutional science, and how these might allow forms of scientific progress by demonstrating the errors of environmental orthodoxies. Institutional forms of explanation may provide a middle ground between the inaccurate generalizations of orthodox science, and the more phenomenological accounts of local environmental problems many scientists fear are relativist and non-generalizable.
- discuss the implications of such alternative forms of explanation for debates about scientific realism and the status of "global" environmental problems. Democratizing environmental explanations in favor of localities does not suggest championing "local" above "global" concerns. Instead, the aim is to acknowledge how existing explanations reflect different framings, and to seek ways of addressing global problems that are more relevant to the concerns of local people.

This chapter therefore contributes to a "critical" political ecology by proposing means by which the tacit politics within scientific analysis may be made more transparent, and by providing ways to engage critically with existing scientific explanations. Chapter 9 builds on this discussion by examining how scientific debate about environment in general may be made more transparent and accessible.

Democratizing explanations

So far, this book has discussed various ways in which environmental science is embedded within social and political practices. But what are the implications of this embedding for seeking better explanations of environmental problems?

The purpose of this chapter is to examine how far practices within environmental science can be reformed to acknowledge social and political framings, and to increase the chance of making explanations more relevant to diverse social groups and localities. In essence, this task is to democratize environmental explanations because it seeks to open environmental science up to greater transparency, and to reflect a wider range of social framings and knowledge sources.

Such a challenge, however, raises a number of dilemmas about democratization and scientific realism that form the basis of this chapter's discussion. As discussed in Chapters 1 and 3, defenders of orthodox science (e.g. Gross and Levitt, 1994) fear that democratizing environmental explanations may result in a slide toward relativism – or the belief (in an extreme form) that all truth claims about environmental change should be taken as equally plausible (see Box 3.5 concerning the differences between realism, relativism, and constructivism). Against this, many philosophers and sociologists of science point out that orthodox scientific "laws" and generalizations can never be taken to be universally accurate because they are based on such selective assumptions and modalities (see Harré, 1993; Tennant, 1997). Indeed, it is important to appreciate that a focus on the "construction" of environmental science does not imply that environmental change is unreal or imagined, but instead indicates an interest in how causal statements about biophysical processes have been made, and with which political influences and impacts.

Some important questions about democratizing environmental explanations include whether increasing transparency and framings of environmental science may allow the evolution of more accurate, universal explanations of environmental problems, or the emergence of several, and co-existing, "plural" explanations of environmental change. A further question is whether such pluralism may imply greater scientific realism in general. Some environmentalists and Critical Theorists have also suggested that "science," by its very nature, is exploitative of people and environment, and hence needs to be reformed in order to reflect more socially aware objectives (see Chapters 1 and 5).

This initial section introduces three key themes in the democratization of environmental explanations. The first theme is the belief of some

Critical Theorists that science needs to be changed in its very nature from exploitation to social emancipation. The second theme is the approach of science studies more generally adopted in this book, of analyzing environmental science in order to reveal tacit politics and assumptions. The third theme is how far such analysis may also allow a more critical engagement with hegemonic science by seeking ways to falsify environmental orthodoxies (or "myths") with explanations that are apparently more accurate. The possibilities of democratizing environmental explanations by both revealing tacit politics, and through some attempts at scientific realism, are then discussed more fully later in the chapter.

Critical Theory and liberatory science

The main objective of this chapter is to examine debates within science studies and new philosophies of science concerning greater awareness of social framings of science. But it is worth referring briefly to some old debates in social theory concerning the ability of science to act as an emancipatory or liberatory force.

As discussed in Chapter 5, some early forms of environmentalism within the new social movements of Europe and North America during the 1960s drew inspiration from the rejection of science and technology as the means through which industrial societies applied instrumental or exploitative rationality. In particular, the Critical Theorists, Habermas and Marcuse, debated how to reform science for better social development and liberation of the human spirit. Specifically, this debate would focus on whether science was problematic because it was used for exploitative purposes, or whether there was something inherently exploitative in the current forms of science adopted. Habermas, for example, argued that scient*ism* – rather than science itself – was the chief problem, and hence sought ways to increase the positive use of science for human needs (see Leiss, 1972; Alford, 1985; Vogel, 1996). Marcuse, on the other hand, sought to reform science itself to incorporate new values and objectives within its methods. As reported in Chapter 5, Marcuse wrote:

> the domination of nature has remained linked to the domination of man [sic] ... If this is the case, then the change in the direction of progress, which might sever this fatal link, would also affect the very structure of science – the scientific project. Its hypotheses, without losing their rational character, would develop in an essentially different epistemological context (that of a pacified world); consequently, science would arrive at essentially different concepts of nature and establish essentially different facts.
>
> (*One Dimensional Man*, 1964: 166–167)

And he later wrote:

[We need] a science and technology released from their service to destruction and exploitation ... [for the] collective practice of creating an environment ... in which the non-aggressive, erotic, receptive faculties of man [sic], in harmony with the consciousness of freedom, strive for the pacification of man and nature.

(*An Essay on Liberation*, 1969: 31; see also Vogel, 1996)

Marcuse was writing specifically about his definition of human nature, or the lost vitality and eroticism characteristic under the oppressive circumstances of industrial society (see Chapter 5). Yet later theorists have adopted similar approaches, particularly from debates in Critical Realism (Bhaskar, 1986; Collier, 1989). Such theorists have also argued that seeking social emancipation – or the escape from social injustices – cannot be achieved by using existing forms of scientific knowledge as a guide to these injustices. As a result, there is a need to reconsider the basis upon which to measure and explain injustice and social agency.

But despite these calls for reform in scientific analysis, there have been some barriers to change. As Morrow (1994) and Vogel (1996) have noted, such comments refer entirely to the practice of science itself, yet few efforts have been made to integrate these concerns with debates about scientific methodology. Indeed, such discussions about "science" as a form of social explanation now look rather vague and general beside the use of natural science as a guide to explaining biophysical change. Indeed, many observers might now consider these two uses of the word "science" to be separate debates, and consequently there is a need to understand the emergence of scientific research as a separate and legitimate institution beyond such criticisms (Jasanoff, 1990).

This separation of debates about exploitative and explanatory uses of science may also be seen in some approaches to environmental research. As noted in Chapter 5, much mainstream environmentalism has adopted a contradictory attitude to science, by first criticizing the instrumental domination of nature by science and technology, yet also using the knowledge created by orthodox natural sciences to describe the alleged ecological crisis without acknowledging the problems with this science (Yearley, 1992). Similarly, Dickens's (1996) study of Critical Realism and environment sought to achieve a more democratic and emancipatory explanation of environmental degradation. Yet this study was based upon many orthodox definitions and explanations of environmental degradation that could also be criticized for not being democratic, and need not be seen to be the only way that Critical Realism can be applied to environment (e.g. see Jackson, 1997; Forsyth, 2001b).

The problem with the discussion of "science" or social oppression as the cause of environmental degradation under early Critical Theory is that it presupposes the nature of exploitation, and then seeks to find political strategies to avoid this problem. As discussed in Chapters 5 and 6, such assumptions about causes of environmental degradation are not always warranted, and reflect wider discourses of opposition to capitalism and modernity rather than a full appreciation of the diverse causes of biophysical change. Furthermore, such an approach separates assumed causes of environmental problems from politics, rather than acknowledging the influence of political debates upon the construction of explanations. An alternative approach is to look at scientific discourse in order to identify the hidden politics and assumptions, and then use this knowledge to assist in social reform.

Revealing the tacit politics of science

A further approach to democratizing environmental explanations is to analyze the discourses and practices of science in order to indicate the tacit – or hidden – implications of science for political debates. This approach is generally the position adopted by Sociology of Scientific Knowledge (SSK). As Wynne summarized:

> A key element of SSK is that it involved identifying (and problematizing the role in knowledge-establishment of) tacit contextual commitments and assumptions, of the kind which have become routinized and taken-for-granted in the prevailing cultural fabric, and which may have shaped accepted "natural knowledge."
>
> (1998: 339)

Much of this book has already summarized the objectives of SSK (see especially Chapter 4) and the purposes of revealing the tacit politics of science. These approaches can assist in the democratization of science by indicating how far "black box" statements about environmental causality have reflected the perceptions and experiences of only limited social groups, or how far scientific boundaries and debates contribute to the coproduction of science and politics. Revealing the tacit assumptions within science portrayed as being universally applicable and politically neutral both weakens the power of that science to influence policy decisions, and empowers social groups not represented in science to challenge existing explanations. Furthermore, deconstructing science in this way does not presuppose causal structures or "real" explanations of environmental degradation in the manner of the discussions in Critical Theory above.

Each chapter of this book seeks to reveal the tacit politics within environmental science, and so it is unnecessary to explain these approaches in this section. Yet it is worth noting that this book also asks how far revealing the tacit politics of science may also allow the construction of more accurate or relevant forms of environmental explanation, especially for social groups not represented in currently hegemonic scientific discourses. Such a question, however, is controversial because it questions some basic assumptions about scientific realism and relativism that are commonly misrepresented and need clarification.

Can environmental "myths" be falsified?

Some observers have expressed frustration at the deconstruction of scientific discourse associated with Sociology of Scientific Knowledge. Perhaps most importantly, critics have suggested that deconstruction alone is insufficient to replace unrepresentative science with something more accurate (Wildavsky, 1995). Another fear is that a discussion of science based upon poststructuralist debates about discourse is anti-empirical (e.g. Bryant and Bailey, 1997: 192), or may be easily delegitimized by natural scientists who see such discussions as impractical and even dangerous to progress (e.g. Gross and Levitt, 1994; see Segerstråle, 2000). Consequently, some scholars have suggested that science studies should seek ways to engage with orthodox falsification in order to replace inaccurate and unjust explanations of environmental degradation.

Such views bubbled over in a debate between two science studies scholars, Hans Radder and Brian Wynne. Criticizing Wynne, Radder wrote:

> Just to say that ... knowledge is not "simply" or not "automatically" falsified may be comfortably vague, but it is also disturbingly unhelpful, especially when we are confronted with difficult and far-reaching political epistemological questions. Can we still believe, and act upon, the knowledge claims of climate scientists, or can't we? ...To put it bluntly, there is *nothing* in the constructivist account of the global warming issues that urges us to be (or indeed, not to be) "more precautionary in practice."
>
> (1998: 329–330, emphasis in original)

To which, Wynne replied:

> Radder argues that the realism–relativism issue is crucial ... by this he appears to mean that such works must take sides on truth claims... I simply say that this is not the point.... I argue that SSK [Sociology of Scientific Knowledge] can help illuminate that problem in a new and meaningful way, but not in any direct and final way. It is a sterile, misleading, false and unhelpful way to define political interaction to suppose that it can be based upon such positive convictions.
>
> (1998: 339, 342)

The crucial point made by Wynne is that orthodox "testing" of conflicting environmental statements requires taking sides on truth claims – or assuming that one side may be more accurate than the other. (A similar process is called "purification," after Latour, as discussed in Chapter 4.) Instead, under more sociological and constructivist approaches, it is important also to ask both how far either side may reflect wider social contexts, and how each side has emerged in conjunction to the other rather than by itself as an independent claim about reality. Radder's argument is that SSK needs to be more aggressive in attacking truth claims made from orthodox science.

This chapter seeks to reconcile both positions by examining ways in which environmental science might be reformed in order to demonstrate the errors of environmental orthodoxies (or dominant, yet questionable explanations), and seek to replace these with more diversified and localized explanations. Such an approach does not, however, suggest that alternative explanations allow "direct and final" falsification of orthodoxies, but a greater illumination of how, and under what conditions, they may be considered accurate or inaccurate. As Latour wrote, such a process represents a notional "Parliament of Things," an exercise that does not in principle reject the general philosophy of scientific progress:

> We want the meticulous sorting of quasi-objects to become possible – no longer unofficially and under the table, but officially and in broad daylight. In this desire to bring to light, to incorporate into language, to make public, we continue to identify with the intuition of the Enlightenment.
>
> (1993: 142)

Such "meticulous sorting" may contribute to the democratization of environmental explanations by providing greater accountability for how scientific "facts" have evolved. This knowledge allows greater ability to criticize hegemonic explanations of environmental degradation, and provides more scope for alternative framings of environmental explanations. This chapter describes some approaches to achieving such a "Parliament" of explanations through reforming scientific techniques. Chapter 9 builds on this by examining political arenas and infrastructure that may empower such analysis further (for example, the approaches of sustainable livelihoods and environmental entitlements, in the section entitled "Environmental Adaptations in the Developing World," pp. 253–258).

Before this discussion, however, it is necessary to examine the concepts of "truth" and falsification, before considering alternative approaches to environmental explanation.

Integrating social framings and scientific realism

So, how is it possible to integrate social framings and scientific realism in order to produce more accurate and relevant alternatives to environmental orthodoxies and "myths"? This section examines some general debates about environmental science and social contexts within scientific methods themselves. (Chapter 9 provides a broader discussion of the democratization of scientific debate in general.)

The democratization of scientific method, however, raises important questions that relate to some key scientific controversies. First, does challenging universalizing environmental explanations imply the adoption of relativism? One of the greatest fears among defenders of orthodox science is that the deconstruction of existing explanations – or analyzing the social and political embeddedness of scientific statements – is that it suggests all truth claims are relative, and that scientific progress is not possible.

Second, does diversifying and localizing environmental explanations imply denying the existence of "global" environmental problems, or championing "local" framings above others? Similarly, does localizing science entail imposing predefined notions of locality or identity onto social groups in ways that are not warranted (as discussed in Chapter 7)?

Third, is it reasonable to seek some form of realism at all, or is this simply a reiteration of processes of purification that may only result in the establishment of alternative hybrid structures of knowledge rather than some ultimately more accurate vision of reality?

This section summarizes new thinking from sociology and Philosophy of Science about integrating social framings and scientific explanations, including themes of semantic and critical realism introduced in Chapter 3. The section first discusses new thinking about environmental "truth" and falsification. It then discusses so-called institutional approaches to explanation, and the implications for debates about scientific realism.

Beyond environmental "truths" and "falsehoods"

One of the most important challenges for integrating social framings and scientific realism is the need to redefine orthodox notions of scientific progress.

As discussed in Chapter 3, the concept of scientific progress is an essential component of orthodox science and refers to the process of producing scientific statements that increasingly represent reality. According to logical positivists (in the early part of the twentieth century), progress lay in "proving" "laws" of nature by verifying these apparent trends in new datasets. Under the critical rationalism of Karl Popper (in the mid to late twentieth century), verification was replaced by falsification, or the testing of theories or hypotheses.

Both verification and falsification sought to achieve scientific progress by seeking evidence for the accuracy or inaccuracy ("verisimilitude") of different "laws" or theories. This form of scientific progress also draws upon the philosophy of truth known as the correspondence theory of truth. Under this theory, a statement – such as the environmental orthodoxy that deforestation causes erosion – may be defined as "true" if it can be shown to correspond to an existing and accepted definition of how it should be (Leplin, 1984; Allen, 1993; Psillos, 1999).

This orthodox approach to falsification and scientific progress, however, has been criticized since the 1960s by a variety of philosophers and sociologists of science. Chapters 3 and 4 have already reviewed much of these criticisms, referring mainly to the ability to generalize so broadly from limited empiricism in socially explicit framings, and the hidden politics and social values within the social regulation of scientific practice itself (see Laudan, 1977, 1990; Harré, 1986, 1993; Latour, 1987; Shapin, 1994). The seminal critiques of Kuhn (1962) and Lakatos (1978) in particular pointed to the social shaping of hypothesis testing within socially determined paradigms or networks of scientists, rather than a uniformly meaningful "tide" of knowledge advancing via science (see Chapter 6).

Further work has underlined the social contexts through which falsification can take place. Larry Laudan's *Progress and its Problems* (1977) rejected the relevance of "truth" as a guiding principle in scientific progress because a long succession of debates in philosophy has illustrated the difficulty of knowing when "truth" has been achieved. Concepts such as "scientific progress," and "conjecture and refutation" may therefore reflect episodes of topical political debate rather than a gradual progression toward scientific realism. Laudan wrote: "if scientific progress consists in a series of theories which represent an even closer approximation to the truth, then science cannot be shown to be progressive" (1977: 126).

Steve Shapin's *A Social History of Truth* (1994) also argued that "truth" and "reality" are defined according to the norms and conventions of specific times and places. His study of social norms and networks of medicinal scientists in seventeenth-century England led him to propose that social norms set the conditions of "truth" that science aimed to satisfy, rather than the orthodox belief that science uncovers truth in essence. Bruno Latour (1987, 1993), of course, has argued that the process of "purification" offered by science in effect only replicates predefined framings of hybrid objects made by historic social relations, rather than establishing real and universally applicable explanations and causal linkages in a socially independent nature (see Chapter 4).

Such criticisms suggest that orthodox approaches to falsification overlook semantic and linguistically driven approaches to truth that present alternatives to the correspondence theory of truth. As introduced in Chapters 3 and 4, semantic and linguistic boundings (cernings) of environmental problems acknowledge the importance of the observer in both identifying and attributing meaning to concepts or objects that may or may not be considered true (Levin, 1984; Norris, 1995). Under such alternative conceptions of truth, conceptualizations of "reality" are seen to be dynamically shaped according to the language, social framings, and agendas of the society and scientists who helped formulate them over time, and such conceptualizations continue to be shaped by further political interaction. Hence, the use of orthodox falsification, through the testing of hypotheses by empirical investigation, may therefore present proof of the existence of particular framings of reality, as reflected in the hypothesis and empiricism, rather than alternative framings relying on different perceptions and the empiricism to match these. Yet equating such scientific practice and findings with "truth" may be to legitimize them as "reality" when the same investigation using different boundings may yield equally valid empirical results.

These kind of arguments, however, have frustrated many defenders of scientific realism, who refuse to believe that all concepts of truth are socially driven, or who still hope for some ability to refer objectively to ecological truth. The co-founder of the Worldwatch Institute, Lester Brown, for example, has urged that we need to revitalize economic policies so that prices and incentives reflect "the ecological truth" (2001: xvii), as though this "truth" is both singular and known. Such matter-of-fact approaches to environmental realism overlook the political contestations over truth, and the semantic and institutional bases on which many environmental "problems" are experienced and discussed.

The tendency to rely only upon the correspondence theory of truth has also beleaguered perennial debates between environmentalists and eco-optimists concerning the "truth" underlying the state of the environment. Such debates have often been associated with specific reports and the production of statistics from both sides. The *Limits to Growth* report of 1972 (Meadows *et al.*, 1972) was one early statistical prediction of ecological degradation, and was followed by *The Global 2000 Report to the President* (USCEQ, 1980), which was also pessimistic about environment. The more optimistic *The Resourceful Earth* (Simon and Kahn, 1984) followed as a further statistical response to suggest that degradation was not so serious. The most recent example of this "tit-for-tat" publication of statistics has been *The Skeptical Environmentalist* by the statistician Björn Lomborg (2001), which again provided optimistic statistical summaries of environmental degradation in different ecosystems or locations (see Box 8.1).

Such publications provide interesting information about the levels of uncertainty about environmental degradation. But they are also couched within the terms of existing environmental discourse and scientific concepts, rather than accepting that these terms impose structures onto complex biophysical and social realities that may not always be best suited by these concepts (see Chapter 2). As such, these publications adopt the correspondence theory of truth, in which the definitions of "problems" such as deforestation, desertification, and erosion are accepted as "real" and measurable. These studies, however, avoid discussion about how far, under what conditions, and for whom these conceptualizations of degradation actually do represent problems. Simply measuring the state of predefined concepts of environmental degradation as a test of the truth of environmental degradation may only add to political contestations of environmental policy.

Integrating social framings of environment with scientific realism therefore requires the rejection of orthodox models of falsification and the correspondence theory of truth. There is little point in seeking to democratize framings of environment if the measurement of environmental degradation uses categories and concepts of degradation that are themselves contested and variable. Instead, there is a need to acknowledge the political factors that lead toward the conditions for truth to be established, rather than simply to assume that "truth" is already defined and agreed upon. So-called institutional approaches to explanations offer ways to integrate scientific realism with such variable framings of truth.

Box 8.1   *The Skeptical Environmentalist* and statistical analyses of environmental degradation

Ever since the publication of the *Limits to Growth* (Meadows *et al.*, 1972), there has been much environmental debate providing statistical claims and counter claims concerning environmental degradation and resource deple-tion. The generally concerned *Global 2000 Report to the President* (USCEQ, 1980), for example, was followed by the optimistic *The Resourceful Earth* (Simon and Kahn, 1984). In 2001, a further book was published, *The Skepti-cal Environmentalist* by the statistician, Björn Lomborg (2001), which pro-vided further optimistic rejections of environmental concern. Indeed, the book included a foreword by Julian Simon, the co-author of *The Resourceful Earth*.

The Skeptical Environmentalist urged readers to be aware of two key concerns. First, much environmental concern is the result of the so-called "Litany" of unnecessarily pessimistic press coverage, sensationalist reports, and doomsaying that appeal to people because they are instantaneously newsworthy. Second, Lomborg urged a reevaluation of the "facts" to measure the "real state" of the world, by providing a variety of statistics that question many projections about environmental problems such as increasing pollution, deforestation, biodiversity, climate change, and water resources. The book's optimistic conclusion is that "children born in the world today – in both the industrialized world and developing countries – will live longer and be healthier, they will get more food, a better education, a higher stan-dard of living, more leisure time and far more possibilities – without the global environment being destroyed. And that is a beautiful world" (2001: 352).

Predictably, the book invoked much immediate criticism. The World Resources Institute (WRI) used its website to list nine reasons to criticize *The Skeptical Environmentalist*. Reasons included alleged pseudo-scholarship because of selective quotation of statistics; the lack of specific environmental training by Lomborg; and a confusion of statistical association with actual causality in many figures cited. WRI also alleged the book con-tained a variety of errors and confusions on specific topics such as the confu-sion of wild and farm fish production, or the avoidance of damage to forests resulting from agricultural conversion. Perhaps most fundamentally, WRI accused Lomborg of confusing the issue of measuring growing human pros-perity without asking if this causes environmental damage. WRI went on to publish its own report on declining forests worldwide to refute many of Lomborg's statements. The co-founder of the Worldwatch Institute, Lester Brown, refused to enter into any face-to-face discussions with Lomborg.

Yet more generally, *The Skeptical Environmentalist*, and its criticisms reveal further insights into the shaping of environmental knowledge. From a Cultural Theory perspective, the book represented a classic example of the "Individualist" myth of nature, and WRI represents the "Egalitarian" myth (see Chapter 4). In both cases, the statistics produced by either side reflect the worldview that nature is either resilient or fragile. Furthermore, the sta-tistical measurement of predefined concepts such as "deforestation" may preclose the meaning of such environmental changes, and therefore deny alternative framings of how such changes may be experienced. Many of the positive scenarios described by Lomborg do not apply to all people. It is also worthwhile noting that, despite Lomborg's criticism of environmental doom-saying for attracting media attention, his book also touted publicity by being so overtly contrarian and dismissive of environmental concern. The implications of the controversy surrounding *The Skeptical Environmentalist* are therefore to place more attention to how each side makes its arguments, and how this influences how we understand environmental change, rather than to accept either side as universally true. Such statistical analyses do not provide an objective image of the "real" world.

Source: Lomborg, 2001; http: //www.wri.org.

"Institutional" forms of explanation

Institutional forms of explanation are approaches to science that acknowledge the social framings and institutions that make explanatory statements possible. Institutions may include semantic framings, transcendental (or phenomenological) experiences of reality, or the networks and conventions that may influence the production of apparently real causal statements (see Chapters 3 and 4). These approaches are called institutional because they acknowledge these institutions openly, although it is worth noting that orthodox "laws" and generalizations are similarly contingent on institutions, but these are not acknowledged under orthodox science.

A variety of philosophers of science have claimed that such institutional approaches to science allow the chance to reconcile scientific realism with sociological and semantic analyses of truth. Aronson *et al.* (1994) write:

> most of the problems associated with fending off … anti-realist attacks result from a failure to separate metaphysical [ontological], semantic and epistemological issues … Most scientific discourse is not about the natural world but about representations of selected aspects of that world. Our conceptions of what nature is are mediated by our representations of nature in models, which … are subject to important constraints. Constraints on our best representations of naturally occurring structures and processes mostly reflect historical conditions for the intelligibility of those representations and the experimental procedures we have devised for manipulating them.
>
> (1994: 2, 4)

Similarly, Nancy Cartwright commented:

> Philosophers distinguish phenomenological from theoretical laws. Phenomenological laws are about appearances; theoretical ones are about the reality behind the appearances. The distinction is rooted in epistemology. Phenomenological laws are about things which we can

at least in principle observe directly, whereas the laws can be known only by indirect influence. Normally for philosophers "phenomenological" and "theoretical" mark the distinction between the observable and the unobservable.

(1983: 1)

Some defenders of orthodox science have argued that such distinctions are already acknowledged within the frameworks of orthodox science (e.g. Levitt, 1999: 357). For example, Bunge wrote: "physicists do not ignore phenomena: on the contrary, they often start with them. But they do not limit themselves to what appears to us" (1991: 267).

But the key difference between this kind of attention to phenomena, and the deeper critique of orthodox science adopted by science critics, is that orthodox scientists do not acknowledge how different social boundings or networks may influence either the construction of truth in philosophical terms, or the practice and enforcing of "truth" within policy. Instead, institutional forms of explanation point to the localized, and semantically bound influences on the creation of apparently truthful explanations and descriptions, as alternatives to explanations or descriptions based upon universalizing, or propositional truth statements characteristic of orthodox science.

Such boundaries of causal statements may be defined in various ways. Searle (1995), for example, distinguished between "brute" and "institutional" facts in order to indicate the influence of predefined purposes for different objects within different social settings (see Chapter 3). Drawing on transcendental, or Critical Realism, Bhaskar (1991) argued that science proceeds through the combination of so-called conventional referring – or denoting terms for hypothetical activities – and then practical referring – or the fixing of this term through official description and measurement (Lewis, 1996). In this sense, science produces "transitive" (or socially constructed and changeable) explanations and descriptions of "intransitive" (or underlying and unchanging) reality.

One implication of such institutional approaches to explanation is that apparent "laws" of nature emerge because they make sense to the society or network that produces them. "Laws" can also exist unchallenged as long as their creators maintain the boundaries around the truth claim, or the members within the institution remain a homogenous unit, even if alternative framings may provide conflicting explanations involving the same subject matter. The continued strength of institutions upholding different truth claims may help to explain how environmental orthodoxies (or generalized beliefs about environmental explanation) continue to be adopted and promoted as true in various institutional settings, despite the increasing occurrence of apparently contradictory evidence.

Figure 8.1 provides a comic illustration of this phenomenon. The figure presents a humorous cartoon of snakes attending a class on "nature," and is based on a well-known joke in England about how English people have

THE ENGLISHMAN GOES TO THE BEACH ONCE
A YEAR TO SHED HIS SKIN .

Figure 8.1 A cartoon representation of the problem of scientific explanations based on empirical evidence framed by people with different experience

Source: Reproduced with permission of Giles Pilbrow.

the tendency to be so excited about experiencing hot weather on holiday that they take few precautions against sunburn. In the cartoon, the snakes explain the perceived reality of the English behavior in terms that the snakes (but not the English) find meaningful. The humor in this cartoon is that the causal statement is apparently real and empirically testable (some difference in verisimilitude may be found, for example, between beaches in colder northern England, and those in sunny Spain). But the resulting explanation of English behavior reflects only the reality of the snakes (who really do have to shed their skins), and not the humans. Perhaps less amusingly, however, the cartoon is also a good representation of how some explanations of environmental problems in many locations are based upon the same combination of empirical evidence and framings imposed by outsiders.

This example illustrates a point made elsewhere in this book that scientific "laws" only hold true under certain conditions and assumptions. There is a need to show caution because they may encourage us to have too much confidence in these laws as reality. As Nancy Cartwright argues in *A Dappled World*: "the impressive empirical successes of our best physics theories may argue for truth of these theories but not for their universality ... Laws, where they do apply, hold only *ceteris paribus*, ..." (1999: 4).

The implications of these arguments are to shift the consideration of environmental realism from the application of blanket, universalizing statements that are either "true" or "false" (as characteristic of orthodox science), toward the appreciation that a variety of statements may be considered true or false according to the strength of the truth conditions that bind them. In essence, it is more important for a "critical" political ecology to consider the political factors underlying truth conditions about environmental change, than to assess predefined notions of ecological truth (see also Aronson *et al.*, 1994).

Truth conditions may take a variety of forms, based upon historical or linguistic framings of environmental change. For instance, dominant beliefs about the loss of wilderness, forestry, and heritage may provide tacit framings about what should, and should not, be considered worthy of protection (Howarth, 1995; Neumann, 1998). Similarly, some debates about watershed management in Asia have tended to assume the only purpose of land-use policies in mountain areas is to ensure provision of water to lowland cities rather than also sustainable agriculture in the uplands (Hamilton, 1988; Alford, 1992; see Chapter 2). Yet changing the truth conditions and boundaries of empirical analysis may reveal empirical results that question the accuracy of previously bounded statements. Indeed, this has occurred in numerous studies (such as concerning the role of upland agriculture in watershed degradation) and has resulted in findings that contradict the previously unchallenged assumptions of the preceding framings. Such studies have often formed the basis of debates about environmental orthodoxies and evidence that challenges them (for example, concerning the belief that erosion and deforestation are always degrading, or that deforestation always causes erosion and water shortages: see Ives and Pitt, 1988; Ives and Messerli, 1989; Calder, 1999).

Kukla (1993) summarized this problem by referring to different kinds of epistemic boundaries in the discussion of truth statements. Epistemic boundaries are constraints on a belief coming from the stock of concepts in a mind. In this sense, the boundaries and resulting truth statements reflect the principles of semantic realism, or the classification of the world according to meaning. In addition, there are also syntactic boundaries, or those referring to the logical ordering of objects. Under syntactic ordering, there may also be abductive boundaries, when alternative concepts never occur because of the configuration of the system or social "machine" that orders them; and implementational boundaries, when a system recognizes a concept as true, but acts as though it is false. Environmental orthodoxies may therefore be examples of abductive boundaries because they give the impression that no alternatives exist; and local adaptations or resistances to land-use policies based on orthodoxies (such as those listed in Chapter 2) are implementational boundaries because people persist in constructing livelihoods despite the continuation of supposedly degrading practices like erosion and deforestation. Using the political theorist James Scott's terminology, universalizing orthodoxies may be examples of top-down development consistent with "seeing like a state"; and the local adaptations of citizens in opposition to these theories are akin to "weapons of the weak" (Scott, 1985, 1998).

These different kinds of boundaries offer ways of assessing how political activism or powerful networks may influence the formulation of environmental explanations. Yet, in addition, this form of analysis also allows the chance of democratizing environmental explanations by falsifying more powerful and generalizing environmental explanations. This possibility arises when truth statements under one set of epistemic boundaries (or truth conditions) are apparently contradicted when empiricism on the same topics is conducted under a different set (for example, as when predictions of degradation from population increase, deforestation, or erosion are contradicted by evidence of local adaptations).

Commonly, under orthodox science, such apparent disagreements are treated in black and white terms of correspondence to predefined truths; or are dismissed as local anomalies; or occasionally are claimed to represent the desire to romanticize "local" agriculture and ethnic groups despite the pressing logic of orthodox environmental science. Yet, in effect, these alternative framings of environmental problems represent a further insight into biophysical reality from a new perspective, and which can provide a more accurate (and usually more socially relevant) explanation of change than the larger-scale generalization.

One approach to explaining this co-existence of different epistemic boundaries is the concept of type hierarchies. Type hierarchies may be defined as the structural relationships between accounts of reality with different semantic structures (Harré, 1993; Aronson *et al.*, 1994: 127). For example, the words "animal," "dog," and "Fido" all refer to natural kinds that adopt different levels of generality when referring to the same object. Similarly, a Ferrari is not a horse, but both are vehicles.

The significance of type hierarchies to environmental science is that scientific statements or explanations are bounded within semantic structures of meaning, and lead to a similar hierarchy of associated statements of causality in different settings. For example, the semantic structures of equilibrium (or "balance of nature") ecology might be associated with notions of lost wilderness, and the need to restore equilibrium in order to avoid catastrophe. Under such framings, it is logical to see interlocking statements linking, for example, deforestation to erosion, or reforestation to global ecological stability. Finding alternative evidence against these "laws" might therefore be seen to be contrary to the semantic structure of the explanations as a whole, and therefore dismissed as inaccurate, or even actively resisted by researchers and policymakers in the dominant system (see Chapter 6), all of which regularly occurs to the evidence and people who challenge orthodoxies (see Chapter 2). As Harré noted:

> laws of nature are always understood as embedded in a background system of kinds, an interlinked ontology of type hierarchies … It is only by virtue of the relevant type-hierarchy that the content of ontological presuppositions can be filled out, and the range of beings relative to which the assessment of the plausibility of the statement of a law can be made.
>
> (1993: 114, 113)

Or, as Aronson *et al.* write in relation to the semantic norms of science:

> A theory is plausible when two conditions are satisfied. It must be capable of yielding more or less correct predictions and retrodictions, the familiar criterion of "empirical adequacy." We could call this a "logical" criterion. But it must also be the case that the content of the theory is based on a model which is type-wise drawn from a chunk of a type-hierarchy which expresses the common ontology accepted by the community. We could call this an "ontological" criterion.
>
> (1994: 191)

Type hierarchies have important implications for a "critical" political ecology, or the analysis of the political conditions that lead to the establishment of environmental "truths." According to the description of type hierarchies above, the continued existence of "laws" of nature also requires the co-existence of a self-enforced community (or indeed, epistemic community) to support it. This co-existence of social structures and scientific explanations indicate how social norms shape scientific statements as "truth"; how such social structures may exclude alternative conceptions of truth; and that reformulating the norms and boundaries of truth statements may reveal alternative – and possibly more accurate – explanations.

The political implications of such self-enforcing communities for the democratization of environmental science are discussed further in Chapter 9. The purpose of this chapter is to demonstrate how these debates about scientific practice might lead to the democratization of environmental explanations. But before some case studies of this kind of research are presented, it is worth asking how far such challenges to orthodox environmental science challenge debates about scientific realism.

Does democratizing science enhance realism?

The previous sections have argued that there are different ways of looking at the world instead of the simple binary distinction of "true" and "false" commonly adopted under the frameworks of orthodox science and the correspondence theory of truth. As an alternative, semantic approaches to truth are particularly useful for environmental themes because so much debate about environment focuses on "problems" that are inherently institutional and semantic in nature. The word "institutional" implies that problems of risk or degradation of complex ecosystems always represent some element of social framings of when, how, and for whom changes or events present problems.

As noted throughout this book (and especially in Chapter 2), concepts such as erosion, forest loss, and pollution – although frequently contributing to environmental degradation or risk – may not always be human induced, and may themselves be metaphors or summaries for associated and cumulative factors such as soil fertility, biodiversity loss, or resilience of affected people and systems. Explaining such institutional factors through the propositional, universalizing "laws" of orthodox science avoids the contextual factors that make such environmental changes meaningful in different ways to different people. Referring instead to the semantic basis of environmental explanations illustrates how they fit into interlocking systems of type hierarchies or associated explanations that reflect overriding social values and experiences within each semantic institution. As Aronson *et al.* noted:

> it is our contention that laws [of nature] are "ontologically localized" to specific types of phenomena. Which law applies to a system depends on the location of the type to which that system belongs in the type-hierarchy that expresses the common ontology of a certain field of phenomena.
>
> (1994: 153–154)

Adopting a contextual and semantic basis to environmental explanation offers the possibility to democratize environmental science in two key ways. First, it strengthens the criticism of dominant environmental explanations (so-called environmental orthodoxies) by showing that contrary evidence (such as the suggestion that more people can produce less erosion) may not be statistical anomalies from generally true trends, but may reflect alternative systems of managing and framing environmental change. Second, conducting research on alternative framings of environmental change may also produce new insights, or surprising refutations of orthodox thinking that question the accuracy of the orthodox explanations (e.g. that erosion or water shortages may not increase, in total, after deforestation). Such insights may be seen as forms of democratization because they indicate how hegemonic environmental explanations reflect culturally specific framings of problems. They also weaken the apparent universality of these explanations. Furthermore, they offer possibilities for building new explanations based on new framings from previously unrepresented social groups.

But do these challenges to orthodox notions of environmental change also mean a step toward greater accuracy and scientific realism? Or do they simply demonstrate that environmental problems may be explained in different ways, using different framings and political priorities? This question has immense practical importance in leading to more informed and biophysically accurate environmental management practices in various locations, and in challenging the authority of many political actors that claim scientific accuracy as a means to increase their credibility and legitimization (see Chapter 6). Claiming a relativistic approach to science – in effect saying that different framings present equally valid explanations – may weaken political criticisms of different environmental management policies or organizations because it suggests that the only deciding factor should be the value systems adopted by each. Yet concluding that altern- ative explanations may also be distinguished on grounds of ultimate accu- racy may enhance political criticisms because it weakens, or even effectively falsifies, the scientific approaches they are based upon.

This question is the root of disagreements between scholars who seek a more realist vision of environmental problems; or those who see the analy- sis and revealing of how statements are made to be criticism enough (such as the debate between Hans Radder and Brian Wynne above, p. 207). In general terms, these different approaches may be labeled as a conflict between scholars favoring generally favorable approaches to environ- mental realism (particularly critical realists), and those looking more to the social institutions and networks influencing the production of scientific "laws" (notably pragmatists or network theorists) (see Rose, 1990; Proctor, 1998).

Under the Critical Realism of Roy Bhaskar (1975, 1991), for example, environmental reality may be compared to three levels of knowledge: empiricism (simple experiences); actualism (experiences, and the events that give rise to experiences); and realism (the underlying ontology and structures that give rise to events and experiences) (see also Chapters 1 and 3). Identifying new explanatory statements that contradict or improve on existing (orthodox) statements may therefore be likened to "peeling an onion" as each layer of underlying reality is revealed. Critical realism therefore supposes a progressive attitude toward scientific progress, by assuming that new insights and framings increase our knowledge of the world's underlying structures. (Yet, as noted in Chapters 1 and 3, critical realists differ from classical Realists by not equating scientific laws with reality.)

Against this, critics of scientific realism have suggested this approach may be too optimistic about indicating underlying ontology. Hannah (1999), for example, has argued that Critical Realism – as defined by Bhaskar – may over-privilege the perceptions and actions of local ob- servers as necessarily indicative of underlying structures rather than social conventions. Similarly, pragmatists such as Richard Rorty (1989a, b) argue that scientific explanations reflect social and political networks rather than underlying reality. As noted in Chapter 4, the philosophy of pragmatism implies three key tenets: the rejection of essentialist concepts of truth; the perception of no epistemological difference between facts, values, morality, and science; and a belief that social networks or solidarities determine scientific inquiry.

The evidence of research that contradicts environmental orthodoxies suggests that both some elements of Critical Realism and pragmatism may be relevant to the democratization of environmental science. Clearly, some of the more overt generalizations of some environmental orthodoxies, such as "deforestation always causes biodiversity loss," or "population increase always increases degradation" (see Chapter 2), are demonstrably inaccurate when applied universally to all locations. Yet the rejection of such orthodox statements is only dependent on the semantic interpretation of all aspects of these statements. For example, the words "degradation," "deforestation," or "erosion" have to be defined according to how these words are meant to represent different ecological changes and the magnitude of each. Any criticism of orthodoxies using institutional approaches to science is hence done on a more nuanced and less binary basis than orthodox falsification based on the correspondence theory of truth and the rejection of propositional hypotheses. Accordingly, the "rejection" of orthodox generalizations also does not mean that, for instance, deforestation may *never* lead to erosion, or population increase may *never* accelerate the exhaustion of resources. It is important to reject the universality and dogmatism of orthodox environmental generalizations. But this can only be done by simultaneously revising what is meant by falsification and scientific progress.

Similarly, the ability to revise generalized environmental explanations has, to date, generally emerged only when research has been conducted on the practices and experiences of different social groups, or – in the case of Cultural Theory – in the analysis of different worldviews (or "myths" of nature) (see Chapter 4). Such alternative framings can only exist where there are social solidarities to support them. Identifying alternative, or localized, framings of environmental explanation *does* challenge the universality of orthodox environmental explanations. But finding such alternatives does not suggest that these alternative framings should be seen as universal replacements for environmental orthodoxies. (For example, the realization that erosion is not necessarily caused only by agriculture, or that declining soil fertility is a more important threat to agricultural productivity than erosion alone, should not imply that erosion should be discounted as a problem in all locations – see Chapter 2.)

As Proctor wrote: "If Critical Realism could be criticized for being rather too epistemologically confident in its reality claims, pragmatism could likewise be criticized for being too epistemologically tentative" (1998: 367). Making environmental science more biophysically accurate and socially relevant requires revealing both the inaccuracies and social framings of existing environmental explanations at the same time. Doing both does not imply creating a new and universally more accurate scientific explanation, but it does challenge the pretensions of orthodox science to achieve universal accuracy.

The following section now presents some case studies of how environmental science may be reformed to integrate social framings with scientific realism. Such approaches help to achieve a "critical" political ecology by demonstrating the social and political bases upon which different explanations of environmental degradation are made.

Approaches to diversifying and localizing environmental science

This section provides some examples of recent approaches to integrating biophysical explanation with social and political contextualizations of environmental problems. The examples cited reflect a variety of positions regarding the acceptance or criticism of attempts to use science in a realist manner.

Degradation syndromes and "science in places"

One initial approach to integrating environmental science with local environmental perceptions is so-called "science in places" or "degradation syndromes" (NRC, 1999). These concepts have been used within the contexts of orthodox environmental science to explain how environmental problems may be region-specific rather than universal (see Kasperson *et al.*, 1995, in Chapter 7). The word "syndrome" is used to imply the co-existence of several forms and causes of environmental degradation within a locality.

Three types of syndrome have been identified. Utilization syndromes refer to locations where land uses have exceeded local ecological criteria for sustainability, such as in the Dust Bowl, the Sahel, or in locations of mass tourism. Development syndromes refer to locations where overt attempts at land development have led to degradation, such as in the Aral Sea, the Asian Tigers, or the agricultural Green Revolution. Sink syndromes are where regions or localities have become despoiled through the disposal of wastes such as by smokestacks or toxic dumping. The term "science in places" refers to the need to tailor-make the application of environmental science to locations where such syndromes are experienced. The NRC wrote:

> Making knowledge more usable means enhancing the capacity of groups around the world not only to *obtain and interpret it, but also to critique it and to adapt it to their own place-specific contexts...* Aggressive and inclusive fostering of local capacity in science and technology must therefore be a centerpiece of any strategy for the sustainability transition.
>
> (1999: 297, emphasis added)

The concepts of degradation syndromes and science in places offer a means to acknowledge the role of locality in the experience and causes of environmental degradation. The statements of the NRC, above, however, suggest that acknowledging the importance of locality should mean adapting existing science to local areas, rather than allowing local areas to create their own framings and explanations. This approach reflects many of the assumptions of orthodox science in assuming it is necessary to understand environmental change first before applying these understandings to localities, rather than acknowledging that science and politics are coproduced. The "fostering of local capacity" discussed in the statement above does suggest some form of local inclusivity. But if this capacity is designed to increase the adoption of predefined scientific explanations, then it may serve to extend scientific networks from outside the locality, rather than support the development of locally framed environmental explanations.

The approaches of degradation syndromes and science in places may therefore be criticized by constructivist environmental scientists for not going far enough in localizing and diversifying environmental science. But in contexts where environmental risks may be agreed to be "global" by all localities, such extension of scientific knowledge may be extremely valid. The following section discusses the problems of distinguishing "local" and "global" environmental problems and science. But first, we assess some alternative means of diversifying and localizing environmental science.

Events ecology

Events ecology is an attempt to reform the frameworks of orthodox ecological science by integrating studies of environmental degradation with the application of a causal–historical analysis of human–environment relations. Events ecology adopts a critical stance to orthodox generalizations, or "laws" of nature, and adopts insights from the new, non-equilibrium ecology, including an awareness of historical changes in landscape, and historical framings in explanations themselves (Vayda, 1996; Vayda and Walters, 1999).

Unlike orthodox approaches to ecological explanation, events ecology seeks to identify individual ecological events and then seek explanations for these by posing a number of open-ended questions. This approach adopts a partly phenomenological attitude by seeking to understand "events" as local changes of significance, rather than as "facts" that can be incorporated into preexisting theories, or "factors" that imply events have causal significance. Researchers ask themselves a variety of counterfactual questions about the event in question, and whether antecedent events would have had influences upon it.

For example, Walters (2001) asked: "Why did mangrove planting emerge and spread in a specific island of the Philippines at a specific time?" Two possible explanations could be investigated by the questions:

"Were there any threats from storms that made mangrove planting advisable?" or "Were there any shortages of local construction materials?" In this example, the answer to the second was affirmative. In some senses, events ecology represents a successful attempt to integrate orthodox ecology with insights from historical anthropology and oral histories.

The objective of events ecology is to place the local historical events underlying ecological change as the primary objectives in seeking explanation, rather than meta-narratives associated with much political ecology such as the role of the state or the influence of global capitalism. Similarly, this approach also questions predefined ecological theories such as succession.

"Hybrid science"

The concept of "hybrid science" also seeks to integrate a variety of social and natural science techniques in the search for a locally framed explanation of environmental problems. Yet unlike events ecology, hybrid science can also be used to show the errors of predefined, orthodox "laws" of nature, or to reveal the institutional bases that different explanations rely upon. It gains its name because it uses a hybrid blend of different knowledge sources, and because it acknowledges Latour's arguments that physical objects are a hybrid blend of society and nature (Forsyth, 1996; Batterbury *et al.*, 1997).

A variety of studies have illustrated the principles of hybrid science. One study in Thailand (Forsyth, 1996), for example, questioned long-standing assumptions about environmental degradation in mountainous zones sometimes called the "Theory of Himalayan Environmental Degradation" (see Chapter 2). According to this "theory," population increase within traditional upland agrarian communities is expected to lead to the cultivation of steeper and steeper slopes, with resulting increases in deforestation and soil erosion, and claims that upland agriculture causes lowland problems of sedimentation and water shortages (see Eckholm, 1976). Much research in Nepal has already questioned these assumptions (see Thompson *et al.*, 1986; Hamilton, 1988; Ives and Messerli, 1989; Metz, 1991).

The study in Thailand sought to test the Himalayan "theory" by assessing what had occurred following the establishment of one village in a mountainous region and the consequent growth in population. The research used a variety of knowledge sources, including detailed oral histories and questioning; historical analysis of land-use, land-cover changes using aerial photographs and a Geographical Information System (GIS); and historic measurements of soil erosion on slopes. The findings revealed, however, that – counter to expectations – local farmers had *not* used steeper slopes more frequently, but instead had realized that these slopes were more vulnerable to erosion and declining soil fertility. Instead, farmers had used relatively less steep slopes (of below 20 percent) more frequently, where declining soil fertility (rather than erosion) was a greater problem. The study also showed – again counter to expectations – that overall forest area in the locality had actually increased since the establishment of a local land-tenure system in the 1970s (although forest quality had declined). Finally, the research suggested that much sedimentation onto the lowlands may not be the result of upland agriculture alone, but instead be caused by deep gullies that preexisted agriculture, and which were characteristic of granite land elsewhere in the tropics.

The implication of this study is that the orthodox beliefs concerning the ecological impacts of upland agriculture are highly overstated. The study also suggests that some proposals to manage degradation in this area – such as controlling erosion on steep slopes, or reforesting large areas – may not address underlying biophysical problems of either sedimentation or declining soil fertility, and may interfere unnecessarily with local livelihoods. (Other work in the region has also suggested that reforestation may not help water shortages: see Alford, 1992.) The use of so-called hybrid science therefore indicated the failings of orthodox generalizations, and gave greater insights into the complexity and different framings of environmental problems.

Other studies have also used principles of hybrid science. Fairhead and Leach's (1996) study of historic changes in the forest–savanna convergence zone in Guinea, for example, used a variety of satellite imagery and local oral histories to question orthodox beliefs about the prevalence and causes of deforestation (see Chapter 2). Sillitoe (1993, 1998) used in-depth chemical testing of soil nutrients to indicate the value of local soil conservation practices in Papua New Guinea. Robbins (1998) used GIS to compare the implications of different forest classification techniques on the identification of "forest" and "forest land." Rocheleau and Edmunds (1997) stressed the need to reframe environmental explanations in terms of local qualitative experiences of problems. Dahlberg and Blaikie (1999) adopted a similar approach to achieve "closure" on controversies about land degradation in Botswana, and, by comparing different accounts of environmental change side-by-side, to establish how and why they differ.

The aim of hybrid science is not to uncover biophysical change in a final and complete realist manner, but to reveal how far hegemonic discourses of degradation may actually match the experience of people within specific localities. The practice of hybrid science (or similar inquiries) may also empower political ecology by showing the political basis necessary for different accounts of ecological "reality" used to justify different policy options, and by giving the means to challenge orthodox environmental explanations that are presented as unassailable "truth."

Hybrid science also advances orthodox explanations of environmental degradation in two other key ways. It challenges fixed approaches to spatial scales of inquiry by conducting research within scales identified by people who experience environmental problems. Second, hybrid science is a form of integrated assessment that examines the social solidarities underlying environmental explanation simultaneously with the study of environmental problems themselves. As discussed in the following section, such dual appreciation of explanations, and the social structures that uphold them, is needed to understand how environmental explanations become dominant.

Implications for the analysis of "local" versus "global" environmental problems

The key arguments of this chapter have been that local alternatives to global environmental generalizations present advances in both local relevance and physical accuracy of environmental explanation; yet that all forms of environmental explanation reflect a wider social framing and solidarity such as a network or community. But do these mean that there are no such things as globally "real" explanations and "laws"? Similarly, do these imply that some global ecological concerns about environmental protection should be forgotten or dismissed because all people do not share them?

The answer proposed by this book to these questions is "no": there is no need to dismiss either the potential existence of globally applicable "laws" of nature, or the value of ethical statements that aim to revise environmental behavior worldwide. Instead, the objective of this book has been to demonstrate that existing environmental explanations are *not* universally accurate representations of reality, and that many political views of ecology are either not shared by all, nor as scientifically justified as commonly presented. For example, the belief of Lester Brown (2001) that it is an urgent ecological need to recover large areas of the planet with trees in order to stop erosion, sequester carbon, or restore biodiversity may actually not be biophysically accurate. Instead, such views reflect the historic co-evolution of science and politics from social and scientific networks that have prioritized trees (with all their implications of wilderness, heritage, and relative cost-effectiveness to alternative means of mitigating climate change, etc.), and which have not accepted alternative framings and empirical contradictions to these statements as acceptable influences. (See also the discussion of carbon-offset forestry in Chapter 6.)

In essence, this book has rejected current definitions of environmental reality, rather than the existence of reality itself. This statement underlies all realist scientific inquiry: the separation of epistemology (of knowledge about environment) and ontology (the underlying generating mechanisms and structures of environmental change). Simply questioning the relevance and applicability of currently dominant explanations of environment does not imply rejecting the principle of realist biophysical mechanisms, or indeed (in the most optimistic case) of the possibility of getting close to knowing them. But it does imply adopting a critical view to how these laws reflect different social and political influences, and – importantly for so many environmental orthodoxies – how they both support and restrict varying options for social and economic activities on the grounds of alleged scientific truth.

As discussed above (pp. 218–222), a crucial question underlying all inquiry is how far resilient explanations based on powerful networks may also coincide with accurate understandings of ecology in a realist sense (or, in Bhaskar's words, how far transitive structures may also reveal intransitive reality). In the global context, this question may mean asking if some apparently universal "truths" (such as perhaps that pure water freezes at 0 °C) are genuinely "brute facts" (Searle, 1995), or "immutable mobiles" – items that are universally accepted in different networks or cultures (Latour, 1987). (Brute facts were discussed in more detail in Chapter 4, and immutable mobiles were discussed in Chapter 7.) The first approach would suggest that such "facts" are reality; the second would stress the universality of social networks that make such a "fact" adopted in a maximum number of locations and cultures. Yet if apparent truth claims are not overtly attached to different social networks, these could represent a form of truth. Latour wrote:

> If they [truth claims about the real world] are not attached, people know exactly what nature is; they are objective; they tell the truth; they do not live in a society or culture that could influence their group of things, they simply group things in themselves; their spokespersons are not "interpreting" phenomena, nature talks through them directly.
> (1987: 206; also in Ward, 1996: 111)

Clearly, this book has argued that there are many political and methodological reasons to explain why certain environmental explanations have been portrayed as universal and unassailable, despite increasing evidence to the contrary. Yet even Critical Realists such as Roy Bhaskar, who have been considered more optimistic about scientific realism than many other critics of science, would still say that apparently universally agreed "facts" (if they do exist) are still reflective of social needs that have made such boundings of environmental processes and impacts appear as "truths." If these needs change, boundings will be different, and new apparent "truths" would emerge.

Moreover, this approach also accounts for the continued adoption of many statements we know to be false because they suit dominant social requirements. For example, most people on Earth talk of, and organize their lives according to, the principle that "the sun rises in the morning." Of course, this statement is patently false in a realist context, but there is no need to question the statement until such time as social needs make the epistemic boundaries of the statement inapplicable (such as engaging in space travel). This is an extreme example, but – as demonstrated by environmental orthodoxies – every day, many people worldwide are faced with receiving environmental advice or policies that bear more resemblance to the epistemic boundaries of the advisers than the recipients of the policies, with frequently severe implications on local livelihoods.

There is consequently a need to evaluate the applicability of environmental explanations at different scales *alongside* the existence of the social networks or solidarities that make these statements meaningful at that scale. Many current analyses of the importance of scale in environmental assessment do not make this link with social solidarities (e.g. Clark, 1985; Rosswall *et al.*, 1988; Wilbanks and Kates, 1999). DeHart and Soulé (2000), for example, conducted research to ask if the $I = PAT$ equation (see Chapter 2) works in local places, and answered in the affirmative. Yet this research failed to acknowledge that the framings of environmental problems were imposed from outside these localities, and consequently could be claimed to have made assumptions about the nature of environmental problems in these areas. Evaluating environmental explanations side-by-side with the different social networks and solidarities that uphold them (as in "hybrid science") may help to avoid the projection of environmental explanations onto people or places that find them irrelevant. Integrating social and physical science in this way may also increase our understanding of environmental change.

Table 8.1 shows a simple classification of environmental explanation according to both "local" and "global" scales, and the brute and institutional facts of Searle (1995). The objective of the table is to suggest a framework for assessing the universality of different environmental explanations according to allegedly universal biophysical properties, and the institutional framings placed upon more general aspects of environmental change by different networks or solidarities. The table is not a rigid definition in realist terms of what is a local or global problem, but a guide to different ways of assessing concepts within environmental discourse according to different claims to be seen as either "local" or "global"

Table 8.1 "Local" and "global" environmental problems defined in constructivist terms

		"Local"		"Global"
"Brute facts"	1	Local physical variations (e.g. *aridity, tectonic uplift, infiltration rates, soil erodibility*)	2	Uniform physical properties (e.g. *freezing points, thresholds of toxic pollution such as Persistent Organophosphate Pollutants*)
"Institutional facts"	3	Local cultural adaptations/problems (e.g. *shifting cultivation, pastoralism, environmental vulnerability*)	4	Globally-identified problems (e.g. *global deforestation, anthropogenic climate change*)

Source: the author.

scientific statements. Under this scheme, universal biophysical properties, for example, may be seen to be global brute facts, and "global environmental problems" (as discussed in Chapter 7), such as anthropogenic climate change, are global institutional facts, representative of framings and social networks rather than indicating a universally real ecological risk for each location worldwide.

For example, using this diagram, it is possible to indicate environmental orthodoxies (or generalized environmental explanations in common discourse) as resting in category 4 (as reflecting much institutional framing), but commonly portrayed as in category 1 (as unassailable universal truths). Category 3 (or local institutionally defined environmental problems or adaptations) should be afforded more credibility. Ideally, under realist approaches to science, information in categories 1 and 2 should be used as practical guidance to environmental policy. But it should be noted that even apparent brute facts reflect social framings and needs, and are also contingent on many assumptions. For example, the example of the "brute fact" that water freezes at $0\,°C$ assumes water is free from impurities, and that the temperature is at sea level, which itself is in flux).

The objective of this table is not to suggest that there are clear environmental problems or properties that can be placed in each column. The aim is to indicate that discussions of how far scientific explanations are accurate at a local or global scale need to acknowledge how (and by whom) these different scales are defined. It is important to note that epistemologizing truth statements in this way does not suggest that local practices should be seen as feasible in all other locations; or that global environmental opinions (such as that forests or wildlife should be preserved for their own right) should be denied as desirable social choices. Instead, there is a need to see the fallibility of many of the alleged scientific truths that are used to legitimize many environmental policies and alleged imperatives. As argued throughout this book, failing to see the shortcomings of environmental science may only serve to restrict local livelihoods and fail to address underlying biophysical causes of environmental problems. The next chapter builds on this analysis of scientific practice, by considering how political reforms and new arenas of debate can empower such new approaches to environmental science, and assist the transparency, accountability, and relevance of environmental explanations.

Summary

This chapter has attempted to answer some of the questions raised in previous chapters concerning the ability to enhance both social relevance and biophysical accuracy of science by focusing on debates concerning scientific method. This discussion contributes to a "critical" political ecology by seeking to democratize environmental science by revealing the tacit politics contained in scientific statements, and by exploring the possibility for challenging and even overturning dominant environmental explanations.

The chapter argued that the universalizing, propositional statements about environmental explanation associated with the frameworks of orthodox science, relying on the correspondence theory of truth, are inadequate to acknowledge the institutional nature of many environmental problems as experienced by different people in a variety of locations and circumstances. As an alternative, the chapter reviewed debates within semantic and Critical Realism that allow some means of integrating social contextualization of environmental change with biophysical accuracy and the possibility of refuting existing, orthodox explanations.

The chapter presented some examples of different methodologies for more realist and socially relevant forms of explanation. These methodologies – such as events ecology and hybrid science – consider local framings of external ecological reality from both the scientist and local citizens' perspectives, with the intention of increasing both our awareness of these framings, and greater relevance of biophysical explanations to different social needs. It was argued that using such techniques allow a greater democratization of scientific practice – because they increase the transparency of explanations, and allow a local approach to explanations on terms determined by people experiencing problems, rather than according to meta-narratives of explanation or fixed spatial scales. The chapter made it clear that democratizing science in this way does not mean rejecting "global" environmental explanations or concerns in principle, but instead the need to understand better how, and by whom, such explanations are claimed to be global or local.

The next chapter now builds on this discussion by examining the political factors underlying the evolution of scientific networks and public access to scientific debate.

9 Democratizing environmental science and networks

This chapter now examines the political institutions and procedures that can increase transparency and public participation in environmental science. If environmental science reflects social and political framings, then how can political debate be reformed to make these framings more visible and relevant to more people? How can people not represented in the framings of environmental explanations be empowered to influence environmental science?

The chapter will:

- discuss the dilemmas of enhancing public participation in environmental science. Some observers have suggested that increasing participation may democratize scientific debates by acknowledging diverse forms of expertise, and by building trust in science. Against this, critics have suggested that scientific consensus and certainty are based upon the enforcement of networks and boundaries that are, by definition, exclusionary.
- examine how environmental assessments and scientific organizations may increase transparency and accountability. Such actions may form new ways of regulating the production of scientific knowledge, and may improve the communication of scientific findings from scientific networks to other groups.
- consider how alternative scientific networks or institutions may be empowered, especially from marginalized social groups or in developing countries. Such networks may often not seek to impose predefined "laws" or explanations of environmental degradation, but build local capacity to achieve inclusive political debate about the management of resources and environmental risks.

In common with Chapter 8, this chapter therefore presents practical means to address some of the problems of environmental science discussed in earlier chapters. The aim of a "critical" political ecology is to conduct environmental politics without using a priori definitions and explanations of environmental degradation. This chapter helps achieve this objective by describing political arenas that allow the discussion of environmental objectives at the same time as acknowledging the political embedding of environmental science.

Scientific expertise and public participation

The earlier chapters of this book listed a variety of problems in using environmental science as a politically neutral basis for environmental policy. The aim of this chapter is to examine how far these problems can be addressed by democratizing the networks and institutions that produce science.

Chapter 8 started the analysis of democratizing environmental science by looking at ways to reform scientific methods themselves. In that chapter, "democratization" was defined as revealing the tacit politics within scientific statements, and in diversifying and localizing universalistic scientific explanations or "laws" of nature. How can similar reforms be made to public debate about the purpose of environmental science, and to scientific networks and institutions?

This objective raises a number of associated questions. First, it is necessary to consider the definition of networks. As discussed in Chapters 4 and 6, networks may refer to the people, actors, and organizations that uphold scientific practices or beliefs. Yet networks may be interpreted formally in the sense of clearly identifiable scientists and institutions, or more complexly in the sense of Actor Network Theory, or the extended translation model of science (Callon, 1995) that refers to the epistemological impacts of different networks on the definition and hybridization of physical objects as part of networks. Democratizing networks may therefore imply not simply changing the participants, but reconsidering the knowledge claims and approaches to biophysical objects in general.

Second, is it more effective to seek to reform existing networks, or to establish alternative, co-existing networks? Many scientific networks depend on the claim to represent expertise. Yet much research in cultural ecology and science studies has illustrated alternative forms of expertise within "lay" people such as farmers (e.g. Batterbury, 1996; Wynne, 1996a). Third, how far does the discussion of "science," as commonly portrayed in popular debates, itself foreclose what is considered to be expertise or legitimate knowledge?

These questions are addressed throughout this chapter. This initial section outlines some general dilemmas in discussing democratization of scientific networks. One crucial problem is in blending public participation with scientific certainty. Will science be considered effective if it includes dissenting voices and diverse opinions? Does environmental science seek to restrict public criticism in order to make its work seem more successful?

This section considers the political barriers to public participation in environmental science, and then examines some dilemmas of using orthodox concepts in environmental politics for analyzing the democratization of science. The following sections then assess means of reforming scientific institutions, and building alternative scientific networks.

Scientific networks and consensus

This book has discussed many of the problems of believing that environmental science can be separated from politics, or that existing explanations of environmental problems can be used as a neutral backcloth for political debate. The most extreme position of this belief can be expressed as a belief that science provides the "facts" for policymakers to use. This belief has been described as science "speaking truth to power" (Price, 1965), or the model of "synoptic rationality" in which decisions are made based on collating "all the facts" (Collingridge and Reeve, 1986: 63).

This book has argued that such approaches are clearly inappropriate for describing the evolution of scientific knowledge and its interface with policy. As discussed in previous chapters, scientific knowledge is clearly coproduced with political activism; the boundaries of scientific networks can be exclusive and related to political interests; and the very construction of scientific explanations of complex biophysical changes rely on social framings often rooted in history and language. How can the interface between science and policy be more effectively explained?

One of the first approaches to complicating the rigid separation of science and politics was the concept of "transcience" (Weinberg, 1972). Transcience can be described as the zone between pure science and pure politics that consist of topics where scientific experiments cannot reduce uncertainty to known levels. "Uncertainty" is commonly defined as the situation where we don't know what we don't know, whereas "risk" is used to define probabilities that can be calculated (e.g. Douglas, 1985). Under "transcience," science cannot provide answers sought by policymakers, and so policy criteria are used to direct the research and models chosen. For example, policy criteria such as protection of public health may dictate the inquiries into nuclear physics in order to identify the risks posed by nuclear power plants (Jasanoff and Wynne, 1998: 9).

The concept of transcience highlighted that certain topics of public policy based on scientific knowledge would remain uncertain, and shaped by topical political concerns, or the actions of influential politicians or scientists (see also Cobb and Elder, 1972; Kingdon, 1984). During the Second World War, for example, the British writer, C.P. Snow expressed concern about the "court politics" between Churchill and his leading scientists, which gave them "more direct power than any scientist in history" (Snow, 1961: 57–63; in Weingart, 1999: 153). The speech of James Hansen of the National Aeronautical and Space Administration to the US Congress in 1988 on the topic of global warming is a further case of a prominent scientist shaping political action regarding a scientifically uncertain topic.

Blurring lines between scientific expertise and political advocacy like this clearly raises questions concerning the alleged separation of science and politics (Jasanoff, 1990). When science advisers become integrated into policy debates, their status within the frameworks of orthodox science as neutral observers may change:

What transforms scientific knowledge into an expert appraisal is its inscription within the dynamics of decision-making. Yet this inscription, at least in the case of scientifically and politically complex questions, immediately leads the scientist to express opinions or convictions which (however scientifically founded) cannot in any way be identified with knowledge in the strict sense which science generally affords this term.

(Roquelpo, 1995: 170; also in Weingart, 1999: 157)

It is also worth asking how certain topics become seen as "uncertain," because these indicate where and how existing expertise is considered insufficient. Funtowicz and Ravetz (1985, 1992, 1993) proposed one model that advanced on the concept of "transcience" by proposing three levels of scientific certainty (see Figure 9.1). The model suggested that most uncertain and intractable policy dilemmas come when high decision stakes coincide with high systems uncertainty.

As shown in Figure 9.1, at the lowest level of uncertainty and decision stakes, the activities of "normal" science (in the terms of Kuhn's paradigms, see Chapter 3) are usually sufficient to provide legitimate information. Beyond this level, new participants and skills need to be consulted in order to resolve policy dilemmas (Funtowicz and Ravetz called this "professional consultancy"). At the highest levels of uncertainty and decision stakes, scientific experts may need to share inquiries with amateurs, stakeholders, or professions quite dissimilar to their own. Such "post-normal science" is seen, optimistically, to generate sufficient reframings in order to reduce uncertainty to a level where professional consultancy or "normal" (or applied) science can operate again.

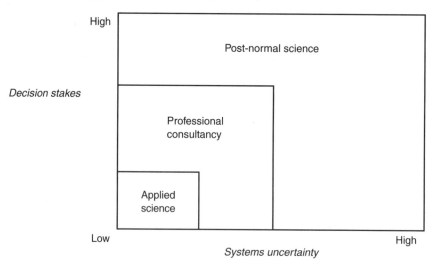

Figure 9.1 Three kinds of science

Source: Funtowicz and Ravetz, 1985, 1992.

The model of "post-normal science" is attractive for explaining political barriers to scientific participation because it shows ways in which divisions may be made between scientific progress and public consultation. But it also may be criticized for a number of reasons. As Jasanoff and Wynne (1998: 12) note, this model may be questioned because it assumed that uncertainty and decision stakes might be independent of each other; or because it implies that reducing uncertainty in post-normal science may simultaneously decrease decision stakes. Furthermore, MacKenzie (1990) and others have claimed, reductions in uncertainty occur if decision stakes are reduced for unrelated reasons. MacKenzie's model of the "certainty trough" (see Figure 9.2) indicates that "uncertainty" is also dependent on access to, and communication with, expert institutions, and that such factors may vary between different social groups.

A more political discussion of scientific uncertainty therefore highlights the political barriers to reaching public consensus about the veracity of specific explanations or scientific findings. "Uncertainty" is not just the statistical probability of successful explanation achieved via science, but also the degree of public access to, and participation in, the production of knowledge. Such factors also account in part for the emergence of institutionalized environmental explanations (or environmental orthodoxies). By definition, such explanations are seen to be unchallengeable "truth," yet in practice contain many aspects of uncertainty or irrelevance to people where they are applied (see Chapter 2).

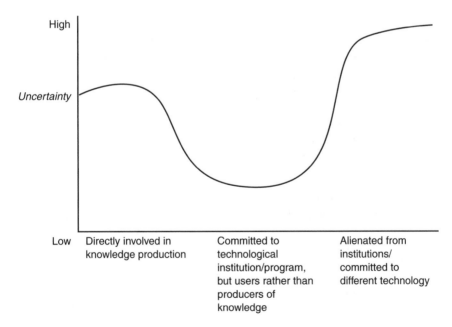

Figure 9.2 The certainty trough

Source: MacKenzie, 1990; Jasanoff and Wynne, 1998: 13.

These political approaches to scientific uncertainty indicate the important role of scientific networks in constructing where uncertainty is seen and not seen. Yet, as discussed in earlier chapters, the emergence of scientific networks and expertise is in part directed by the political attention to different problems, and who participates in the analysis and regulation of findings. Consequently, some sociologists of science have argued that risk analysis is often far less reliable than alleged because regulatory scientists and government officials tend to depict risk assessments as if they are fully determinate, and understate both the complexity of biophysical factors and their own ignorance about them (Wynne, 1992; van Zwanenberg and Millstone, 2000: 262). Such problems have also been associated with monopolistic science or expert institutions – such as state science bodies – where such criticisms are resisted (de Jong, 1999). Accordingly, the optimistic "post-normal science" of Funtowicz and Ravetz may not occur because practitioners working within "normal science" often exclude dissenting or worried voices. As Collingridge and Reeve noted, the enforcement of consensus in this way may have severe implications on the ability for science to challenge the status quo:

> When research is directed from outside on problems where disciplinary distinctions are blurred, and where any proposed solution will have a high error cost, consensus is quite impossible. The price of a super-efficient normal science is the impossibility of scientific research exerting any significant influence on policy directions.
>
> (1986: 147)

Such statements have important implications for the inclusivity of science when "uncertainty" is perceived to be a topic of political concern. The search for a clearly agreed solution, within certain networks, may dissuade transdisciplinary research (or "hybrid science," see Chapter 8). Furthermore, the boundaries between what is considered science, or acceptable science, may be further enforced, with the further implication that expert institutions working on reducing uncertainty become important boundary organizations (see Chapter 6). As Jasanoff, in her study of US science advisers as policymakers, noted:

> By drawing seemingly sharp boundaries between science and policy, scientists in effect post "keep out" signs to prevent nonscientists from challenging or reinterpreting claims labeled as "science." The creation of such boundaries seems crucial to the political acceptability of advice... Curiously, however, the most politically successful examples of boundary work are those that leave some room for agencies and their advisers to negotiate the location and meaning of the boundaries.
>
> (1990: 236)

Scientific uncertainty, therefore, is not simply a matter of calculated risk, but is also a function of public participation in the generation and dissemination of knowledge. Yet, the desire to reduce apparent uncertainty by asserting the role of professional (or applied) science may paradoxically increase that uncertainty by reinforcing barriers between scientists and lay people. Furthermore, such reassertion of barriers may also result in privileging forms of knowledge produced by scientists relating to universal properties and inference, rather than local contextualization and meaning of such general biophysical changes gathered by consulting with local people. Ironically, such reassertion of the boundaries of orthodox science contributes to the production of universal "laws" – or environmental orthodoxies – that do not acknowledge local contexts or vulnerabilities to biophysical changes.

So, how can social concerns about environmental risk and uncertainty be communicated to scientific networks in ways that do not reiterate the boundaries of orthodox science? Also, how can such communication take place in ways that acknowledge the coproduction of environmental science and politics? These questions are addressed throughout this chapter. The next section, however, considers some further dilemmas for democratizing environmental science based on challenging some conventional approaches to ecological "community" and "rationality" that have significant implications for the production of more democratic environmental science.

Challenging orthodox concepts of environmental democratization

Democratizing environmental science networks may therefore face a number of political barriers. Yet there is also a need to reconsider some orthodox concepts through which environmental democratization may be achieved. This section briefly summarizes some problems with these concepts when seeking to democratize environmental science.

Much discussion of environmental democratization has focused on the adoption of predefined notions of ecological rationality. For example, Mason wrote:

> Environmental democracy is defined as a participatory and ecologically rational form of collective decision-making: it prioritizes judgments based on long-term generalizable interests, facilitated by communicative political procedures and a radicalization of existing liberal rights.
>
> (1999: 1)

This statement correctly draws attention to the tacit political models of democratization contained within environmentalism, yet does not draw equal attention to the tacit assumptions about ecology (a point Mason acknowledges). As discussed in Chapter 5, much discussion of ecological rationality as a liberatory force against the instrumental rationality of oppressive industrial and state regimes (e.g. Eckersley, 1992) may be associated with a coproduced form of ecology that essentializes economic growth with environmental degradation and lost wilderness. This approach to ecological rationality may avoid alternative approaches from groups not represented in the construction of this rationality. Furthermore, such an approach may even restrict the livelihoods of unrepresented groups if it considers their actions to be environmentally damaging.

Second, the concept of "community" has also been used to indicate a sense of belonging and locality associated with democratic governance. Ecologists have used the concept to also refer to collections of biological species, or for groupings that lie between the individuals and entire populations. As reported in Chapter 1, such "community"-based models of ecology encouraged some of the early political ecologists to urge the limitation of individual human actions because of the impact on the community. Eugene Odum (1964: 15), for example, wrote: "[ecology] deals with the structure and function of levels of organization beyond that of the individual and species." Similarly, Paul Sears, writing in the same volume, commented that "by its very nature, ecology affords a continuing critique of man's [sic] operations within the ecosystem" (1964: 12).

These uses of the concept of "community," however, contain a variety of assumptions about togetherness that can be challenged (see Leach *et al.*, 1997; Agrawal and Gibson, 1999). Such statements suggest that human activities may be assessed in the aggregate and within ecological limits. But they do not acknowledge the ways "communities" can contain various subcategories, or how such limits are defined. The concept of "community" can be differentiated biophysically in relation to the institutional scales within which explanatory statements can be made (such scales refer to the semantic or epistemic boundaries that control the production of apparently true statements, see Chapters 3 and 8). "Community" can also be differentiated socially in reference to divisions such as gender, age, race, or class that make shared values less easy to predict (see Chapter 4). Indeed, as discussed in Chapter 7, the notion of "global" ecology that was enhanced by the first photographs of the Earth from space does not necessarily indicate a unified environmental perception, or the existence of global limits that are equally present or meaningful to all people (Yearly, 1996; Jasanoff, 2001). As discussed in Chapter 6, the words "institution" or "network" may be more useful as means to describe unified norms or experiences than "community," as these words indicate the contingent nature of what is shared by individuals rather than the automatic assumption that a community may really exist (Berry, 1989; O'Riordan and Jordan, 1999).

Third, notions of environmental "expertise" may also be questioned on a number of grounds. Clearly, as discussed above (pp. 233–237), the frameworks of orthodox science portrayed a distinction between "expert" and "lay" knowledge that can be challenged when inquiry concerns topics of local practice or perceptions of risk not shared by people formally classified by experts (see Wynne, 1996a; Fischer, 2000; Tesh, 2000). Indeed, much research in developing countries has highlighted how many farmers in supposedly threatened locations have adopted practices that maintain environmental protection despite rising population and affluence (Batterbury and Forsyth, 1999). There is a need to acknowledge a greater role of expertise among people historically identified as "lay." Such acknowledgment of expertise does not, of course, imply that all individuals have equal knowledge or training about specific risks, but that the experience and framing of risks by "lay" people is crucial for the understanding of risk (see debate between Daly, 1991 and Turnbull, 1991; Irwin, 1995).

These challenges to orthodox concepts of environmental expertise may also be directed at some attempts to redress the balance between "experts" and "lay" people. Some authors – and particularly the eco-anarchists such as Murray Bookchin – have argued that bureaucratic politics and instrumental industrial growth has caused a rationalization of expertise within the state apparatus, and a disturbing distance between experts and lay people. (Such arguments are clearly influenced by the sociologist Max Weber.) As a response, some observers have suggested that environmental decision-making should be diversified to the local level in order to reverse such rationalization, and to integrate environmental management with local experiences (see Eckersley, 1992; Murphy, 1994; Smith, 1996).

There is a need, however, to question how far such calls for devolution are also based on a perception of ecology as a response to instrumental rationality rather than a more contextualized assessment of different knowledge claims about complex biophysical processes and risks. It may be reasonable, for example, to see entrenched scientific expertise as characteristic of liberal democratic government (e.g. Ezrahi, 1990). But to see such entrenched government as itself the cause of ecological degradation (e.g. Beck, 1995) may be to essentialize instrumental rationality with environmental degradation in ways that overlook the complexity of biophysical change and our experiences of it (see Chapter 5).

Fourth, a common approach to environmental democratization has discussed the need to reform the "public sphere" by communicating new ecological discourses. This generally Habermasian approach defines the public sphere as a formal or informal arena in which environmental norms and policies may be discussed and agreement reached (see Habermas, 1987; Dryzek, 1990; Calhoun, 1992). In particular, some authors have pointed to the role of environmental social movements and activism as means of revising environmental policy in favor of groups seeking to regulate dominating state or industrial interests (e.g. Eder, 1996; Mol, 1996; Blowers, 1997; Brulle, 2000).

Such generally optimistic statements about the effects of public debate have been criticized by more constructivist writers who suggest these approaches overlook the role of communicative institutions such as courts, regulatory agencies, expert bodies, and news media in creating, and shaping multiple "public" spheres (Jasanoff and Wynne, 1998: 27;

Edwards, 1999). Indeed, theorists influenced more by Foucault have pointed out how the language adopted in such so-called public spheres – such as cost–benefit analysis by economists, or legal arguments in courts – may present further barriers to participation, even in inquiries that are claimed to be "public" (Tewdyr-Jones and Allmendinger, 1998). Indeed, under such circumstances, expert power and state power may be inseparable (Turner, 2001). The well-known Habermasian task of rationalizing the public sphere to enable greater public participation may therefore overlook the numerous barriers to participation, and the deliberate creation of co-existing public spheres by various organizations to establish different forms of legitimization and support for various objectives.

The general implication of these challenges to orthodox approaches to environmental democratization is to move debate away from "one" pre-defined notion of ecological rationality for a supposed unified "community," toward the acknowledgment that environment may be perceived and valued in various forms, from different people, in diverse political arenas. Such plurality of perspectives does not imply that all perceptions are equally powerful, or that each perspective may co-exist without contradictions. As discussed in Chapter 8, the recognition of diverse perspectives does not imply cultural relativism (or the belief that all are equally valid). Indeed, the cross-comparison of some perspectives may allow a form of scientific progress (see also de Jong, 1999: 198; van Zwanenberg and Millstone, 2000). Plural institutions and networks may exist at a variety of scales, and may exist in formal or informal political settings such as within state bureaucracy or *de facto* adopted by farmers or citizens regardless of official recognition. Yet, as argued by sociologists such as Giddens (1990), Beck (1995), and Seligman (1997), for example, the perception and experience of risks may be increasingly individuated because of the growing multiplicity of roles and tasks performed by individuals rather than groups.

A further implication is – as discussed in Chapter 7 – environmental risk cannot be explained simply in terms of linear causality or biophysical change alone. Instead, it is necessary to acknowledge the institutional shaping, communication, and response to perceived environmental threats, of which addressing perceived biophysical events is important, but not exclusive. As Jasanoff wrote:

> The social sciences have deeply altered our understanding of what "risk" means – from something real and physical if hard to measure, and accessible only to experts, to something constructed out of history and experience by experts and laypeople alike ... Trying to assess risk is therefore necessarily a social and political exercise, even when the methods employed are the seemingly technical routines of quantitative risk assessment ... it makes very little sense to regulate risk on the basis of centralized institutional authority, insulation from public demands, and claims to superior expertise.
>
> (1999: 150)

Table 9.1 Models of risk perception for public policy

Model	Epistemology	Location of authority	Policy prescription	
			Style	*Mechanism*
Realist	*realist*	*expert communities*	*managerial*	*expert advice*
Constructivist	*constructivist*	*social/interest groups*	*pluralist*	*public participation*
Discursive	*constructivist*	*professional discourses*	*critical*	*social movement*

Source: Jasanoff, 1998.

Indeed, Jasanoff (1998) has proposed that risk perception may be classified into three basic models which draw attention to different modes of analysis and expertise (see also Fischer, 2000). The Realist (or positivist) model lends most emphasis to identifying the underlying biophysical cause of risk, communicated via expert advice from expert communities. The constructivist model acknowledges the constructed and plural nature of risks, but seeks to achieve understanding via social inclusion and public participation. The discursive model is also constructivist in epistemology, but acknowledges the political barriers to participation posed by the "vernacular" or language of risk assessment permissible within specified (public) arenas (see Table 9.1).

The objectives of a "critical" political ecology are to assess how far political practices may lead to alternative explanations of environmental reality. According to these objectives, the constructivist and discursive models of risk analysis are most applicable for understanding the evolution of environmental science. Yet these models also imply adopting critical stances toward orthodox concepts of community, rationality, and expertise to see how far such concepts contribute to the coproduction of environmental science and politics.

The following sections of this chapter now assess the challenges in building more locally relevant alternatives to the universalizing "laws" adopted under the Realist model, by either reforming powerful scientific institutions, or by empowering the emergence of alternative institutions from less powerful sectors of society.

Regulating scientific institutions

The preceding discussion listed a number of challenges to the democratization of environmental science, and to orthodox approaches in environmental politics to achieving "environmental democratization." How can environmental politics acknowledge these concerns in order to achieve an effective reform of environmental science and networks?

Perhaps the most obvious means of reforming environmental science is to seek change within the institutions and networks of orthodox environmental science. Changes might include seeking greater reflexivity or transparency within scientific and expert organizations, in order to acknowledge underlying social and political values, or to show the decision-making process through which advocacy statements are made. As discussed in Chapters 3 and 6, Mertonian norms of regulating scientific knowledge (based on principles such as sharing knowledge, or inviting and responding to criticism) may be considered neither accurate nor sufficient in terms of describing the evolution of scientific statements. Acknowledging public concerns about environmental science – or enhancing public participation in the framing of inquiry – may improve both the perceived relevance of science, and public trust in scientific institutions.

Part of the problem also lies in reconsidering the role of science itself. Historically, the word "science" was enough to denote the source of respectable "truth." For example, the future US President, Woodrow Wilson, spoke in 1896 of "calm Science seated there, recluse, ascetic, like a nun" (Wilson, 1896). Yet today, public trust in science as an institution is much less, caused in part by the increasing diversity and individualization in roles performed by people, and what this means for receiving knowledge that can assist with these roles (Giddens, 1990; Seligman, 1997). Indeed, according to Ulrich Beck, such individuated experiences of risks, and declining levels of public trust are key aspects of "Risk Society." As Beck wrote: "[science is] more and more necessary, but at the same time, less and less sufficient for the socially binding definition of truth" (1992: 156).

The challenge facing public organizations under such conditions is to ensure that decisions are made in the public interest, or with sufficient ability for worried citizens to voice concerns and feel they have been listened to (Reich, 1990; Fischer, 2000). Those seeking to reform scientific institutions in this optimistic way, however, may also encounter some significant obstacles. First, the desire to enhance participation in scientific inquiry often implies seeking to soften, or diversify the boundaries around scientific inquiry and networks of scientists. Yet, as discussed above, the achievement of consensus and meaningful causal statements can often depend on the existence and enforcement of epistemic boundaries, and the status of perceived "experts" within orthodox, "normal" science. Second, some challenges to the credibility of scientific practice run the risk of causing great offence to professional scientists, who see such criticisms as inaccurate and unfair, especially if made by people untrained in physical science. Third, the definition and adoption of "participation," "transparency," and "accountability" are themselves contested, and will reflect different models of democracy or justice that should themselves be worthy of analysis (Mason, 1999).

This section considers means of reforming existing (and often orthodox) scientific institutions. The discussion forms part of wider debates about public understanding of science, but there is no attempt to review all of these debates in this book (see Irwin, 1995; Irwin and Wynne, 1996; Dierkes and van Grote, 2000). Instead, this section focuses on institutional structures through which expert organizations can become more reflexive about political framings of science, and more accountable to public concerns about how science has been conducted.

The "science wars" and direct challenges to science

Perhaps the most direct form of public criticism of orthodox scientific institutions in recent years was the number of intellectual exchanges during the 1990s known as the "science wars" (see also Chapter 1; Jasanoff, 2000; Segerstråle, 2000). The "science wars" went beyond the expected level of "conjecture and refutation" associated with orthodox science because it was a discussion about science in general, and its accountability, governance, and claims to represent truth. Indeed, the implications of the "wars" are still felt today in debates about large-scale environmental explanation (such as Global Climate Models, or Land-Use-Cover-Change measurements), and challenges from more localized and contextualized approaches such as in this book.

It is commonly reported that the "science wars" began after the decision to terminate public funding for a superconducting supercollider particle accelerator during the 1992–1993 session of the US Congress. This decision quickly led to wider debates about the purpose of public funding for science, especially after the end of the Cold War, and the ability for the public to influence such decisions. One memorable phrase from this time was coined by Daryl Chubin, a senior analyst at the US Office of Technology Assessment, who described the self-serving arguments used by physicists to justify public funding for equipment such as the supercollider as "quark barreling" (see Fuller, 2000: 134). The debate led onto wider questions of sociology and politics of scientific knowledge, and an angry backlash from orthodox scientists who considered such questions as irrational. These debates comprised a variety of viewpoints and levels of disagreement, ranging from the generally uncontroversial to the overtly confrontational. For example, Mary Midgley wrote: "Science education is now so narrowly defined that many scientists simply do not know that there is any systematic way of thinking besides their own" (1992; in Nader, 1996: 13).

Barnes *et al.* wrote:

> The scientific profession possesses considerable authority in modern societies, and indeed wherever "science" is identified and designated as such, the implication is that something especially trustworthy or reliable is being described. Such authority is of course of inestimable value to individual scientists, and they have a vested interest in its maintenance. They can be expected to police the existing boundaries of science, to avoid the intrusion of whatever may detract from its reputation, and to seek to expel anything potentially disreputable which arises within.
>
> (1996: 140)

And Jasanoff:

> Like the strict constructionists of the Constitution, the critics of
> science studies ascribe an almost mystical primacy to the original
> intent of scientific authors, from Bacon to Einstein to figures of our
> own day – an intent, moreover, that only other scientists are licensed
> to decipher safely. The critics' constant fear is that science studies mis-
> represents the words and works of scientists, citing them out of
> context or distorting them through unnatural juxtapositions. Not for a
> moment do they share the humanist's sense of the fluidity and ambigu-
> ity of language – even scientific language – let alone of texts, artifacts,
> and agents being connected in complex webs of meaning. It is no
> wonder that, from this standpoint, humanistic readings of science are
> so readily construed as assaults on the truth.
>
> (1999: 498)

A common response of some ardent defenders of orthodox science was to
reiterate the boundaries of scientific networks, and to question the motives
of the critics. Most famously, Gross and Levitt wrote:

> To put it bluntly, the academic left dislikes science ... Within the
> academic left, hostility extends to the social structures through which
> science is institutionalized, to the system of education by which profes-
> sional scientists are produced, and to a mentality that is taken, rightly
> or wrongly, as characteristic of scientists. Most surprisingly, there is
> open hostility toward the *actual content* of scientific knowledge and
> toward the assumption, which one might have supposed universal
> among educated people, that scientific knowledge is reasonably reli-
> able and rests on a sound methodology.
>
> (1994: 2, emphasis in original)

This defense of scientific knowledge and methodology also led to criticisms of many other political concerns that the science defenders saw to be
based on ideology rather than on science. Environmental concern reflecting deep-green (or ecocentric) beliefs, for example, was criticized because:
"in part it [environmentalism] is an act of ritual abasement before a per-
sonified "nature" rather than a program of practicable measures for
dealing with concrete environmental dangers" (Levitt, 1999: 132).

Environmentalism is therefore criticized because it may lead to the
knee-jerk rejection of technologies such as nuclear power that could
potentially reduce the causes of anthropogenic climate change. Furthermore, Levitt noted:

> Environmentalism has no credentializing process. Scientific acumen is
> not a requisite for participation or even leadership, and considerable
> prominence has been given to figures whose scientific competence is

nearly non-existent ... in this atmosphere it is difficult to weigh choices in the light of the facts that scrupulous science provides.

(ibid.: 133)

This book, of course, might concur with the need to question the use of scientific statements by different environmentalists in order to legitimize environmental policy on highly contested themes (see Chapters 2 and 5). Yet, unlike Levitt and other science warriors, this book would still question the ability for environmental science to produce socially neutral facts.

These criticisms of environmentalism, however, have been easily associated with "brownlash," or the attempts to discredit environmental concern by businesses unwilling to adopt environmental regulation, and has led to further responses from pro-environmentalists. Paul and Jane Ehrlich, for example, published the pro-environmentalist book, *Betrayal of Science and Reason: How Anti-Environmental Rhetoric Threatens our Future* (1996) partly as a way to demonstrate that environmentalism may also adopt the frameworks of orthodox science that mathematicians such as Levitt sought to protect. Other authors have similarly written to protect the concept of "nature" as a scientifically legitimate basis from which to formulate environmental policy (Soulé and Lease, 1995; see Chapter 1).

As discussed throughout this book such approaches to science and nature overlook diverse problems in how far these concepts can represent politically neutral representations of biophysical reality. Furthermore, under such discussions of environmentalism, much debate from science studies has been inappropriately cast as attempts to legitimize bodies that seek to avoid environmental regulation. The objectives of science studies, instead, are to highlight how tacit framings and institutions shape, and are shaped by, the practice of science. This statement can be made about the so-called "facts that scrupulous science provides" or the alleged "science and reason" claimed by both critics and defenders of (deep-green, or ecocentric) environmentalism.

The debates of the "science wars" helped to publicize some of the concerns of science studies about entrenched orthodox scientific networks. Yet these debates have also encouraged stereotypical images of "pro" and "anti" science, such as the mistaken belief that all professional scientists are uncritical of scientific procedures, or that science critics are opposed to science in any form. Alternative approaches have sought to reform procedures within specific organizations or assessments.

Reforming national and international organizations

Further attempts to reform environmental science have focused on increasing the transparency and participation within existing scientific or policy organizations. Such organizations may be at national or international levels, and include power over research funding, or dissemination of environmental expertise.

Fuller (2000) lists three ways to govern science at the stage where public funding is allocated to competing research proposals. The first model, of "finalization" (Schaefer, 1983), assumes that mature science will be governed through a combination of inertia within science's own networks and the overseeing directions imposed by the state. The second model, "cross-disciplinary relevance," was proposed by Alvin Weinberg (1963) (of transcience fame), which optimistically urged that public funding be prioritized for research providing maximum benefit for the widest selection of disciplines. The third model is "epistemic fungibility" (Fuller, 1993), which acknowledges that cross-disciplinary relevance often does not influence research funding because grant applications are often made to each discipline's own peer group (see also Collingridge and Reeve, 1986).

The model of epistemic fungibility also points out – in common with some debates in science studies – that scientific disciplines are also networks of people who are experts within their discipline, but are lay people outside disciplines. Achieving epistemic fungibility therefore implies acknowledging the different constituencies who control decisions affecting the allocation of research funding. Yet this objective may be challenged by the vested interests of closed networks. Fuller wrote:

> Were disciplinary communities made to be routinely accessible to each other, then much of the aura of expertise and esoteric knowledge that continues to keep the public at a respectful distance from scientists would be removed.
>
> (2000: 142)

Environmental organizations may also be examined for transparency and accountability in the formulation of environmental expertise. Chapter 6 began some discussion of the role of "boundary organizations" as institutions that control the coproduction of science by being accountable to different networks in science and policy. This discussion may be extended to organizations' responses to criticism from environmental groups.

One study by Landy *et al.* (1994), for example, focused on the United States Environmental Protection Agency (EPA). Landy *et al.* argued that the EPA was characterized by a system of "interest group liberalism," or "policy entrepreneurship" that effectively turned administrators into advocates of competing policy proposals (in Fischer, 2000: 228; also Landy, 1995). Under this system, Landy argued that policymakers had to engage in a variety of strategies to portray policies in positive ways to different constituencies such as journalists, legislators, judges, and the public, in ways that can easily be seen by the public to be manipulative and untrustworthy.

The solution to this problem, they argued, lies in increasing transparency and inclusivity of policy decision-making processes within the EPA. Furthermore, the EPA's role in decision-making should be shifted from a position of policy entrepreneurship to providing public information about technological, legal, or financial feasibility of different options. The implication of this kind of study is not to suggest that administrators within the EPA may be corruptly promoting selected policy options, but to indicate the influences of hitherto unacknowledged organizational culture on how policy debate is conducted.

Similar analyses have been conducted on international organization. For example, the World Bank has commonly been criticized for avoiding environmental concerns by supporting large-scale infrastructure projects such as dams and highways; and for being apparently unaccountable in decision-making (e.g. Rich, 1994; Mehta, 2001). The Bank has responded to concerns with a variety of measures: in 1987, it upgraded its environmental office to full departmental level, and in 1989 introduced Environmental Impact Assessments (EIA) for projects. But most attention has been given to the decision to suspend funding for the Narmada dam in India in 1991, and then the decision of the incoming president, James Wolfensohn, to cancel the Arun 3 dam in Nepal in 1995 because of potential environmental and social impacts. Both dams were the subject of much public controversy and activism. As Bank staff are instructed in training sessions: "Don't get zapped by the Narmada effect, do your EIAs!" (in Goldman, 2001: 200).

Yet critics have suggested such responses by the Bank may still be problematic because they have emerged to avoid the most overt criticism rather than assessing environmental principles in general. Furthermore, new environmental codes adopted by the Bank still hide a variety of simplifications in environmental practice, and the simultaneous actions ensure new codes are seen as legitimate.

Goldman (2001) discussed these concerns in relation to the World Bank assessment of the proposed Nam Theun 2 dam in Laos. In particular, Goldman noted that the Bank had contracted research to the International Union for the Conservation of Nature (IUCN), but that the IUCN suppressed anthropological work that highlighted the impacts of the dam on diverse indigenous peoples in the region, and therefore problematized the IUCN's intention to develop parts of Laos into a National Biodiversity Conservation Area. The dam investors' consortium then hired a consultant from Norway who concluded that all peoples in the location of the dam could be resettled without harm, and who also argued that such groups could be described as one ethnic group, with few differences from other peoples in Laos. All such claims are contested by Goldman, who argued that the dam project framed environmental management in the eyes of the lowland state, international trade, and outsiders' visions of nature. Such criticisms echo comments made in Chapter 7 concerning the projection of risk onto regions and remote people. He wrote:

> The new authoritative logic of eco-zone management that is carving up Laos is designed to ensure that there will be "sustainable" hardwood supplies for export, watersheds for dams, and biodiversity

preservation for pharmaceutical firms and eco-tourists. This world-view represents most small producers as ecologically destructive and backward.

(2001: 207)

In addition, the production of environmental knowledge and assessments about the dam were accompanied by actions by the World Bank and IUCN to represent such information as legitimate and credible. In particular, the Bank sought to demonstrate the need for outside environmental guidance by claiming, for example, that the government of Laos showed no environmental awareness. The IUCN even went so far as to say that *no* conservation practices existed in Laos, and by once suggesting that the word does not exist in the Lao language. Such statements are demonstrably wrong when faced with the diversity of words (in various languages within Laos) that can mean "conservation," and by the abundance of information about conservation practices adopted by shifting cultivators (e.g. Fox *et al.*, 2000). Yet the purposes of such statements are to legitimize the intervention by these organizations, and to suggest that the framings and approaches adopted by them are the only options.

There are many ways to assess international organizations such as the World Bank for environmental policy and transparency, including awareness of gender, local participation, and the influence of internal management cultures (see Wade, 1997; Kurian, 2000). There is insufficient space in this chapter to review all of these concerns, but the implications of the example from Laos is that environmental reforms need not necessarily lead to more democratic approaches to environmental expertise and science. Superficially, the Bank adopted environmental expertise from organizations – such as the IUCN – that have important reputations for offering environmental expertise. But this expertise was framed according to predefined, and highly contested, development objectives, and did not seek to challenge the underlying assumptions about environmental causality, or seek to include significant participation at the sub-state level.

The lesson of this discussion is that successful democratization of environmental science and networks needs to assess the process by which expertise is formulated and legitimized, rather than simply accepting "environmental" practices in principle. Simply claiming to be "environmental" need not democratize environmental science and networks if the environmental principles and assumptions are constructed in selective and predefined ways. This conundrum indicates the need to consider the constructivist and discursive models of risk (discussed above, p. 241) in order to indicate how far discourses of environmental reform may actually democratize or reinforce existing networks. The examples of the EPA and approaches to research funding discussed earlier suggest that procedures can be reformed to increase transparency. But the example of the Nam Theun 2 dam suggests that the World Bank and IUCN have opportunistically used apparent environmental reform as a further arena in which to conduct predefined, and highly contested, development objectives. Critics must be diligent to identify when such "reforms" are being used to reinforce, rather than democratize, environmental networks. If discussion seems to be resisted by the networks, then outright opposition or the empowerment of alternative networks may be more effective.

Reforming environmental assessments

Finally, the practice of environmental assessments may also be reformed. Assessments are the means by which scientific networks may communicate findings to policymakers (see Social Learning Group, 2000a, b; Farrell *et al.*, 2001). Commonly, however, assessors have adopted the classic "science speaks to power" model, discussed above (p.233), which has resulted in some important failings.

For example, the Global Biodiversity Assessment (GBA) was undertaken by the United Nations Environment Program (UNEP) at the second conference of the parties to the Convention on Biological Diversity (CBD) in 1995. The report aimed to provide a comprehensive survey of scientific and policy dilemmas concerning biodiversity, and aimed to revolutionize thinking on biodiversity in the same way as the IPCC had done for anthropogenic climate change (Cash and Clark, 2001). The assessment was more than 1,000 pages long, and had included the participation of more than 1,500 scientists from more than 80 countries. The survey was also designed to be free from political criticism by being conducted independently of the CBD. Indeed, the GBA drew upon the expertise and organization skills of UNEP, the World Conservation Union (WCU), and the World Resources Institute (WRI); by including a wide variety of scientists from both developed and developing countries; and by adopting a comprehensive peer review process of its findings. Yet despite these actions, and initial positive reviews of the assessment, parties to the CBD largely ignored the GBA; it did not shape the political agenda for biodiversity conservation; nor was it used as a source book by individual nations for furthering domestic biodiversity policies. Indeed, the GBA was viewed with suspicion by some countries, particularly in the developing world (see Raustilia and Victor, 1996; Reid, 1997). In the words of one participating scientist: "it [the GBA] sank like a lead balloon" (Kaiser, 2000: 1677; in Cash and Clark, 2001: 2).

Cash and Clark (2001) have suggested the failure of the GBA resulted from four errors specifically associated with that assessment. First, the GBA did not acknowledge the political context of the negotiations on biodiversity conservation. The assessment was initiated between the first and second conference of the parties to the CBD, and accordingly was produced at a time when different parties were lobbying for different policy outcomes. The assessment was easy to portray, or delegitimize, as partial to different political viewpoints rather than as a basis from which to establish policy. Indeed, the comparatively more authoritative assessments of ozone depletion in 1985, and the IPCC initial reports in the late 1980s, pre-ceded the establishment of conventions on these topics. Second, the GBA failed to address the needs of its users by focusing too much on technical, "state of the art" scientific measurement, rather than on how biodiversity is perceived and experienced in many countries. Third, the GBA failed to treat assessment as a communication process between assessors and users. Part of the reason for this lack of communication was the desire to demon-strate impartiality in its findings, but the result was perceived irrelevance. Fourth, the GBA failed to connect global and local levels of assessment, by overlooking means to acknowledge local concerns and exposure to risks, or local capacity to deal with them.

These failings in a high profile and expensive environmental assessment indicate both the attractions and contradictions of portraying science to be independent of politics. In order to make the GBA seem legitimate in the eyes of users, the assessors sought to put distance between the assessment and users. Yet by not communicating with users, the assessment lost credi-bility and relevance, and could not help being criticized by some develop-ing country representatives as reflecting only the framings and practices of the Convention's more powerful parties.

Because of these kinds of experiences with environmental assessments, Farrell *et al.* (2001) suggested reforms to four under-appreciated elements of assessment design. First, it is important to acknowledge the assessment initiation and context, or the hidden framings that make an assessment appear necessary. Who called for the assessment and why? In the case of the GBA, one critic was quoted as saying, "the scientific community just decided we needed this and did it" (in Kaiser, 2000: 1677; Cash and Clark, 2001: 3). As discussed widely in this book, such framings have important epistemological implications on the research findings and causal state-ments. Second, what are the science–policy interactions of the assessment? How far are scientists isolated from policymakers? Third, who participates in assessments, and under what conditions? And fourth, what are the dif-ferent assessment capacities available to ensure effective assessments in different contexts? Assessment capacity refers to the ability of relevant organizations, actors, and political arenas to ensure participation in assess-ments, and successful communication between different parties.

The purpose of reconsidering these four elements of assessment design is to increase the appreciation of environmental assessments as dynamic and social processes. Yet there are still important questions about how these suggestions can result in the democratization of the scientific net-works adopted by assessments. First, many assessments still enforce boundaries between the formulation of scientific advice, and then its appli-cation to policy. This is perhaps shown most in the case of the IPCC, where scientists involved in the research of climate change are excluded from writing the associated summaries for policymakers, which are written by political representatives. Similarly, the Ozone Transport Assessment Group of 1995–1997 separated "technical" and "political" issues. Such separation in these cases still indicates a belief that scientific practice itself may not reflect social and political framings, or that "science" can be conducted in political neutrality, and then be communicated to "power."

Second, a further question concerns the extent of public participation within environmental assessments. As discussed at the beginning of this chapter, it is not clear if achieving scientific certainty about environmental problems may simply mean obtaining consensus among a controlled network of participants. Some commentaries about environmental assessments have approached participation warily because it might prevent the achievement of consensus. For example, Farrell *et al.* wrote:

> Expanding participation does not necessarily benefit the assessment process – particularly in the short term. It can reduce the assessment's quality, make the assessment logistically unmanageable and/or increase the difficulty of reaching consensus.
>
> (2001: 330)

Participation may therefore be controlled in order to reduce the potentially disruptive influence on apparently successful assessments. Farrell *et al.* claim that increasing participation from developing countries in the IPCC was a source of potential disruption, but resulted in the positive outcomes of providing more attention to development-oriented aspects of climate change. Against this, however, critics have suggested such participation has not democratized the IPCC enough. Increased participation has been mostly at the level of inter-state negotiation, rather than at the sub-state level. Furthermore, the IPCC still holds on to the emphasis on atmospheric greenhouse gas concentrations as a guide to environmental risk, rather than understanding more contextual analyses of vulnerability (e.g. Dowlatabadi, 1997; Kates, 2000b; Demeritt, 2001; see Chapter 7). In this sense, increased participation may have reinforced the existing Realist and linear model of risk adopted within the IPCC, and not have resulted in a more diversified reframing of how risk is presented.

Increased participation in some environmental assessments may therefore have limited impacts on the democratization of science and networks. As Farrell *et al.* note, it is important to ensure that participation includes aspects of culture and perception of environmental problems, rather than be restricted to the nominal inclusion of "token" representatives of unrepresented groups. It should also be noted that there are many types of assessment, with different contexts and possibilities for change. Despite such statements, however, it is not clear how far some environmental assessments, predicated on the belief that science itself can be separated from political framings, can actually adopt such greater participation in meaningful terms. According to this belief, participation cannot, by definition, democratize science. Yet as this book has sought to demonstrate, the very identification of environmental problems and causal links are indeed determined by social and political factors.

This section has discussed a variety of means by which existing scientific networks may be democratized or reformed in order to increase transparency, and increase the possibility for people outside the network to influence them. But have these measures successfully challenged the boundaries of networks, or increased widespread participation? One alternative is to empower alternative networks as a way to challenge the authority of existing, more powerful systems of expertise. The next section now considers these alternative networks.

Empowering alternative networks

So, if attempting to reform existing scientific institutions may prove problematic, is it possible to develop alternative networks that may address the needs of less represented groups more successfully?

This next section now examines the emergence of alternative institutions or networks of science that may exist in parallel to, or outside of, the formal boundaries of orthodox scientific institutions. Such alternative networks may be considered a form of democratization because they allow greater localization and diversification of environmental explanation. They may also increase transparency and public participation in local environmental science.

The alternative networks discussed in this chapter are not simply those associated with social movements. As discussed in Chapter 5, many discussions of ecology associated with "new" social movements have adopted ecological discourses that essentialize environmental degradation with social oppression and instrumental rationality in modern societies. Indeed, Richard Harvey Brown (1998) has discussed the possibility for a "democratic science" based on improved social communication to a public sphere in order to avoid such instrumentalism. This book has argued that such discourses themselves need to be democratized in order to understand how – and with whose participation – such assumptions were made. Similarly, as discussed in Chapter 6, other social movements may often harness or replicate existing environmental discourses in order to achieve political success.

Instead, this section considers institutions that acknowledge the coproduction of political activism and environmental science, and which seek to acknowledge the political boundaries associated with public participation. Such institutions need not reject interfaces with formal scientific knowledge from expertise from other institutions. But they may seek to advance the constructivist and discursive models of risk discussed above, rather than simply provide capacity for the implementation of science constructed under the Realist model.

The discussions in this section are all related in various ways to wider debates concerning so-called Deliberative and Inclusionary Processes in Environmental Policymaking (DIPS) and participatory approaches to environmental policy (see Dryzek, 1990; Button and Madson, 1999;

Holmes and Scoones, 2000). This chapter cannot summarize all of these debates, but focuses mostly on the ways in which debates within political ecology can engage with diversifying the local negotiation of environmental science.

Environmental adaptations in developing countries

The concept of environmental adaptations was introduced in Chapter 2 to refer to environmental practices and livelihood strategies that allow the protection of resources despite the existence of poverty or increasing populations (see Netting, 1993; Batterbury and Forsyth, 1999). Examples of adaptations include soil conservation measures such as *diguettes* (or lines of stone) to prevent erosion, or soil mounds to enhance agricultural fertility (e.g. Tiffen and Mortimore, 1994; Batterbury, 1996; Sillitoe, 1998), or gradual transformation of forest–savanna landscapes to enhance the production of specific tree species valuable for local livelihoods (e.g. Fairhead and Leach, 1996; Schmidt-Vogt, 1998). Environmental adaptations are often considered forms of so-called community-based natural resource management (CBNRM) (e.g. Leach *et al.*, 1997).

Environmental adaptations indicate two important lessons for empowering alternative scientific networks. First, adaptations may be considered to be a form of alternative network because they offer exceptions to generalized predictions of environmental degradation such as the $I = PAT$ equation (see Chapter 2). Second, adaptations also indicate the importance of local, and often unpredictable, factors of culture and social organization that are not always included in rational choice, or positivist approaches to common-property resource theory (e.g. Ostrom, 1990). As a result, some authors have argued that studying the institutional bases of environmental adaptations, and transferring these to new locations, may be effective ways of diversifying and localizing environmental management (see Mehta *et al.*, 1999, 2001).

Empowering and transferring adaptations, however, imply a number of difficulties. First, there are problems in identifying how far practices are "local." As discussed in Chapter 7, there is a need to assess how far conceptions of local practices may reflect outside constructions of locality or indigenous people that may hinder locally determined development. It is also increasingly difficult to identify groups of people or environmental practices that are not in some way connected to regional or global networks of trade, investment, or migration (Bebbington and Batterbury, 2001). Such increasing global integration both affects the institutional basis of environmental adaptations, and the causes of environmental degradation. Murton (1999), for example, found that Tiffen and Mortimore's (1994) originally positive findings that "more people" may mean "less erosion" in Machakos, were increasingly less apparent because local farmers were spending time away from soil conservation, and instead were engaging in sporadic migration to cities for waged employment.

Second, there is also a need to consider the role of "global" environmental risks alongside "local" environmental problems. Some environmental adaptations may be effective against locally defined problems, but may be insufficient to mitigate risks that may occur more globally, or practices that have environmental impacts outside localities. For example, Chapter 6 described the example of the International Center for Research in Agroforestry (ICRAF) framing research on shifting cultivation in Southeast Asia in terms of regional impacts on declining biodiversity or regional haze from fires. Such concerns also have to be matched with local framings of problems, and how far policies suggested from a regional perspective (such as restricting shifting cultivation) might actually impact negatively on local livelihoods.

Third, it is sometimes difficult to separate the concept of environmental adaptations from the underlying securities – such as land tenure, health, education, or access to resources – that allow adaptations to succeed. Clearly, some adaptations – such as building terraces, installing *diguettes*, or shaping forest islands – are only attractive to farmers if they are confident they can reap the rewards in the future. Enforced resettlement, or appropriation of land by state or investors, or during times of political unrest, may therefore undermine the adoption of environmental adaptations. Similarly, all members of the locality are unlikely to benefit from adaptations in the same way. As discussed above (p. 238), the concept of "community" frequently hides a variety of social divisions along lines of gender, caste, age, etc. that may differ in access to underlying securities such as land tenure or education. (Indeed, this criticism may also be applied to the concept of "social capital," which has also been used in generally positive terms about local development.)

Because of these problems, some observers have argued that more attention should be given to the means by which local environmental governance may be achieved, rather than the imposition of predefined "laws" about environmental degradation. Box 9.1 describes some potential institutional forms that may allow the successful transfer of experience of environmental adaptations to new locations. These approaches differ to orthodox approaches to environmental management or environmental politics by allowing the local framing of environmental problems, and by acknowledging that concepts of "community" include a variety of conflicts and social divisions that may be constantly experienced and negotiated. Together, they form suggestions for how local environmental governance may be achieved, which may also include constructive engagement with expert knowledge from outside localities.

The concepts of sustainable livelihoods and environmental entitlements focus on building environmental adaptations at the micro level. Sustainable livelihoods is a more general term referring to the means of establishing capabilities and assets that may enable social groupings (such as individuals, households, or localities) to maintain reliable sources of income despite resource scarcities (see Chambers and Conway, 1992;

Box 9.1 Some possible institutional forms for integrating environmental governance and learning

Sustainable livelihoods

The concept of sustainable livelihoods is a framework for integrating environmental management with local livelihood strategies. In simple terms, a "livelihood" may be defined as the capabilities, resources, and other assets and activities required for making a living. A "sustainable livelihood" may be defined as one that

> can cope with and recover from stress and shocks, maintain and enhance its capabilities and assets and provide sustainable livelihood opportunities for the next generation; and which contributes net benefits to other livelihoods at the local and global level and in the short and long term.
>
> (Chambers and Conway, 1992: 1)

Sustainable livelihoods differ from orthodox approaches to environmental management in two key ways. First, they reject the assumption that there is an inescapable link between poverty and environmental degradation, and instead seek to empower local strategies to conserve resources. Second, they allow local people to frame environmental problems and resource conservation in terms that are necessary for their livelihoods.

Environmental entitlements

The environmental entitlements debate seeks to apply the "entitlements" approach of Amartya Sen (see Chapter 7) to the means by which individuals or social groups can gain access to, and protect, environmental resources. The approach seeks to highlight the "endowments" or "entitlements" that may allow different people to use resources, and the varied way in which social institutions – or shared norms of behavior or environmental perception – may influence this access (Leach *et al.*, 1999). Institutions may exist at micro, meso, or macro scales, involving negotiations within households, villages, or regions, and between different actors. Environmental entitlements also help avoid a simplistic approach to "community"-based natural resource management by highlighting the social divisions that exist within so-called communities.

Adaptive management

Adaptive management is a form of resource management that allows local people to negotiate and reframe official advice on environmental management from orthodox science. It aims to understand the potential for different management techniques by looking at the responses to management itself from local people who use scientific information (Berkes *et al.*, 1998). Adaptive management has been used in contexts where resources are subject to a variety of local and regional demands on land use, such as in areas of protected forest, or in the evolution of community forestry (Robbins, 2000; Klooster, 2002). It provides a means by which management techniques may be adapted dynamically in order to reflect the demands made by a wide group of users as the result of regular communication between resource managers from different user groups.

Islands of sustainability
The concept of islands of sustainability refers to a region or locality that has adopted forms of economic cooperation and environmental protection that allow it to integrate economic success with environmental conservation (Wallner *et al.*, 1996; Bebbington, 1997). Similar to sustainable livelihoods, "islands of sustainability" commonly exist in zones considered to be subject to widespread environmental degradation and poverty. "Islands" allow local livelihoods to continue through the co-creation of political unity between different villages or farms, common economic strategies, and the establishment of trading links with other localities. The establishment of common links between different parties in each "island" is crucial to successful integration.
Sources: Chambers and Conway, 1992; Wallner *et al.*, 1996; Bebbington, 1997, 1999; Berkes *et al.*, 1998; Carney, 1998; Scoones, 1998; Batterbury and Forsyth, 1999; Leach *et al.*, 1999; Robbins, 2000; Klooster, 2002.

Scoones, 1998). Some authors have referred to this as *bricolage*, or the ability to adopt a flexible and interconnecting range of income sources and environmental activities to reduce vulnerability to changes (e.g. Batterbury, 2001; Cleaver, 2001). Often, sustainable livelihoods are based on a combination of three key actions: agricultural intensification, income diversification, or short- or long-term migration by some or all members of a household. For example, in the Sahel of West Africa, Mossi farmers from Burkina Faso may seek to overcome long-term problems of drought and declining agricultural productivity by seeking short-term employment in cities. In Papua New Guinea, Wola shifting cultivators have maintained soil fertility despite growing populations by innovating with soil mounds and the adoption of sweet potato that thrives on such mounds (Sillitoe, 1998).

The concept of environmental entitlements is a similar approach but focuses more on the institutional controls of access to resources. The approach reflects Amartya Sen's entitlements approach to food security discussed in Chapter 7, and seeks to indicate the different institutions – or shared behavior and expectations – through which individuals or social groupings may gain access to resources, often in variable or short supply (Leach *et al.*, 1999). For example, in the semi-arid Indian province of Rajasthan, water management is crucial for irrigation and for urban sanitation. The underlying biophysical variation in groundwater leads to variable supplies of water for either boreholes or local surface water supplies. According to the approach of environmental entitlements, *endowments* for water may be defined as the private arable and pasturelands occupied by farmers, and the water rights that enable access to communal water supplies. *Entitlements* for water supply, on the other hand, include irrigation water, crops, and income from marketed products, and these are influenced by collective action among owners of contiguous plots, or communal repair work on gullies and canals. A variety of institutional controls influence endowments and entitlements. Access to water endowments are influenced by micro institutions such as inheritance of land, labor contributions to agriculture, and macro-scale institutions such as interactions between the governments of India and Rajasthan concerning watershed development policy, and land laws. At the meso scale, entitlements are influenced by market forces and credit institutions. The result of these interactions is a supply of water to large farmers, marginal farmers, and livestock rearers (Ahluwalia, 1997).

Local strategies may also allow more overt resistance to environmental changes or controls that are imposed from outside. In the Dominican Republic, for example, Rocheleau *et al.* (2001) noted how local *bricolage* by different farmers allowed resistance to the introduction of *Acacia* trees as both cash crops and carbon-offset forestry. Such local resistance has also been noted against state-led soil or forest conservation policies often dating back to colonial science and management objectives in South Africa (Driver, 1999); India (Jewitt, 1995; Srivaramakrishnan, 2000); West Africa (Fairhead and Leach, 1996; Batterbury, 2001); and in Thailand (Johnson and Forsyth, 2002). A less confrontational approach may be adaptive management (Berkes *et al.*, 1998), or the mutual shaping of external environmental management plans by formal scientists and local people. This approach has been praised for developing models of community forestry in Mexico, for example, because it allows the integration of different framings and experiences of forests from different users (Klooster, 2002). Yet such negotiations may still be controversial. In India, for example, Robbins (2000) noted that the resulting consensus within such negotiations still reflected powerful groups, and that the consequent environmental assumptions about the impacts of forest use could still be questioned from other perspectives.

The concept of "islands of sustainability" (Wallner *et al.*, 1996) also proposes a large-scale application of sustainable livelihoods for integrating economic competitiveness and environmental sustainability, often in regions where orthodox thinking would assume widespread environmental degradation and poverty. In the rural Andes, for example, a combination of action by NGOs, agricultural producers, and local governments have succeeded in intensifying agriculture, and in increasing investment in new, high-value products such as horticultural crops (Bebbington, 1997). In such cases, the establishment of local trade associations, with coordinated agricultural and environmental practices (including both indigenous and imported techniques) may increase prosperity and agricultural production, despite orthodox expectations that such regions may experience downward cycles of poverty and environmental degradation.

The formation of localized zones of environmental governance that can resist wider forces of economic and political control is, of course, one of the key objectives of all cultural and political ecology that focuses on social justice and environment in the developing world. The attention to how far local governance can also influence the scientific assumptions underlying environmental management, and the means to achieve such governance, are crucial elements in ensuring that environmental science is included in such decentralized political control.

Marginalized social groups and environmental science

The formation of alternative scientific networks for environmental explanation may also be conducted in locations that are not necessarily associated with developing countries. Many social groups around the world may be considered marginalized or under-represented in hegemonic science. As discussed in Chapter 4, it is common to assume that women, ethnic minorities, and people in lower economic classes may be less represented in common scientific discourse. Indeed, feminist analysts of science have observed that some women have considered "science" to exist outside of their day-to-day experiences, and consequently see science to be both irrelevant and unapproachable (Harding, 1986; Schiebinger, 1993; Lederman and Bartsch, 2001).

Empowering alternative scientific networks for marginalized groups, however, raises important dilemmas. First, it is important to identify under-represented groups in a critical and comprehensive manner. Simply assuming that, for instance, "women" or "ethnic minorities" or "children" are necessarily under-represented may overlook how such groups have succeeded in gaining recognition, and may also essentialize marginalization with these categories. Second, it is important to appreciate that the objective is not to get groups such as women into science (as it currently exists), but to reframe science itself in order to better reflect the needs and concerns of unrepresented people. These two problems exist simultaneously:

> The gendered character of scientific knowledge means that women's location always begins from outside science ... It is difficult indeed for any woman to become "inside" the practices and authority of orthodox science. It is even more difficult if she is not white or middle class.
> (Barr and Birke, 1998: 78)

A third problem is whether seeking to empower alternative networks may imply having to adopt the language of dominant networks in order to gain credibility. If they do adopt similar language, how far does this make them lose their alternative status? Indeed, the need for environmentalists to use orthodox science in order to gain credibility, when some have argued against the principles of science and technology, has been well recorded by critics of environmentalism (Yearley, 1992). Similarly, as discussed in Chapter 6, the problem of attempting to reform environmental policy through local social movements – or the "Liberation Ecologies" approach of Peet and Watts (1996) – may also experience the problem of needing to use existing environmental discourses rather than introduce new themes.

The empowerment of alternative scientific networks may therefore undertake both the reframing of scientific inquiry to reflect the concerns of marginalized groups (under the constructivist model of risk), and the development of new, and more inclusive forms of measuring and discussing risk (under the discursive model). Empowering networks themselves may include activism on behalf of marginalized groups, and the establishment of new arenas for scientific debate and dissemination of scientific knowledge.

Chapter 6 already reported one apparently successful example of reframing AIDS research in the USA through an alliance of scientists and people with AIDS (Epstein, 1996). Epstein's work indicated that the alliance of patients and scientists with links to medical research establishments succeeded in creating a change of emphasis toward the treatment of symptoms of AIDS, and a more sensitive approach to patients in general. This activism was helped in part by campaign objectives that sought to modify, rather than overthrow existing science networks, by seeking other ways to address the risk posed by AIDS rather than seeking ways to avoid transmission of the virus alone.

Yet such reframing of scientific research groups may also be done by formal intervention in processes of research funding and dissemination. In the USA in 1992, for example, the Carnegie Commission deliberately changed its support for research in order to link science and technology to societal goals, and particularly to less wealthy people (Carnegie Commission, 1992). The concept of "science shops" has also been used to promote scientific needs and findings within urban neighborhoods. Science shops act as brokers between community groups and university researchers on themes of concern defined by the lay groups rather than researchers (for instance, concerning the origin of local pollution). In the Netherlands, for example, individuals can approach science shops for information, and if this is not available, they are then put in contact with interested researchers (Barr and Birke, 1998: 16, 138).

Another well-known example is the Kerala Sastra Sahitya Parishad (KSSP) organization of India. The KSSP was established in 1962 as the result of a number of scientists and social activists who feared that scientific information was inaccessible to most people. After some years of translating scientific books from English into the local language of Malayalam, in 1972, the organization adopted the motto "Science for Social Revolution," and sought to make local development more oriented to local concerns. The organization in particular opposed the construction of the "Silent Valley" dam in 1984, and then used local volunteers to assist the central government campaign to increase local literacy in the area. The KSSP was awarded the "Alternative Nobel Prize" in 1996 (see Fischer, 2000: 162).

Yet, in some cases, environmental problems may also be experienced in ways that are not commonly discussed, or where there are no overt causal links between environmental causes and the problems experienced. One particularly emotive example is the case of lost pregnancies, which, as Linda Layne (1990: 69) suggests is associated with a "veil of silence," indeed, so silent that it is one common bereavement for which there are no Hallmark cards available. Disturbingly, it is estimated that some 31–43 percent of all pregnancies in the USA end in miscarriage (Layne, 2001: 25). Yet the links of lost pregnancies and environmental factors – such as chemical toxins – is poorly researched.

Layne (2001) studied three cities in locations close to high concentrations of chemical toxins in the USA (Woburn, Massachusetts; Love Canal, New York; and Alsea, Oregon). Layne found that few women who had lost pregnancies were also willing to link these to exposure to toxins. Indeed, many women sought to blame themselves for miscarriages. For example, one woman from Alsea believed she caused her own lost pregnancy by her "own stupidity" in taking a strenuous hike, rather than her documented exposure to dioxin-containing herbicides (ibid.: 42). In cases like these, the public taboo concerning the loss of pregnancy, and stereotypical social expectations that women need to succeed as mothers, have meant that women often seek explanations for miscarriages that focus on their own presumed role rather than on external factors. As Layne noted: "In our culture, we deal with events like unsuccessful pregnancies, which challenge our cherished narratives of linear progress and the cultural mandate to be always happy, primarily by pretending they don't happen" (2001: 25)

Layne's study has important implications for empowering alternative networks of scientific explanation. The tendency for women to blame themselves rather than toxins (or other external causes) for lost pregnancies suggests the existence of a "storyline" (see Chapter 4) about the role of individuals in causing miscarriages. Yet clearly, such an explanation is likely to be highly simplistic and unnecessarily blameworthy of women concerned. Challenging this trend requires creating a new public arena in which potential alternative causes for miscarriages may be discussed. Layne sought to achieve this by calling for "an agenda for a feminist discourse of pregnancy loss" (1997). Creating a new arena for discussing the problem may therefore increase the search for potential causes for miscarriages, and enhance support for men and especially women who have been affected by them.

Much political ecology, of course, has focused on addressing the environmental concerns of marginalized social groups or environmental problems that have been overlooked by official policies. Assessing the institutional forms or social solidarities that allow greater analysis and questioning of assumed scientific causes of risk might enhance this process. Indeed, seeking further political attention for the environmental risks experienced by less powerful groups is a key requirement of democratizing environmental concern in the years to come.

Participatory environmental assessments

Finally, it is worth considering the techniques by which environmental research itself may highlight alternative conceptualizations of environmental problems. This chapter has already discussed how the languages and techniques of risk assessment may themselves be a barrier to public participation (under the discursive model of risk). Diversifying the languages and arenas through which environmental risks are discussed or defined may further increase the democratization of environmental science and networks.

Participatory environmental assessment may be defined as forms of research that allow maximum opportunity for people under research to define and express their thoughts about environmental problems in terms of their own choosing. The aim of such research is to reduce as many influences from outside agendas, networks, or assumptions as possible (often such assumptions are held by researchers themselves). Ideally, such research avoids the problems of "speaking on behalf of others" or essentializing "local" knowledge discussed in Chapter 7. Participatory assessments also allow ways to reform formal environmental assessments fundamentally by allowing participants to frame the purpose of environmental research themselves, rather than by simply allowing participants to discuss the research findings alone (see above, pp. 249–251).

There are many forms of participatory environmental assessment, and a full discussion of all techniques is beyond the scope of this book (see Anderson and Jaeger, 1999; Fischer, 1999; Holmes and Scoones, 2000). Box 9.2 lists some possible methods for encouraging participation in policy discussions. These techniques may be divided into different categories. Legislative theatre and community video, for example, offer ways for local people to express different elements of support or concern for different policy options or perceived problems. Focus groups, citizen juries, and deliberative polling, for example, offer means for citizens themselves to engage in discussions about the nature of problems, and to express concerns both individually and collectively. Other techniques such as multi-criteria mapping or participatory scenario building aim to highlight the diversity of different evaluations and policy options available. In all cases, such participatory assessment techniques seek to demonstrate the complexity of local concerns about environmental problems or propositions. By so doing, they also move away from the uniformity of orthodox science and "laws" of nature that suggest a priori conceptualizations of causes and effects of environmental changes, or the black-box statements of cost–benefit analysis (CBA) and contingency valuation conducted by environmental economists (see Lohmann, 1998).

Political ecologists need not necessarily engage closely in the undertaking of participatory environmental assessments. But if political ecologists are to engage critically with contested notions of environmental problems and the potentially damaging political impacts of orthodox explanations and associated policies, then they have to be aware of how assessment

---

Box 9.2   Some possible methods for encouraging participation in policy processes

- Participatory appraisal and priority assessment.
- Multicriteria mapping.
- Citizen juries.
- Standing panels.
- Focus groups.
- Community issue groups.
- Community video.
- Legislative theatre.
- Participatory scenario building.
- Future search workshops.
- Citizen foresight panels.
- Visioning exercises.
- Deliberative polling.
- Consensus conferences.
- Stakeholder decision analysis.
    Source: Holmes and Scoones, 2000: see also Anderson and Jaeger, 1999; Durant, 1999; Dürrenberger *et al.*, 1999; Fischer, 1999; Hörning, 1999.

---

methods may overlook the diversity and complexity of local environmental perceptions. Participatory methods allow some means to indicate such diversity. Ideally, such assessments have to be matched by the existence of political arenas that acknowledge their findings, and display transparency when discussing the different evaluations. Yet, as discussed above (pp. 239–240), such optimistic notions of a public sphere may never exist, and it may be more realistic to expect to see different evaluations being counted or discounted in a variety of formal and informal arenas, including law courts, newspapers, street marches, or websites. Yet the degree of participation included in different environmental assessments may be grounds by which different knowledge claims may be themselves assessed.

This section has summarized some methods by which alternative scientific networks may be empowered in order to reduce the influence of hegemonic environmental assessments and assumptions. Yet before this chapter ends, it is necessary to discuss some important implications of democratizing environmental science and networks for debates concerning the political transparency and participation in environmental science.

Implications for integrating environmental governance and learning

This chapter has sought to identify ways to democratize environmental science and its networks by reforming existing science networks or empowering the emergence of alternative networks seeking to represent the perspectives of under-represented people.

However, it is important to acknowledge that democratizing environmental science does not imply the replacement of one set of explanations with another in a final way. Instead, as discussed throughout this book, different environmental explanations are contingent upon a number of framings, objectives, and their associated boundaries. Revising orthodox and unquestioned environmental science toward more locally accurate and relevant explanations consequently reflects an evolving debate about what science should achieve, and for whom. Such different approaches to science may therefore reflect different ideologies of social justice and democracy that may themselves constitute networks, and accordingly be acknowledged and discussed as such.

The awareness and criticism of one's own institutional assumptions – or "institutional reflexivity" – is therefore a key requirement in ensuring that challenges to hegemonic environmental science may also be called democratization. According to Cultural Theory (see Chapter 4), each institution may be located within one of the myths of nature, and as such, institutional reflexivity implies recognizing the limitations of world visions associated with just one myth (e.g. Thompson, 1993). More poststructuralist approaches such as narratives or storylines (see Chapter 4), are less rigidly linked to the different myths, and instead see reflexivity as a critical analysis of how such understandings or ideologies have emerged over time, and how different future framings and public participation may result in different epistemologies.

One concept that may allow greater integration of ideologies and resulting science is the so-called "virtuous circle of facts and norms" (Kearns, 1998). Adopting many insights of Critical Theory (and especially Habermas), the "virtuous circle" refers to the ability to integrate learning about environmental reality with the constant reshaping of ideologies and perspectives through which such inquiry is framed. Kearns argued that the influence of such reshaping of environmental history has been demonstrated in the case of research on the history of the western USA. Environmentalist writers such as William Cronon (1991) and Donald Worster (1977, 1979) can be compared with Richard White (1980) and Patricia Limerick (1991), who gave more attention to questions of social justice, and who also sought to understand environmental change from the perspectives of ethnic groups and classes not referred to in more ecologically minded histories. Kearns wrote:

> An understanding of the "other's" point of view entails recognizing both the specific differences that frame worldviews and the particular context in which those framings take place ... Only the *voice* of the other can adequately alert us to plurality and difference ... If historical and geographical writings can build on this work ... and continue to attend to the sets of agenda of those groups in subjugated positions, then the circle between facts and values will indeed have turned virtuous.
>
> (1998: 404; emphasis in original)

Such integrated environmental explanation and social learning has been reflected in various other discussions of democratizing environmental science in this book (e.g. Collier, 1989; Bhaskar, 1991; Kukla, 1993, in Chapter 8). It is also similar to the concept of discursive democracy (Dryzek, 1990), or Robert Chambers' (1997) question, "Whose reality counts?" Such views point to the need to seek democratic challenges to institutionalized ideological or scientific statements – but also to the need to constantly reassess the values and knowledge guiding them. This book has argued that environmental science and politics are coproduced. Democratizing environmental science also means making the democratizing process dynamic, transparent, and inclusive, but also self-critical.

Summary

This chapter has brought the book's substantive discussions to a close (prior to the Conclusion) by examining political means to increase transparency and public participation in the formulation of environmental science. The chapter builds upon the discussion of democratizing environmental scientific techniques in Chapter 8. Together, these chapters also suggest practical ways to carry out research and debate under a "critical" political ecology.

The chapter argued that scientific uncertainty cannot be understood without acknowledging the extent of public participation or observation of scientific inquiry. Yet increasing public participation may also mean challenging the status of established scientific organizations as sole providers of expertise. Similarly, reinforcing orthodox scientific networks may also mean reiterating linear models of risk, which emphasize projected biophysical changes as causes of risk, rather than contextual factors underlying the interpretation or vulnerability to such changes. Furthermore, the language or techniques through which environmental risk and science are discussed also form effective barriers to public participation.

The chapter discussed two main ways of revising orthodox or hegemonic scientific networks in favor of more decentralized, or less powerful viewpoints. The first way is to reform existing scientific institutions. The chapter discussed direct attacks on orthodox science (such as during the "science wars"); and different approaches to reforming scientific organizations such as the World Bank, and formal environmental assessments. These approaches have had limited success, but have sometimes led to the evolution of new approaches of making scientific expertise appear legitimate rather than including public participation in the early framings of environmental science.

Second, the chapter outlined various ways in which different scientific networks could be empowered as alternatives to orthodox scientific institutions. This section looked at environmental adaptations in developing countries, and particularly institutional approaches such as environmental entitlements or sustainable livelihoods as means to increase local governance over environmental management. This section also considered the emergence of environmental science relevant for marginalized people and subjects, such as the causes of lost pregnancies, and a variety of participatory environmental assessment techniques that can enhance local framings of environmental science.

But despite these optimistic proposals for reframing and governing science, the chapter noted two remaining problems. All scientific explanation relies in part on the establishment of communities or networks of explanation that require boundaries in order to be meaningful and credible. Democratizing environmental science and networks depends ultimately on a guiding ideology about the nature of social justice. Such ideologies need to be acknowledged and questioned for their potential impacts on learning about environmental change.

Finally, as discussed throughout this book, concepts of "science" are commonly used to support and legitimize different political strategies. It should not be surprising that certain networks or organizations do not wish to listen to criticism, or change strategies because of greater public participation. Learning to identify successful, rather than superficial, forms of scientific participation and governance may therefore become an important new theme of environmental democratization in the future.

10 Conclusion

"Critical" political ecology and environmental science

This book has discussed various ways in which environmental science and political processes are mutually embedded. The book has drawn upon a wide range of debates to show how scientific statements are made; how social movements and international organizations shape science; and how greater public participation may be allowed in the formulation of scientific statements. This final chapter now seeks to consider the implications of this book for debates in political ecology and for future approaches to environmental science. How does a "critical" political ecology differ from other types of political ecology? How can this book influence debates about the formation and implementation of environmental policy?

The chapter begins with a summary of the book's key arguments, and then goes on to discuss the book's implications for other debates in political ecology and environmental policy. The key themes addressed by this chapter are the relationship of ecology as a science, and ecologism as an ideology; theoretical approaches to explaining the political structures and causes of environmental degradation; and means of incorporating a more politicized approach to science within environmental debate and decision-making.

Summary of the book's arguments

The chief purpose of this book has been to challenge many existing beliefs about the separation of environmental science and politics. Many environmental scientists, political activists, and political ecologists have suggested that lines may be drawn between the explanation of environmental problems as a scientific project on one hand, and the discussion of environmental policy as a political project on the other. Instead, this book has argued that environmental science and politics should be seen as *coproduced* – or as mutually reinforcing at every stage. Politics are not merely stimulated by scientific findings but are prevalent in the shaping and dissemination of environmental science. Politics are also influential in the strategies used to present different environmental explanations as legitimate bases for policy.

This book has advanced many existing discussions of science and politics (e.g. Rouse, 1987; Aronowitz, 1988; Ward, 1996) by focusing explicitly on environmental science, and on demonstrating the epistemological linkages between physical-science studies of, for instance, erosion and deforestation, and wider political and social debates about ecological decline and opposition to state and industrial practices. Furthermore, this book has sought to advance debates in political ecology by placing more attention onto political factors underlying "ecology," or the assumptions political ecologists use to discuss environmental degradation. A "critical" political ecology is one that eschews meta-narratives or received wisdom about environmental degradation, and instead adopts a critical attitude to how such supposedly neutral explanations of ecological reality were made. There is little point in conducting political analysis of environmental degradation or studying the political allocation of risks if concepts of degradation and risk are themselves political and not acknowledged as such.

As discussed in Chapter 1, this approach reflects discussions in political ecology that go back to the British Political Ecology Research Group of the 1970s, and have been discussed by a variety of political ecologists since (e.g. Peet and Watts, 1996; Vayda and Walters, 1999; Mukta and Hardiman, 2000). Moreover, a "critical" political ecology is rooted within three other intellectual debates of Critical Theory (concerning the analysis of political oppression and construction of knowledge); Critical (or "skeptical") Realism (regarding the discussion of biophysical reality through the guise of social knowledge); and Critical Science (or the criticism of orthodox scientific practice in constructive ways). This book does not seek to find a better form of science, which can then be applied *post-hoc* to more democratic fora. Instead, this book seeks to show how science and politics are coproduced, and the power this realization gives for revealing the covert use of science for political objectives, and the ability to devolve environmental scientific governance within diverse social groupings and for locally determined purposes at various time and space scales.

These themes, however, are relevant to a variety of topical concerns about the status of ecology as a guiding principle for environmental policy; the relationship between contemporary environmental concerns and the production of science; and the possibility to reform scientific practices by enhancing transparency and public participation in environmental science. These topics are now dealt with in turn.

Ecology and ecologism

As noted throughout this book, these are controversial times for writing about environmental science and politics. The growth in academic debates about the relationship of science and politics has emerged as more and more companies or industrial interests seek to discredit environmental concerns by publishing information that questions environmentalist objectives. In response, ecologists are increasingly turning to the defense of orthodox science as a means to protect environmentalism against criticism. Such trends are apparent in the suggestions of Paul and Anne Ehrlich (1996) that environmentalism represents "science and reason," or of Lester Brown (2001) that the economy should be changed to reflect the "ecological truth."

This book has argued against such uses of orthodox science, but in no way seems to undermine an informed debate about environmental concern and regulation. The book has argued that many existing approaches to ecology, when presented as unproblematized universal truths, have caused a variety of problems in manners not acknowledged by ecologists such as Brown or the Ehrlichs. One of the clearest dilemmas stated in this book concerns the confused status of "ecology" in environmental debates as an allegedly accurate science, yet also as an ideological statement about how the world is meant to be. Such confusion was apparent in the writings of many environmentalists and political ecologists from the early calls to make ecology the "subversive science" (Sears, 1964). Yet the failure to acknowledge this mixture of scientific prediction and ideological ecologism – or ecocentrism – has led to a variety of environmental explanations and policies that do not address the biophysical complexity of many long-term environmental changes in many locations around the world, or the diverse institutional bases in which environmental problems are experienced. Many scientific prescriptions based on orthodox explanations have been ineffective against environmental problems such as declining soil fertility or so-called desertification. Environmental policies based on such explanations have overlooked actions taken by poor people to lessen degradation. In the worst cases, policies have even restricted livelihoods and increased the vulnerability of people by reducing land available for agriculture or restricting other economic options. These problems indicate that the mechanisms for environmental explanation under ecology and ecologism are problematic, and possibly deeply flawed: "If we disentangle environmental discourse, we find a complex medley of ethical and epistemological issues nowhere more confused than in the ecocentrist appeal to nature as a privileged source of invariant meaning" (Gandy, 1997: 237).

Instead, this book suggests that ideologies and science need to be seen as co-constructed, and specific environmental explanations as contingent upon social and political framings. Such comments do not deny the existence of environmental degradation, but illustrate the inadequacy of concepts used to define it, and particularly when such concepts are transferred uncritically between different contexts. Criticizing the Universal Soil Loss Equation (USLE), for example (see Chapter 2), does not imply that erosion is never a problem, but means that more attention should be given to alternative causes of declining soil fertility, than simply to assume that erosion, or the Equation, are as universally problematic as the title suggests.

Many insights to the difficulties of explaining ecological change have also been achieved within the variety of debates known as the "non-equilibrium," or non-linear ecology (Botkin, 1990; see Chapter 3). This book has drawn attention to these debates, but urges that current work be supplemented by more attention to the social and political influences behind the identification of different time and space scales for environmental explanation, or on what is considered "normal" in ecological terms (see also Robbins, 1998; Zimmerer, 2000). It is tempting to refer to forest fires or floods as crises, but such events need to be contextualized according to landscape histories, and the distribution of impacts on different people. Attempting to restrict all such events may overlook their historic role in creating landscapes, or the impacts of such policies on different users of the landscape (Leach and Mearns, 1996; Adams, 1997).

Perhaps the most significant current example of social framings of environmental policy is the widespread assumption among many environmental activists that reforestation presents a panacea for a wide selection of environmental problems ranging from loss of biodiversity and wilderness, water shortages and erosion, or the mitigation of climate change (e.g. Brown, 2001, see Chapter 2). Much research about reforestation, and particularly plantation forestry, for these purposes may now form a new paradigm of "normal" science (Kuhn, 1962). This book in no way argues against the desire to protect wilderness as one environmental choice, or the many potentially beneficial ways that trees or selective reforestation can support a variety of environmental concerns. Yet this book has also illustrated various ways that such approaches to reforestation have been questioned by research on watershed degradation, or how carbon-offset forestry may provide a variety of negative impacts on people and ecosystems (e.g. Hamilton, 1988; Howarth, 1995; Rocheleau and Ross, 1995; Cullet and Kameri-Mbote, 1998; Fairhead and Leach, 1998; Calder, 1999; see Chapters 2, 6, and 7). There needs to be greater critical awareness of how far environmental beliefs in management policies such as reforestation represent black-box statements in which controversies, potential negative impacts, and alternative solutions are hidden.

This book has shown a variety of ways in which such black-box statements of environmental causality can be made more transparent, or how such statements are reinforced by wider political debates and trends in society. Such analysis is inherently political, yet it also engages closely with the techniques and inference mechanisms of science itself. Box 10.1 lists some key differences between so-called "orthodox" approaches to ecology, and newer, "critical" insights associated with this book. These approaches do not suggest that environmental explanations can exist outside of social and political framings, or that environmental protection should not form a guiding principle for a variety of policies. But they do show that many popular broad-brush statements about ecological degradation and fragility can be criticized for simplicity. A "critical" political ecology allows an engagement with both the social framings and the predictive capacity of ecology in order to show the coproduction of environmental politics and science within different contexts.

Box 10.1   Main contrasts between "orthodox" and "critical" thinking about environment

"Orthodox" approach

Stability and equilibrium within ecosystems; a "balance of nature" which could be disrupted by human activities.

Gradual, linear change within ecosystems.

Homeostatic regulation of systems Environmental change (degradation) may be inferred from "snapshots" or short-term processes.
Assessment and statistics produced and cited by major agencies – national and global – assumed to be authoritative; left unquestioned.

Science and its methods in assessing and modeling environmental change assumed to be neutral and value-free.

There is an aggregate environment to which the "population," "society," or "community" relates.

There are uniformly agreed principles of environmental protection broadly summarized by the attention to impacts resulting from population growth, in conjunction with affluence and technology.

"Critical" approach

Non-equilibrium perspectives; importance of variability over space and time, and of social influences on how environmental processes are scaled
Punctuated changes and contingencies; importance of historical influences on current dynamics; "path-dependency"
Open, "chaotic" systems
Attention to historical sources and the reconstruction of actual change using time–series data
Critique of influential statistics and "scientific" method on the basis of other data sources, including "local" knowledge and "citizen science"
A number of perspectives on a particular environmental issue can co-exist, upheld by different people or institutions, and representing different social and political values or positions
Social groups may be differentiated in many ways; people use and value environment in different ways, and may define differently what is meant by "degradation"
General beliefs about the causes of environmental degradation have to be re-examined in order to see how far they represent framings from different social groups; accepting predefined environmental explanations may not address underlying causes of apparent problems, and may lead to social injustices

Sources adapted from Forsyth, Leach and Scoones, 1998;
Leach and Mearns, 1996.

Political ecology, structure, and agency

This book has therefore called for greater attention to the hidden politics within the scientific discourses of ecology. Yet, a further theme of debate within political ecology concerns whether focusing on scientific discourse might take attention away from more deeply set structural causes of injustice and environmental degradation. In particular, some political ecologists have feared that deconstructing scientific discourses may imply weakening criticism of industry or capitalism as the causes of environmental degradation. Indeed, as Watts and McCarthy wrote (also reported in Chapter 1):

> A compelling and liberatory political ecology must begin with an accurate understanding of capitalist dynamics for the simple and profound reason that they lie at the roots of most problems with which political ecology concerns itself.
>
> (1997: 85)

As discussed in this book, the focus on capitalism has influenced other trends in political ecology. For example, some political ecologists have sought to explain environmental struggles in the context of the opposition of society (commonly comprising grassroots, or non-governmental organizations) against oppressive actions undertaken by industrial or state concerns (e.g. Bunker, 1985; Taylor, 1995; Wapner, 1995; Bryant and Bailey, 1997). It is also common for environmental theorists to present environmental degradation as resulting from social oppression associated with capitalism or the instrumental reason of modernity (e.g. Eckersley, 1992; O'Connor, 1996; Wallerstein, 1999). Accordingly, many approaches to environmental politics have adopted a Habermasian vein, where social movements and criticism may reduce the instrumental reason of oppressive states and industry, and lead to a more socially representative environmental policy (e.g. Eder, 1996; Mol, 1996; Brulle, 2000).

This book has reiterated such calls for social justice. Yet influenced by Foucault, the book has argued for a more critical understanding of what is considered either environmental degradation or oppression, rather than by adopting explanations of environmental degradation based upon the operation of capitalism alone. Similarly, this book also suggests that the classification of political actors into divisions of state, society, and economy should be conducted alongside how far interactions between each may reinforce or challenge existing discourses of environmental degradation. Not to do so may risk seeking a form of environmental democratization that does not challenge the political basis of environmental concerns as well. Ecological discourses form important structuring devices for environmental politics, as well as opposition between state, society, and economy. Accordingly, specific "actors" such as NGOs, state agencies, or transnational companies may not act autonomously to create ecological oppression or liberation, but may themselves – as well as their critics – be acting within structures defined by environmental discourses or storylines (Hajer, 1995; Harré *et al.*, 1999; see Chapters 5 and 6).

The use of linguistic and discourse analysis to question some basic assumptions about how we understand environmental degradation has been criticized by some writers for avoiding the genuine causes of environmental exploitation under capitalism (for example, see the criticism of Leach and Mearns, 1996, by Bernstein and Woodhouse, 2001). Such criticisms, however, may overlook the importance of integrating local framings and experience of ecological change into models of environmental explanation, and the potential negative impacts of relying only on explanations that reflect meta-narratives of causality between capitalism and environment. Much environmental critique of capitalism within some social sciences still reflect the broad-based concerns of Marcuse (e.g. 1964) about the domination of human nature by science, technology, and industrialism developed during the "new" social movements of Europe and North America (e.g. Luke, 1999; Lipietz, 2000). Indeed, the evaluations of environmental degradation often referred to under critiques of capitalism have frequently, and contradictorily, reflected framings of wilderness and balance-of-nature often (and perhaps stereotypically) associated with urban middle-classes (Enzensberger, 1974; Guha and Martinez-Allier, 1997). Increasingly, scholars are questioning whether these simple associations of development and the "domination of nature" are fair (Foster, 2000; see Chapter 5). Explanations also need to acknowledge other social divisions such as gender, caste, and age, although with concern to reflect diverse meanings attached to these (Rocheleau and Edmunds, 1997).

The discussion of social justice under political ecology should therefore acknowledge the different contexts through which justice is defined and used (Collier, 1989; Low and Gleeson, 1998). Analyzing the hidden politics in environmental discourse and explanations is not an abandonment of social justice as a purpose of research, but the acknowledgment of the simplifications and cultural specificities contained in ecological critiques. As Wynne noted, "It has been recognized for some time that sociological deconstruction of knowledge may find itself in unwelcome company, politically speaking" (1996b: 363). Yet, as the Mexican scholar, Enrique Leff noted, deconstruction – and reconstruction – of nature need not lose sight of socially just development:

> There is a need to establish a concept of nature that is appropriate for the building of socialism based on the social use and democratic and participatory management of the environment viewed as a resource base, means of production, and condition of existence, which in turn determines different production life-style patterns.
>
> (1995: 143)

This book has discussed different approaches in constructivist social science to consider how environmental or scientific discourses have been constructed, and with the agency of different actors. Cultural Theory offers a strong criticism of political analysis that counter-poses actors from "state," "society," and "economy" against each other, because such positions represent different "myths" of nature. These myths will be present in all debates and the resulting discussions will result in the production of environmental knowledge aimed at supporting these opposing positions rather than final and unchallengeable explanations of environmental fragility. More poststructuralist theorists instead suggest that these "myths" are too reductionist and uniform. Alternatively, concepts such as narratives, storylines, and Actor Network Theory show more culturally and historically situated accounts of how different environmental discourses and political forces have emerged (see Chapter 4). This book has argued that these approaches offer greater flexibility in showing how different political actors interact to produce environmental explanations and conceptualizations of environmental problems that are now seen as "fact."

However, the potentially damaging impacts of narratives also need to be acknowledged. There is a risk that attractive stories concerning exceptions to environmental orthodoxies – or generalized and inaccurate explanations of environmental degradation – may become "seductive siren calls" (Joerges, 1999) used to illustrate wider points, rather than to show the factors behind local successes. Also, narratives might impose other hidden structures on environmental knowledge and the construction of political actors, such as small farmers against powerful states and multinationals, or of local action against global threats (see Harré *et al.*, 1999). Representations of the Chipko movement in India have been accused of co-opting local livelihood struggles into wider arguments about regionalism or ecofeminism (Jackson, 1995; Bandyopadhyay, 1999; Rangan, 2000). Other well-known cases of environmental adaptations such as Tiffen and Mortimore's (1994) account of *More People, Less Erosion* in Machakos, Kenya, may also be used out of context to suggest that erosion *per se* might not occur; or that opposition to the $I = PAT$ equation is case specific; or that environmental problems do not exist. This book does not support such universal conclusions, but instead seeks to indicate the institutional factors that underlie the predictions of environmental change (such as the $I = PAT$ or USLE equations), but also the experiences and responses to environmental changes at the local level.

Under a "critical" political ecology, the complex interrelationship between structure and agency in environmental problems and explanations is acknowledged. As noted in Box 10.1, there is no "single" environment to which "society" relates, and environmental impacts are a complex result of both social and biophysical interactions. "Critical" political ecology, as discussed in this book, has approached environmental explanation from the perspective of achieving social justice in environmental policy, especially in the developing world. But it seeks to diversify and question the philosophical bases upon which such justice – and consequently different explanations – are established.

Rethinking science and realism

In addition to fighting for social justice, political ecology has also sought to demonstrate the political factors underlying environmental degradation and risk. Yet mixing environmental science with politics has led to further concerns. Much debate has focused on whether integrating science and politics might mean the delegitimization of science as a source of authoritative knowledge. Critics such as Levitt (1999) have also feared that acknowledging social and political framings of science might imply relativism, or the reduction of the truth-value of science to the influence of social structures alone.

This book has challenged these concerns in a variety of ways. First, this book has argued that, contrary to much orthodox scientific thought, so-called "laws" of nature are not accurate representations of environmental problems as experienced in diverse contexts, and consequently the claims for orthodox science to predict reality are clearly flawed (see Chapter 2). Second, it is also important to acknowledge that many constructivist criticisms of environmental science do not aim to dismiss notions of an externally-real world, but instead aim to improve biophysical explanations of complex and diverse environmental problems (see Chapter 3). Simply acknowledging social constructions does not imply the rejection of belief in a "real world," or the criticism of scientists working critically and reflexively within orthodox scientific institutions.

This book has argued that orthodox approaches to environmental explanation fail to acknowledge the institutional basis – such as language, problem closure, or culture – through which environmental problems are experienced, and then how such institutional factors are replicated in scientific "laws" and explanations (see Chapters 3, 4, and 7). Such factors have meant that environmental explanations have often been seen to be acceptable in the circumstances where they were developed, yet have also been called "myths" because they fail to acknowledge the semantic or institutional contexts where they have been applied. As Rouse noted, "science sometimes 'works' only if we change the world to suit it" (1987: 118).

Furthermore, the supposed rigor associated with orthodox, or positivist, science – with its claims to political neutrality about how research is conducted – has been clearly inapplicable (see, for example, Chapter 6). As Nancy Cartwright noted in her book, *How the Laws of Physics Lie*:

> The picture of science that I present ... lacks the purity of positivism. It is a jumble of unobservable entities, causal processes, and phenomenological laws. But it shows one deep positivist criticism: there is no better reality beside the reality we have to hand.
>
> (1983: 19)

Hence, philosophers and sociologists of science have proposed that scientific explanations need to be readjusted in order to acknowledge, rather than deny, the influence of social framings:

If we are to rescue realism, we must abandon our logicist ways and think of language, including scientific language, in a new light... Instead of redefining scientific realism in such a way as to avoid truth, the nicest strategy would be to redefine truth in a way that does justice to the notion and fits in with scientific realism.

(Aronson et al., 1994: 8, 124)

This book has attempted such alternative forms of science by using insights from so-called "institutional" approaches to explanation. These approaches, influenced by philosophical debates such as pragmatism, critical (or "skeptical") realism, and semantic realism, aim to acknowledge the social boundaries and assumptions that give rise to apparently real explanations (see Chapter 8). Such approaches have a number of advantages over the adoption of orthodox conceptions of ecology. They show the various means through which social, linguistic, and semantic contexts may shape complex biophysical events into identifiable "problems" and "processes." They also move debate beyond the simplistic discussion of environmental "truths" and "falsehoods" as demonstrated by measurement of predefined indications of environment. In this sense, they offer a more sophisticated analysis of environmental degradation than the simple statistical measurements of the "truths" behind apparent environmental problems as conducted by Björn Lomborg's (2001) *The Skeptical Environmentalist*. Simply measuring environmental concepts such as "deforestation" or "pollution" without indicating how and by whom such changes are considered problematic is to replicate predefined, and commonly flawed, concepts of degradation.

Institutional approaches to explanation also challenge the interpretation of environmental "myths" as falsehoods, and instead suggest means to see these concepts as self-sustaining truths, narratives, or storylines, or indeed scientific paradigms within the constraints of so-called "normal" science (Funtowicz and Ravetz, 1992). As such, "myths" may continue unchallenged because they serve political purposes, and are upheld by institutions and networks that support them (see Chapters 6 and 9). As stated at the start of this book: "A truth is the kind of error without which a certain species of life could not live" (Nietzsche, *The Will to Power,* 1901: 493).

Revealing the institutional basis of truth claims helps to show how particular environmental explanations may be used to support political objectives when they are presented as non-negotiable forms of truth. In addition, institutional approaches to explanation also allow the possibility for reorienting environmental explanations to the criteria and social framings or boundings relevant to a wider diversity of spatial scales and social divisions. Some environmental assessments, such as the IPCC or approaches to land-use cover and change, have already suggested ways to consider local vulnerability to climate change (e.g. NRC, 1999; Wilbanks and Kates, 1999; Kasperson and Kasperson, 2001). This book, however, suggests such assessments may go several steps further by localizing the phenomenological and semantic framings of environmental risk that give rise to fundamentally revised environmental explanations, rather than simply the communication of preexisting and universalist environmental science to localities. Such localization and diversification of environmental explanations also need not be relativist in the sense implied by critics because it can allow a form of scientific progress by demonstrating the flaws in orthodox approaches to science, and because they are grounded in empirical knowledge of biophysical changes (see Forsyth, 1996; Fairhead and Leach, 1998; Robbins, 1998; Sillitoe, 1998; see Chapter 8).

A "critical" political ecology may seek to adopt such institutional approaches to explanation in order to integrate political analysis with the formation of different explanations of ecological reality. This book has argued throughout that the ecological "laws" and principles that underlie much environmental political debate also need to be considered part of environmental politics. Using institutional approaches to environmental explanations achieves this objective by showing how, and for whom, different statements of environmental causality may be seen to be true.

Table 10.1 lists some of the forms of institutional explanations mentioned in this book, which may be used in varying ways to assist political ecology. The table is entitled "Varieties of institutional realism" (after Harré, 1986; Harré and Krausz, 1996) in order to indicate the importance of different institutional contexts on the production of environmental knowledge. The approaches listed include institutions relevant to people both experiencing environmental problems (such as semantic or transcendental realism), and those who seek to study them (such as paradigms and narratives), although all may influence each other. Such approaches offer alternatives to orthodox (positivist, or critical rationalist) approaches to environmental science. All may be considered various forms of coproduction, or the simultaneous production of knowledge and social order (see Jasanoff, 1996b).

The objectives of these alternative approaches to environmental knowledge creation are to challenge a priori or black-box statements about environmental degradation, and instead reveal how such statements reflect wider social and political framings. By discussing such themes, political ecology may seek to demonstrate the hidden politics within different uses of the word "ecology," and contribute to more locally determined forms of environmental management.

A new agenda for political ecology

Finally, it is worth discussing the implications of this book for political ecology itself.

This book has discussed a wide variety of problems that result from the separation of science and politics within environmental policy. The book has been called *"Critical" Political Ecology* because it seeks to enhance a

Table 10.1 Varieties of institutional realism for environmental debates

Variety of realism	Sample references	Epistemology	Influences on institutional boundaries
Paradigmatic/ programmatic	Kuhn, 1962; Lakatos, 1978	Environmental science is organized into distinct paradigms or programs that provide objectives for research.	Scientists and research funders organize research into paradigms and sequences of "normal" science.
Semantic	Dummett, 1978; Searle, 1995; Tennant, 1997	Environmental processes are framed in sequences according to perceived causes and effects from the perspective of the viewer.	Perspectives depend on social context, and the meaning apparent to different individuals or groups.
Transcendental	Bhaskar, 1975	Environmental "problems" and events are perceived only as a result of social needs that make these meaningful.	All individuals and social groupings experience reality at levels of "empirical," "actual," and "real."
Pragmatic	Rorty, 1989a, b	Environmental explanations emerge as the function of the social solidarities that uphold them.	Social solidarities may be found in various contexts, including scientific networks.
Cultural Theory	Thompson *et al.*, 1990	Environmental perceptions and research findings are organized according to dominant "myths" of nature.	Myths represent different worldviews of egalitarian, individualist, fatalist, and hierarchical, and their associated actors.
Narrative/ Actor Network	Callon, 1986; Litfin, 1994; Hajer, 1995	Environmental explanations are organized into "storylines" that have emerged over time as the result of interfaces between political actors and how they frame biophysical reality.	Narratives emerge historically from discourse coalitions; multivalencies between different actors; or acts of identifying and enrolling objects into networks.
Communal	Wynne, 1996a; Leach *et al.*, 1999	Environmental problems and management are organized according to the participation and needs of people within a communal area such as a village or locality.	Local institutions may arbitrate access to common property resources, or resource endowments and entitlements.
Organizational	Jasanoff, 1990; Guston, 2001	Environmental knowledge is shaped by organizational cultures and the implications of liaising with constituencies.	"Boundary" organizations and science advisers influence epistemological linkages between science and policy networks, and provide legitimacy to science–policy recommendations.

Source: the author.

more critical approach to the unquestioned use of existing environmental science or meta-narratives of ecology. But this name does not imply that this book has a monopoly on criticism, or that political ecology in general needs to be criticized.

This book has sought to contribute most to political ecology by suggesting means to integrate political analysis with the formulation and dissemination of environmental science itself. As discussed in Chapter 1, much discussion of political ecology has overtly reiterated the approach in which "science" provides a neutral backcloth for "politics." This book seeks to challenge this approach. A "critical" political ecology seeks to avoid the separation of science and politics by making the political framings of environmental science more transparent, and by offering the possibility to reformulate environmental explanations in ways that are more relevant to locally determined environmental problems and development objectives.

This book therefore suggests Ulrich Beck might be justified when he wrote, "ecological blindness is a congenital defect of sociologists" (1995: 41). Yet this book also urges a critical engagement with what is meant by "ecology." Many social theorists discuss notions of ecology as forms of unproblematized scientific truth, but these notions need to be analyzed in order to reveal their hidden politics and applicability to different environmental problems in various contexts.

Consequently, this book may also be a way to address differences between approaches to cultural and political ecology that traditionally have focused either on physical impacts of land-use-cover changes, or alternatively at the marginalizing impacts of economic development and political oppression on environment and people (Chapter 1; see also Escobar, 1999; Turner, 2002). This book, however, may be considered constructively critical of both approaches on one hand for overlooking the hidden politics of orthodox scientific methods in much land-use-cover change work, or in the simplifications of many meta-narratives underlying ecological critiques of capitalism on the other. Perhaps more importantly, this book has also argued that approaches to political ecology that adopt no critical engagement with the meaning of "ecology" – or which use ecology as a metaphor for the connectivity of political actors – run the risk of reinforcing currently dominant explanations of environmental problems without assessing how far such explanations reinforce historic power relations and selective experiences of environment.

If "political ecology" is to be worthy of its name, it has to be more than another term for "environmental politics," and instead should seek to conduct critical analysis of the political factors that underlie competing definitions and explanations of environmental reality. Such objectives mean increasing public debate about existing environmental assumptions, and increasing capacity for the development of alternative approaches. There is a need for greater public participation in the *formulation* of environmental science, rather than in simply the *access* to science. Under a diversified, critical approach to ecology, more attention will be given to the transparency, legitimacy, and participation in environmental science than to the enforcement of predefined notions of risk and assumed causes of environmental degradation. The demonstration in this book that environmental science is not an a priori basis for environmental politics may be an important step toward these objectives.

Bibliography

Adams, W.M. (1997) "Rationalization and conservation: ecology and the management of nature in the United Kingdom," *Transactions of the Institute of British Geographers* NS 22: 277–291.

Adams, W.M. (2001) *Green Development: Environment and Sustainability in the Third World*, 2nd edn, London: Routledge.

Adger, W. (2000) "Social and ecological resilience: are they related?" *Progress in Human Geography* 24: 3, 347–364.

Adger, W., Benjaminsen, T., Brown, K., and Svarstad, H. (2001) "Advancing a political ecology of global environmental discourses," *Development and Change* 32: 4, 687–715.

Agarwal, A. and Narain, S. (1991) *Global Warming in an Unequal World*, New Delhi: Center for Science and Environment.

Agger, B. (1992) *The Discourse of Domination: from the Frankfurt School to Postmodernism*, Evanston, IL: Northwestern University Press.

Agrawal, A. (1995) "Dismantling the divide between indigenous and scientific knowledge," *Development and Change* 26: 3, 413–439.

Agrawal, A. and Gibson, C. (1999) "Enchantment and disenchantment: the role of community in natural resource conservation," *World Development* 27: 4, 629–649.

Ahluwalia, M. (1997) "Representing communities: the case of a community-based watershed management project in Rajasthan, India," *IDS Bulletin* 28: 4, 23–26.

Alford, C. (1985) *Science and the Revenge of Nature*, Gainesville, FL: University Presses of Florida.

Alford, D. (1992) "Streamflow and sediment transport from mountain watersheds of the Chao Phraya basin, northern Thailand: a reconnaissance study," *Mountain Research and Development* 12: 3, 257–268.

Allen, B. (1993) *Truth in Philosophy*, Cambridge, MA: Harvard University Press.

Andersen, I. and Jaeger, B. (1999) "Scenario and workshops and consensus conferences: towards more democratic decision-making," *Science and Public Policy* 26: 5, 331–340.

Anderson, D. (1984) "Depression, Dust Bowl, demography and drought: the colonial state and soil conservation in East Africa during the 1930s," *African Affairs* 83: 332.

Anderson, J. (1968) "Cultural adaptation to threatened disaster," *Human Organization* 27: 298–307.

Anderson, L. (1994) *The Political Ecology of the Modern Peasant: Calculation and Community*, Baltimore, MD: Johns Hopkins University Press.

Angelsen, A. (1995) "Shifting cultivation and deforestation: a study from Indonesia," *World Development* 23: 1713–1729.

Angelsen, A. and Kaimowitz, D. (1999) *Rethinking the Causes of Deforestation: Lessons from Economic Models*, Washington, DC: World Bank.

Archer, M., Bhaskar, R., Collier, A., Lawson, T., and Norrie, A. (eds) (1998) *Critical Realism: Essential Readings*, London: Routledge.

Arnold, D. (1996) *The Problem of Nature*, Oxford: Blackwell.

Aronowitz, S. (1988) *Science as Power: Discourse and Ideology in Modern Society*, Basingstoke: Macmillan Press.

Aronson, J., Harré, R., and Cornell Way, E. (1994) *Realism Rescued: How Scientific Progress is Possible*, London: Duckworth.

Atkinson, A. (1991) *Principles of Political Ecology*, London: Belhaven.

Baarschers, W. (1996) *Eco-facts and Eco-fiction: Understanding the Environmental Debate*, London: Routledge.

Baden, J. (ed.) (1994) *Environmental Gore: a Constructive Response to "Earth in the Balance,"* San Francisco, CA: Pacific Research Institute for Public Policy.

Bandyopadhyay, J. (1999) "Chipko movement: of floated myths and flouted realities," *Economic and Political Weekly* 34: 15, Apr, 10–16.

Bankoff, G. (1999) "A history of poverty: the politics of natural disasters in the Philippines, 1985–95," *Pacific Review* 12: 3, 381–420.

Bankoff, G. (2001) "Rendering the world unsafe: 'vulnerability' as western discourse," *Disasters* 25: 1, 19–35.

Banuri, T. (1990) "Development and the politics of knowledge: a critical interpretation of the social role of modernization theories in the development of the Third World," in Marglin, F. and Marglin, S. (eds) *Dominating Knowledge: Development Culture and Resistance*, Oxford: Clarendon.

Barnes, B., Bloor, D., and Henry, J. (eds) (1996) *Scientific Knowledge: a Sociological Analysis*, London: Athlone Press.

Barr, J. and Birke, L. (1998) *Common Science: Women, Science and Knowledge*, Bloomington and Indianapolis, IN: Indiana University Press.

Barraclough, S. and Ghimire, K. (1996) "Deforestation in Tanzania: beyond simplistic generalizations," *The Ecologist* 26: 3, 104–107.

Bass, T. (1990) *Camping with the Prince, and Other Tales of Science in Africa*, Cambridge: Latterworth Press.

Bassett, T. and Zuéli, K. (2000) "Environmental discourses and the Ivorian Savanna," *Annals of the Association of American Geographers* 90: 1, 67–95.

Batterbury, S. (1996) "Planners or performers? Reflections on indigenous dryland farming in northern Burkina Faso," *Agricultural and Human Values* 13: 3, 12–22.

Batterbury, S. (2001) "Landscapes of diversity: a local political ecology of livelihood diversification in south-western Niger," *Ecumene* 8: 4, 438–464.

Batterbury, S. and Bebbington, A. (eds) (1999) "Environmental histories, access to resources and landscape change: an introduction," Special edition of *Land Degradation and Development* 10: 4, 279–288.

Batterbury, S. and Forsyth, T. (1999) "Fighting back: human adaptations in marginal environments," *Environment* 41: 6, 6–11, 25–29.

Batterbury, S., Forsyth, T., and Thomson, K. (1997) "Environmental transformations in developing countries: hybrid research and democratic policy," *Geographical Journal* 163: 2, 126–132.

Batterbury, S. and Warren, A. (eds) (2001) "The African Sahel 25 years after the great drought: assessing progress and moving towards new agendas and approaches," Special edition of *Global Environmental Change* 11: 1, 1–8.

Bebbington, A. (1997) "Social capital and rural intensification: local organizations and islands of sustainability in the rural Andes," *Geographical Journal* 163: 2.

Bebbington, A. (1999) "Capitals and capabilities: a framework for analyzing peasant viability, rural livelihoods and poverty," *World Development* 27: 12, 2021–2044.

Bebbington, A. and Batterbury, S. (2001) "Transnational livelihoods and landscapes: political ecologies of globalization," *Ecumene* 8: 4, 369–380.

Beck, U. (1992) *Risk Society: Towards a New Modernity*, Cambridge: Polity.

Beck, U. (1995) *Ecological Politics in an Age of Risk*, Cambridge: Polity.

Benedick, R. (1991) *Ozone Diplomacy: New Directions in Safeguarding the Planet*, Cambridge, MA: Harvard University Press.

Benhabib, S. (ed.) (1996) *Democracy and Difference: Contesting the Boundaries of the Political*, Princeton, NJ: Princeton University Press.

Benjaminsen, T. (2001) "The population–environment nexus in the Malian cotton zone," *Global Environmental Change* 11: 4, 283–295.

Benton, T. (1996) "Marxism and natural limits: an ecological critique and reconstruction," in Benton, T. (ed.) *The Greening of Marxism*, New York: Guilford Press, pp. 157–186.

Berger, P. (1987) *The Capitalist Revolution*, Aldershot: Wildwood House.

Berkes, F., Folke, C., and Colding, J. (eds) (1998) *Linking Social and Ecological Systems: Management Practices and Social Mechanisms for Building Resilience*, Cambridge: Cambridge University Press.

Bernstein, H. and Woodhouse, P. (2001) "Telling environmental change like it is? Reflections on a study in Sub-Saharan Africa," *Journal of Agrarian Change* 1: 2, 283–324.

Berry, S. (1989) "Social institutions and access to resources," *Africa* 59: 1, 41–55.

Bhaskar, R. (1975) *A Realist Theory of Science*, Leeds: Leeds Books.

Bhaskar, R. (1986) *Scientific Realism and Human Emancipation*, London: Macmillan.

Bhaskar, R. (1991) *Philosophy and the Idea of Freedom*, Oxford: Blackwell.

Biot, Y. (1995) *Rethinking Research on Land Degradation in Developing Countries*, Washington, DC: World Bank.

Blaikie, P. (1985) *The Political Economy of Soil Erosion in Developing Countries*, London: Longman Development Series.

Blaikie, P. (1995) "Understanding environmental issues," in Morse, S. and Stocking, M. (eds) *People and Environment*, London: University College London Press, pp. 1–30.

Blaikie, P. and Brookfield, H.C. (ed.) (1987) *Land Degradation and Society*, London: Methuen.

Blaikie, P., Cannon, T., Davis, I., and Wisner, B. (1994) *At Risk: Natural Hazards, People's Vulnerability and Disasters*, London: Routledge.

Blowers, A. (1997) "Environmental policy: ecological modernization or the risk society?" *Urban Studies* 34: 5/6.

Boehmer-Christiansen, S. (1994) "Global climate protection policy: the limits of scientific advice part II," *Global Environmental Change* 4: 3, 845–871.

Botkin, D. (1990) *Discordant Harmonies: A New Ecology for the Twenty-First Century*, New York: Oxford University Press.

Braun, B. and Castree, N. (eds) (1998) *Remaking Reality: Nature at the Millennium*, London: Routledge.

Brechin, S. (1997) *Planting Trees in the Developing World: A Sociology of International Organizations*, Baltimore, MD: Johns Hopkins University.

Brookfield, H., Potter, L., and Byron, Y. (1995) *In Place of the Forest: Environmental and Socio-Economic Transformation in Borneo and the Eastern Malay Peninsula*, Tokyo, Paris, New York: United Nations University Press.

Brown, L. (2001) *Eco-economy: Building an Economy for the Earth*, London and Washington, DC: Earthscan and Earth Policy Institute.

Brown, R. (1998) *Toward a Democratic Science: Scientific Narration and Civic Communication*, New Haven, CT: Yale University Press.

Brulle, R. (2000) *Agency, Democracy, and Nature: The US Environmental Movement from a Critical Theory Perspective*, Cambridge, MA: MIT Press.

Bryant, R. (1992) "Political ecology: an emerging research agenda in Third-World studies," *Political Geography* 11: 1, 12–36.

Bryant, R. (1997a) *The Political Ecology of Forestry in Burma 1824–1994*, London: Hurst.

Bryant, R. (1997b) "Beyond the impasse: the power of political ecology in Third World environmental research," *Area* 29: 1–15.

Bryant, R. (1998) "Power, knowledge and political ecology in the Third World: a review," *Progress in Physical Geography* 22: 1, 79–94.

Bryant, R. and Bailey, S. (1997) *Third-World Political Ecology*, London: Routledge.

Bunge, M. (1991) "What is science? Does it matter to distinguish from pseudoscience? A reply to my commentators," *New Ideas in Psychology* 9: 2, 245–283.

Bunker, S. (1985) *Underdeveloping the Amazon: Extraction, Unequal Exchange and the Failure of the Modern State,* Chicago, IL: University of Illinois Press.

Button, M. and Madson, K. (1999) "Deliberative democracy in practice: challenges and prospects for civic deliberation," *Polity* 31: 4, 609–637.

Calder, I. (1999) *The Blue Revolution: Land Use and Integrated Water Resources*, London: Earthscan.

Calder, I. and Aylward, B. (2002) *Forests and Floods: Perspectives on Watershed Management and Integrated Flood Management*, Rome: FAO, and Newcastle, University of Newcastle.

Calhoun, C. (ed.) (1992) *Habermas and the Public Sphere*, Cambridge, MA: MIT Press.

Callon, M. (1986) "Some elements of a sociology of translation: domestication of the scallops and the fishermen of Saint Brieuc Bay," in Law, J. (ed.) *Power, Action and Belief: A New Sociology of Knowledge? Sociological Review Monograph,* Volume 32, London: Routledge and Kegan Paul, pp. 196–233.

Callon, M. (1995) "Four models for the dynamics of science," in Jasanoff, S., Markle, G., Petersen, J., and Pinch, T. (eds) *Handbook of Science and Technology Studies*, Thousand Oaks, CA: Sage, pp. 29–63.

Carnegie Commissions on Science, Technology, and Government (1992) *Enabling the Future: Linking Science and Technology to Societal Goals*, New York: Carnegie Commission.

Carney, D. (1998) *Sustainable Rural Livelihoods: What Contribution Can We Make?* London: Department for International Development.

Carson, R. (1962) *Silent Spring*, Boston, MA: Houghton Mifflin.

Cartwright, N. (1983) *How the Laws of Physics Lie*, Oxford: Clarendon.

Cartwright, N. (1999) *The Dappled World: a Study of the Boundaries of Science*, Cambridge: Cambridge University Press.

Cash, D. and Clark, W. (2001) "From science to policy: assessing the assessment process," *John F. Kennedy School of Government, Harvard University, Faculty Research Working Paper Services, RWP01–045*, Cambridge, MA: Kennedy School of Government.

Casman, E. and Dowlatabadi, H. (eds) (2002) *Contextual Determinants of Malaria*, Washington, DC: RFF Press.

Castree, N. (1995) "The nature of produced nature: materiality and knowledge construction in Marxism," *Antipode* 27: 1, 12–48.

Castree, N. (ed.) (2001) *Social Nature*, Oxford: Blackwell.

Castree, N. and Braun, B. (1998) "The construction of nature and the nature of construction: analytical and political tools for building survivable futures," in Braun, B. and Castree, N. (eds) *Remaking Reality: Nature at the Millennium*, London: Routledge, pp. 3–42.

Chambers, R. (1997) *Whose Reality Counts? Putting the First Last*, London: IT Books.

Chambers, R. and Conway, G. (1992) *Sustainable Rural Livelihoods: Practical Concepts for the 21st Century*, IDS Discussion Paper 296: Brighton, UK: IDS.

Chapman, G. and Thompson, M. (eds) (1995) *Water and the Quest for Sustainable Development in the Ganges Valley*, London: Cassell.

Chomitz, K. and Kumari, K. (1996) "The domestic benefits of tropical forests: a critical review emphasizing hydrological functions," *World Bank Policy Research Working Paper 160*, World Bank, Policy Research Department, Washington, DC.

Clark, W. (1985) "Scales of climate impacts," *Climatic Change* 7: 5–27.

Cleaver, F. (2001) "Institutional bricolage, conflict and cooperation in Usangu, Tanzania," *IDS Bulletin* 32: 4, 26–35.

Cline-Cole, R. and Madge, C. (eds) (2000) *Contesting Forestry in West Africa*, Aldershot: Ashgate.

Cobb, R. and Elder, C. (1972) *Participation in American Politics: the Dynamics of Agenda Building*, Baltimore, MA: Johns Hopkins University Press.

Cockburn, A. and Ridgeway, J. (eds) (1979) *Political Ecology: an Activist's Reader on Energy, Land, Food, Technology, Health, and the Economics and Politics of Social Change*, New York: Times Books.

Cohen, E. (1989) "'Primitive and remote': hill tribe trekking in Thailand," *Annals of Tourism Research* 16: 1, 30–61.

Collier, A. (1989) *Scientific Realism and Socialist Thought*, Hemel Hempstead: Harvester Wheatsheaf.

Collier, A. (1994) *Critical Realism: The Work of Roy Bhaskar*, London: Verso.

Collingridge, D. and Reeve, C. (1986) *Science Speaks to Power: the Role of Experts in Policy Making*, London: Pinter.

Conklin, H. (1954) "An ethnoecological approach to shifting agriculture," *Transactions of the New York Academy of Sciences* 77: 133–142.

Conley, A. (1996) *Ecopolitics: the Environment in Poststructuralist Thought*, London: Routledge.

Corbridge, S. (1986) *Capitalist World Development: a Critique of Radical Development Geography*, Basingstoke: Macmillan.

Correll, E. (1999) *The Negotiable Desert: Expert Knowledge in the Negotiations of the Convention to Combat Desertification*, Linköping Studies in Arts and Science, no. 191, Department of Water and Environmental Studies, Sweden: Linköping University.

Costanza, R., Kemp, M., and Boynton, W. (1995) "Scale and biodiversity in estuarine ecosystems," in Perrings, C., Mäler, K., Holling, C., and Jansson, B. (eds) *Biodiversity Loss: Economic and Sociological Issues*, Cambridge: Cambridge University Press, pp. 84–125.

Covey, J. (1995) "Accountability and effectiveness in NGO policy alliances," in Edwards, M. and Hulme, D. (eds) *Non-Governmental Organisations – Performance and Accountability: Beyond the Magic Bullet*, London: Earthscan, pp. 167–182.

Crapper, P. (1962) *Land Requirements for the Papua New Guinea Population*, Internal Report, CSIRO Land Research Series no. 1, Melbourne.

Cronin, E. (1979) *The Arun: A Natural History of the World's Deepest Valley*, Boston, MA: Houghton Mifflin.

Cronon, W. (1991) *Nature's Metropolis: Chicago and the Great West*, New York: Norton.

Cronon, W. (1992) "A place for stories: nature, history and narrative," *Journal of American History* 78: 1347–1376.

Cronon, W. (1996) "The trouble with wilderness: a response," *Environmental History* 1: 1, 47–55.

Cullet, P. and Kameri-Mbote, P. (1998) "Joint implementation and forestry projects: conceptual and operational fallacies," *International Affairs* 74: 2, 393–408.

Custodio, E. (2000) *The Complex Concept of Overexploited Aquifer*, Papeles del Proyecto Aguas Subterráneas, Serie A: Uso intensiob del las agues subterráneas, aspectos ecológicos, tecnolólogicos y éticos, Madrid: Fundación Marcelino Botín.

Cutajar, M.Z. (2001) "Notes for closing session," *Global Change Open Science Conference* (IGBP–IHDP–WCRP), Amsterdam, 10–13 July 2001, Bonn: International Human Dimensions Program on Global Environmental Change.

Cutter, S. (1995) "The forgotten casualties: women, children and environmental change," *Global Environmental Change* 5: 3, 181–194.

Dahlberg, A. and Blaikie, P. (1999) "Changes in landscape or in interpretation? Reflections based on the environmental and socio-economic history of a village in northeast Botswana," *Environment and History* 5: 2, 127–174.

Daly, J. (1991) "Does a constructivist view require epistemological relativism? A response to Turnbull," *Social Studies of Science* 21: 3, 568–571.

Davies, B. and Harré, R. (1990) "Positioning: the discursive production of selves," *Journal for the Theory of Social Behavior* 20: 1, 43–63.

de Jong (1999) "Institutionalized criticism: the demonopolization of scientific advising," *Science and Public Policy* 26: 3, 193–199.

DeHart, J. and Soulé, P. (2000) "Does I = PAT work in local places?" *Professional Geographer* 52: 1, 1–10.

Demeritt, D. (1994) "The nature of metaphors in cultural geography and environmental history," *Progress in Human Geography* 18: 2, 163–185.

Demeritt, D. (1998) "Science, social constructivism and nature," in Braun, B. and Castree, N. (eds) *Remaking Reality: Nature at the Millennium*, London: Routledge, pp. 173–193.

Demeritt, D. (2001) "The construction of global warming and the politics of science," *Annals of the Association of American Geographers* 91: 2, 307–337.

Denevan, W. (1989) "The fragile lands of Latin America," in Browder, J. (ed.) *The Fragile Lands of Latin America: Strategies for Sustainable Development*, Boulder, CO: Westview Press, pp. 3–25.

DFID (Department for International Development, Great Britain) (2002) *Poverty and Environment*, information booklet, March, London: DFID.

Dickens, P. (1996) *Reconstructing Nature: Alienation, Emancipation and the Division of Labour*, London: Routledge.

Dierkes, M. and van Grote, C. (2000) (eds) *Between Understanding and Trust: the Public, Science and Technology*, Amsterdam: Harwood.

Dietz, T. and Rosa, E. (1994) "Rethinking the environmental impacts of population, affluence and technology," *Human Ecology Review* 1: 2, 277–300.

Donovan, D. (1981) "Fuelwood: how much do we need?' *Newsletter* (DGD 14), Hanover, NH: Institute of Current World Affairs.

Dougill, A., Thomas, D., and Heathwaite, A. (1999) "Environmental change in the Kalahari: integrated land degradation studies for nonequilibrium dryland environments," *Annals of the Association of American Geographers* 89: 3, 420–442.

Douglas, M. (1978) *Purity and Danger*, London and New York: Routledge.

Douglas, M. (1985) *Risk Acceptability According to the Social Sciences*, London: Routledge and Kegan Paul.

Douglas, M. (1987) *How Institutions Think*, London and New York: Routledge and Kegan Paul.

Douglas, M. (1993) "The depoliticization of risk," in Ellis, R. and Thompson, M. (eds) *Culture Matters: Essays in Honor of Aaron Wildavsky*, Boulder, CO: Westview Press, pp. 121–132.

Dove, M. (1992) "Foresters' belief about farmers: a priority for social science research in social forestry," *Agroforestry Systems* 17: 13–41.

Dowlatabadi, H. (1997) "Assessing the health impacts of climate change: an editorial essay," *Climatic Change* 35: 2, 137–144.

Dregne, H. (1985) "Aridity and land degradation," *Environment* 27: 8, 16–20, 28–33.

Drèze, J. and Sen, A. (1990) *The Political Economy of Hunger: Vol. 1, Entitlement and Well-Being*, Oxford: Clarendon.

Driver, T. (1999) "Anti-erosion policies in the mountain areas of Lesotho: The South African connection," *Environment and History* 5: 1, 1–25.

Dryzek, J. (1990) *Discursive Democracy: Politics, Policy and Political Science*, Cambridge: Cambridge University Press.

Dryzek, J. (1997) *The Politics of the Earth*, Oxford: Oxford University Press.

Dryzek, J. (1998 [1995]) "Political and ecological communication," in Dryzek, J. and Schlosberg, D. (eds) *Debating the Earth: The Environmental Politics Reader*, Oxford: Oxford University Press, pp. 584–598.

Dubos, R. (1964) "Environmental biology," *Bioscience* 14: 1, 11–14.

Dummett, M. (1978) *Truth and Other Enigmas*, London: Duckworth.

Dunne, T. and Black, R. (1970) "Partial area contributions to storm runoff in a small New England watershed," *Water Resources Research* 6: 5, 1296–1311.

Dunne, T. and Leopold, L. (1978) *Water in Environmental Planning*, San Francisco, CA: W.H. Freeman and Company.

Durant, J. (1999) "Participatory technology assessment and the democratic model of the public understanding of science," *Science and Public Policy* 25: 5, 313–319.

Dürrenberger, G., Kastenholz, H., and Behringer, J. (1999) "Integrated assessment focus groups: bridging the gap between science and policy?' *Science and Public Policy* 25: 5, 341–349.

Dye, C. and Reiter, P. (2000) "Temperatures without fevers?' *Science* 289: 1697–1698.

Eckersley, R. (1992) *Environmentalism and Political Theory: Towards an Ecocentric Approach*, London: University College London Press.

Eckholm, E. (1976) *Losing Ground: Environmental Stress and Food Problems*, New York: W.W. Norton.

Eckley, N. (2001) "Designing effective assessments: the role of participation, science and governance, and focus," *European Environment Agency Environmental Issue Report no. 26*, Copenhagen: Office for Official Publications for the European Communities.

Economist, The (1996) "Science and technology: the science of sexual discrimination," June, 22: 97–99.

Eder, K. (1996) "The institutionalization of environmentalism: ecological discourse and the second transformation of the public sphere," in Lash, S., Szersynski, B., and Wynne, B. (eds) *Risk, Environment and Modernity*, London: Sage, pp. 203–223.

Edwards, A. (1999) "Scientific expertise and policy-making: the intermediary role of the pubic sphere," *Science and Public Policy* 26: 3, 163–170.

Edwards, P. and Schneider, S. (2001) "Self governance and peer review in science for policy: the case of the IPCC Second Assessment Report," in Miller, C. and Edwards, P. (eds) *Changing the Atmosphere: Expert Knowledge and Environmental Governance*, Cambridge, MA: MIT Press, pp. 219–246.

Ehrlich, P. and Ehrlich, A. (1991) *Healing the Planet: Strategies for Resolving the Environmental Crisis,* Cambridge, MA: Perseus.

Ehrlich, P. and Ehrlich, A. (1996) *Betrayal of Science and Reason: How Anti-Environmental Rhetoric Threatens our Future*, Washington, DC and Covelo, CA: Island Books.

Ehrlich, P. and Holdren, J. (1974) "The impact of population growth," *Science* 171: 1212–1217.

Enzensberger, H. (1974) "A critique of political ecology," *New Left Review* 84: 3–31.

Epstein, S. (1996) *Impure Science: AIDS Activism and the Politics of Knowledge*, Los Angeles, CA: UCLA Press.

Escobar, A. (1996) "Constructing nature: elements for a poststructural political ecology," in Peet, R. and Watts, M. (eds) *Liberation Ecologies: Environment, Development, Social Movements*, London: Routledge, pp. 46–68.

Escobar, A. (1998) "Whose knowledge, whose nature? Biodiversity, conservation, and the political ecology of social movements," *Journal of Political Ecology* 5: 53–82.

Escobar, A. (1999) "Steps to an antiessentialist political ecology," *Current Anthropology* 40: 1, 1–30.

Escobar, E. (1995) *Encountering Development: the Making and Unmaking of the Third World*, Princeton, NJ: Princeton University Press.

Eyerman, R. and Jamison, A. (1991) *Social Movements: A Cognitive Approach*, Cambridge: Polity.

Ezrahi, Y. (1990) *The Descent of Icarus: Science and the Transformation of Contemporary Democracy,* Cambridge, MA: Harvard University Press.

Fairhead, J. and Leach, M. (1996) *Misreading the African Landscape: Society and Ecology in a Forest–Savanna Mosaic*, Cambridge: Cambridge University Press.

Fairhead, J. and Leach, M. (1998) *Reframing Deforestation: Global Analysis and Local Realities: Studies in West Africa*, London: Routledge.

Farrell, A., VanDeveer, S., and Jäger, J. (2001) "Environmental assessments: four under-appreciated elements of design," *Global Environmental Change* 11: 4, 311–333.

Ferguson, J. (1990) *The Anti-Politics Machine: "Development," Depoliticization and Bureaucratic Power in Lesotho*, Minneapolis, MN and London: University of Minnesota Press.

Fetzer, J. and Almeder, R. (1993) *Glossary of Epistemology/Philosophy of Science*, New York: Paragon House.

Fischer, F. (1999) "Technological deliberation in a democratic society: the case for participatory inquiry," *Science and Public Policy* 25: 5, 294–302.

Fischer, F. (2000) *Citizens, Experts and the Environment: the Politics of Local Knowledge,* Durham and London: Duke University Press.

Fogel, C. (2001) "From invisibility to...? Indigenous peoples, traditional ecological knowledge, and the Kyoto Protocol," Paper presented at the workshop, *Localizing and Globalizing: Knowledge Cultures of Environment and Development,* Kennedy School of Government, Harvard University, April 5–7.

Fogel, C. (2002). "Perspectives on 'carbon sinks': the construction of knowledge and policy to abate global climate change, 1980–2001." Ph.D. Dissertation, University of California at Santa Cruz.

Forsyth, T. (1994) "Shut up or shut down: how a Thai health agency was forced to close when it challenged a major foreign investor," *Asia, Inc.* 3: 4, April, 30–37.

Forsyth, T. (1996) "Science, myth and knowledge: testing Himalayan environmental degradation in Thailand," *Geoforum* 27: 3, 375–392.

Forsyth, T. (1998a) "Mountain myths revisited: integrating natural and social environmental science," *Mountain Research and Development* 18: 2, 126–139.

Forsyth, T. (1998b) "The politics of environmental health: suspected industrial poisoning in Thailand," in Hirsch, P. and Warren, C. (eds) *The Politics of Environment in Southeast Asia: Resources and Resistance*, Routledge, London, pp. 210–226.

Forsyth, T. (1999a) *International Investment and Climate Change: Energy Technologies for Developing Countries,* London: Earthscan and the Royal Institute of International Affairs.

Forsyth, T. (1999b) "Flexible mechanisms of climate technology transfer," *Journal of Environment and Development* 8: 3, 238–257.

Forsyth, T. (1999c) "Environmental activism and the construction of risk: implications for NGO alliances," *Journal of International Development* 11: 5, 687–700.

Forsyth, T. (2001a) "Environmental social movements in Thailand: how important is class?" *Asian Journal of Social Sciences* 29: 1, 35–51.

Forsyth, T. (2001b) "Political ecology and critical realism," in Stainer, A. and Lopez, G. (eds) *After Postmodernism: Critical Realism?* London: Athlone Press, pp. 146–154.

Forsyth, T. (2002) "What happened on *The Beach*? Social movements and governance of tourism in Thailand," *International Journal of Sustainable Development* 5: 3, 325–336.

Forsyth, T., Leach, M., and Scoones, I. (1998) *Poverty and Environment: Research and Policy Priorities*, UNDP and European Commission.

Foster, J. Bellamy (2000) *Marx's Ecology: Materialism and Nature*, New York: Monthly Review Press.

Foucault, M. (1980) "Two Lectures," in *Power/Knowledge. Selected Interviews and Other Writings 1972–1977* (ed. C. Gordon), London: Harvester, pp. 78–102.

Fox J., Dao Minh Truong, Rambo, A., Nghiem Phuong Tuyen, Le Trong Cuc, and

Stephen Leisz (2000) "Shifting cultivation: a new old paradigm for managing tropical forests," *BioScience* 50: 6, 521–528.

Frankfurt, H. (1978) *On the Importance of What We Care About*, Cambridge: Cambridge University Press.

Fuller, S. (1993) *Philosophy, Rhetoric and the End of Knowledge: the Coming of Science and Technology Studies,* Madison, WI: University of Wisconsin Press.

Fuller, S. (2000) *The Governance of Science*, Buckingham and Philadelphia, PA: Open University Press.

Funtowicz, S. and Ravetz, J. (1985) "Three types of risk assessment: a methodological analysis," in Whipple, C. and Covello, V. (eds) *Risk Analysis in the Private Sector*, New York: Plenum.

Funtowicz, S. and Ravetz, J. (1992) "Three types of risk assessment and the emergence of post normal science," in Krimsky, S. and Golding, D. (eds) *Social Theories of Risk*, Westport, CT: Praeger.

Funtowicz, S. and Ravetz, J. (1993) "Science for the post-normal age," *Futures* 26: September, 739–756.

Gandy, M. (1997) "Ecology, modernity and the intellectual legacy of the Frankfurt School," in Light, A. and Smith, J. (eds) *Space, Place and Environmental Ethics: Philosophy and Geography I*, London: Rowman and Littlefield, pp. 231–254.

Geertz, C. (1963) *Agricultural Involution: the Processes of Ecological Change in Indonesia*, Berkeley, CA: The University of California Press.

Gelbspan, R. (1997) *The Heat is On: The High Stakes Battle over Earth's Threatened Climate*, New York: Addison-Wesley.

Giddens, A. (1973) *The Class Structure of the Advanced Societies*, London: Hutchinson University Library, New York: Harper and Row.

Giddens, A. (1990) *Consequences of Modernity*, London: Polity Press.

Giddens, A. (1994) *Beyond Left and Right: the Future of Radical Politics*, Cambridge: Polity.

Gieryn, T. (1999) *Cultural Boundaries of Science: Credibility on the Line*, Chicago, IL: University of Chicago Press.

Goldman, M. (2001) "The birth of a discipline: producing authoritative green knowledge, World Bank style," *Ethnography* 2: 2, 191–217.

Gore, A. (1992) *Earth in the Balance: Forging a New Common Purpose,* revised edn, London: Earthscan.

Gorz, A. (1983 [1975]) *Ecology as Politics*, London: Pluto Press.

Gorz, A. (1994 [1991]) *Capitalism, Socialism, Ecology*, London: Verso.

Goudie, A. (1990) *The Human Impact on the Natural Environment*, 3rd edn, Oxford: Blackwell.

Grainger, A. (1990) *The Threatening Desert: Controlling Desertification*, London: Earthscan.

Greenberg, J. and Park, T. (1994) "Political ecology," *Journal of Political Ecology* 1: 1–12.

Grimes, P. (1999) "The horsemen and the killing fields: the final contradiction of capitalism," in Goldfrank, W., Goodman, D., and Szasz, A. (eds) *Ecology and the World System*, Westport, CT: Greenwood Press, pp. 13–42.

Gross, P. and Levitt, N. (1994) *Higher Superstition: the Academic Left and its Quarrels With Science*, Baltimore, MA: Johns Hopkins Press.

Grossman, L. (1998) *The Political Ecology of Bananas: Contract Farming, Peasants and Agrarian Change in the Eastern Caribbean*, London and Chapel Hill: University of North Carolina Press.

Grove, R. (1995) *Green Imperialism: Colonial Expansion, Tropical Island Edens and the Origins of Environmentalism, 1660–1860*, Cambridge: Cambridge University Press.

Grove-White, R. (1999) "Afterword: On 'sound science,' the environment, and political authority," *Environmental Values* 8: 277–282.

Grove-White, R., Macnaghten, P., Mayer, S., and Wynne, B. (1997) *Uncertain World: Genetically Modified Organisms, Food and Public Attitudes in Britain*, Lancaster: Centre for Study of Environmental Change, Lancaster University.

Grubb, M., Brack, D., and Vrolijk, C. (1999) *The Kyoto Protocol, a Guide and Assessment*, London: Earthscan and the Royal Institute of International Affairs.

Grubb, M., Koch, M., Munson, A., Sullivan, F., and Thomson, K. (1993) *The Earth Summit Agreements: a Guide and Assessment*, London: Earthscan and RIIA.

Guha, R. and Martinez-Allier, J. (1997) *Varieties of Environmentalism*, London: Earthscan.

Gupta, J. (1997). *The Climate Change Convention and Developing Countries: from Conflict to Consensus?* London, Dordrecht, Boston: Kluwer.

Guston, D. (2001) "Boundary organizations in environmental policy and science: an introduction," *Science, Technology and Human Values* 26: 4, 399–408.

Gyawali, D. (2000) *Water in Nepal*, Kathmandu: Himal Books.

Haas, P. (1992) "Introduction: epistemic communities and international policy coordination," *International Organization* 46: 1, 1–35.

Habermas, J. (1974) *Theory and Practice*, translated by John Viertel, Boston, MA: Beacon Press.

Habermas, J. (1981) "New social movements," *Telos* 49: 33–37.

Habermas, J. (1987 [1985]) *The Theory of Communicative Action, Vol. 2*, translated by Thomas McCarthy, Boston, MA: Beacon Press.

Hagel, Chuck, Senator (2001) Testimony to Senate Committee on Commerce, Science and Transportation, in *Notes from Senate Committee on Commerce, Science and Transportation hearing on Intergovernmental Panel on Climate Change (IPCC) Third Assessment Report, May 1*, Washington, DC: NOAA Office of Legislative Affairs.

Hajer, M. (1993) "The politics of environmental discourse: a study of the acid rain controversy in Great Britain and the Netherlands." Unpublished Ph.D. thesis, University of Oxford.

Hajer, M. (1995) *The Politics of Environmental Discourse*, Oxford: Clarendon.

Hallsworth, E. (1987) *Anatomy, Physiology and Psychology of Erosion*, Chichester: Wiley.

Hamilton, L. (1987) "What are the impacts of deforestation in the Himalayas on the Ganges–Brahmaputra lowlands and delta? Relations between assumptions and facts," *Mountain Research and Development* 7: 256–263.

Hamilton, L. (1988) "Forestry and watershed management," in Ives, J. and Pitt, D. (eds) *Deforestation: Social Dynamics in Watershed and Mountain Ecosystems*, London, New York: Routledge, pp. 99–131.

Hamilton, L. and Pearce, A. (1988) "Soil and water impacts of deforestation," in Ives, J. and Pitt, D. (eds) *Deforestation: Social Dynamics in Watershed and Mountain Ecosystems*, London: Routledge, pp. 75–98.

Hammond, A., Rodenburg, E., and Moomaw, W. (1991) "Calculating national accountability for climate change," *Environment* 33: 1, 11–15, 33–35.

Hanbury-Tenison, R. (2001) "The Greens must not be allowed to ruin our planet," *The Daily Telegraph*, July 19, 26.

Hannah, M. (1999) "Skeptical realism: from either/or to both-and," *Environment and Planning D: Society and Space* 17: 17–34.

Haraway, D. (1991) *Simians, Cyborgs, and Women: the Reinvention of Nature*, London: Routledge.

Hardin, G. (1968) "The tragedy of the commons," *Science* 162: 1243–1248.

Harding, S. (1986) *The Science Question in Feminism*, Ithaca, NY: Cornell University Press.

Hardoy, J., Mitlin, D., and Satterthwaite, D. (2001) *Environmental Problems in an Urbanizing World*, London: Earthscan.

Harré, R. (1986) *Varieties of Realism*, Oxford: Blackwell.

Harré, R. (1993) *Laws of Nature*, London: Duckworth.

Harré, R., Brockmeier, J., and Mühlhäusler, P. (1999) *Greenspeak: a Study of Environmental Discourse*, Thousand Oaks, CA: Sage.

Harré, R. and Krausz, M. (1996) *Varieties of Relativism*, Oxford: Blackwell.

Harris, D. (ed.) (1980) *Human Ecology in Savanna Environments*, London, New York: Academic Press.

Harrison, M. (1996) "'The tender frame of man': disease, climate, and racial difference in India and the West Indies, 1760–1860," *Bulletin of the History of Medicine* 70: 68–93.

Harvey, D. (1996) *Justice, Nature, and the Geography of Difference*, Oxford: Blackwell.

Hecht, S. and Cockburn, A. (1989) *The Fate of the Forest: Developers, Destroyers and Defenders of the Amazon*, London: Verso.

Hess, D. (1997) *Science Studies: an Advanced Introduction*, New York and London: New York University Press.

Hewitt, K. (ed.) (1983) *Interpretations of Calamity: from the Viewpoint of Human Ecology,* Boston, MA: Allen and Unwin.

Hicks, E. and van Rossum, W. (eds) (1991) Policy development and big science, special edition of *Verhandelingen der Koninklijke Nederlandse*, vol. 147.

Hoben, A. (1995) "Paradigms and politics: the cultural construction of environmental policy in Ethiopia," *World Development* 23: 6, 1007–1021.

Höfer, T. (1993) "Himalayan deforestation, changing river discharge, and increasing floods: myth or reality?" *Mountain Research and Development* 13: 3, 213–233.

Hohenemser, C., Kasperson, R., and Kates, R. (1985) "Causal structure," in Kates, R., Hohenemser, C., and Kasperson, J. (eds) *Perilous Progress: Managing the Hazards of Technology*, Boulder, CO: Westview Press, pp. 43–66.

Holden, C. (1996) "Social science: researchers find feminization a two-edged sword," *Science* 271: 5257, 1919–1920.

Holling, C. (1973) "Resilience and stability of ecological systems," *Annual Review of Ecology and Systematics* 4: 1–23.

Holling, C. (1979) "Myths of ecological stability: resilience and the problem of failure," in Smart, G. and Standbury, W. (eds) *Studies on Crisis Management*, Montreal: Butterworth, pp. 93–106.

Holling, C., Schindler, D., Walker, B., and Roughgarden, J. (1995) "Biodiversity in the functioning of ecosystems: an ecological synthesis," in Perrings, C., Mäler, K., Holling, C., and Jansson, B. (eds) *Biodiversity Loss: Economic and Sociological Issues*, Cambridge: Cambridge University Press, pp. 44–83.

Holmes, T. and Scoones, I. (2000) *Participatory Environmental Policy Processes: Experiences from North and South*, IDS Working Paper 113, Falmer: Institute of Development Studies.

Holton, G. (1993) *Science and Anti-Science*, Cambridge, MA: Harvard University Press.

Hörning, G. (1999) "Citizens' panels as a form of deliberative technology assessment," *Science and Public Policy* 26: 5, 351–359.

Howarth, J. (1995) "Ecology: modern hero or post-modern villain: from scientific trees to phenomenological wood," *Biodiversity and Conservation* 4: 786–797.

Hutchings, J. (2001) "Kaitiaki," *Genetics Forum*, May/June 2001 (http://www. geneticsforum.org.uk/).

Huxley, A. (1963) *The Politics of Ecology: the Question of Survival*, occasional paper of the Free Society, Center for the Study of Democratic Institutions, Santa Barbara, CA.

Huxley, J. (1947) *Conservation of Nature in England and Wales*, Cmd. 7122, London: HMSO.

Hynes, H. (1993) *Taking Population Out of the Equation: Reformulating I = PAT*, North Amherst, MA: Institute on Women and Technology.

ICRAF (International Center for Agroforestry Research) (1999) *ICRAF in Southeast Asia*, publicity brochure, Bogor: ICRAF.

IPCC (Intergovernmental Panel on Climate Change) (1996) *Second Assessment Report* Watson, R., Zinyowera, M., and Moss, R. (eds). New York: Cambridge University Press.

IPCC (Intergovernmental Panel on Climate Change) (2000) *IPCC Special Report on Land Use, Land-Use Change and Forestry*, Bonn: IPCC (http://www.grida.no/ climate/ipcc/land_use/index.htm).

IPCC (Intergovernmental Panel on Climate Change) (2001) *Third Assessment Report of the IPCC: Summary for Policymakers: Climate Change 2001: Impacts, Adaptation and Vulnerability* (http: //www.ipcc.ch).

Irwin, A. (1995) *Citizen Science: a Study of People, Expertise, and Sustainable Development*, London: Routledge.

Irwin, A., Simmons, P., and Walker, G. (1999) "Faulty environments and risk reasoning: the local understanding of industrial hazards," *Environment and Planning A* 31: 1311–1326.

Irwin, A. and Wynne, B. (eds) (1996) *Misunderstanding Science*, Cambridge: Cambridge University Press.

Ives, J. and Messerli, B. (1989) *The Himalayan Dilemma: Reconciling Conservation and Development*, London: Routledge/UNU.

Ives, J. and Pitt, D. (eds) (1988) *Deforestation: Social Dynamics in Watershed and Mountain Ecosystems*, London, New York: Routledge.

Jackson, C. (1995) "Radical environmental myths: a gender perspective," *New Left Review* 210: 124–140.

Jackson, C. (1997) "Women in critical realist environmentalism: subaltern to the species?' *Economy and Society* 26: 1, 62–80.

Jamison, A. (1996) "The shaping of the global environmental agenda: the role of non-governmental organisations," in Lash, S., Szerszynski, B., and Wynne, B. (eds) *Risk, Environment and Modernity*, London: Sage, pp. 224–245.

Jasanoff, S. (1987) "Contested boundaries in policy-relevant science," *Social Studies of Science* 17: 195–230.

Jasanoff, S. (1990) *The Fifth Branch: Science Advisers as Policymakers*, Cambridge, MA: Harvard University Press.

Jasanoff, S. (1996a) "Science and norms in global environmental regimes," in

Hampson, F. and Reppy, J. (eds) *Earthly Goods: Environmental Change and Social Justice*, Ithaca, NY and London: Cornell University Press, pp. 173–197.

Jasanoff, S. (1996b) "Beyond epistemology: relativism and engagement in the politics of science," *Social Studies of Science* 26: 2, 393–418.

Jasanoff, S. (1998) "The political science of risk perception," *Reliability Engineering and System Safety* 59: 91–99.

Jasanoff, S. (1999) "The songlines of risk," *Environmental Values* 8: 2, 135–152.

Jasanoff, S. (2000) "The "Science Wars" and American politics," in Dierkes, M. and van Grote, C. (eds) *Between Understanding and Trust: the Public, Science and Technology,* Amsterdam: Harwood, pp. 39–60.

Jasanoff, S. (2001) "Image and imagination: the formation of global environmental consciousness," in Miller, C. and Edwards, P. (eds) *Changing the Atmosphere: Expert Knowledge and Environmental Governance*, Cambridge, MA: MIT Press, pp. 309–338.

Jasanoff, S., Markle, C., Petersen, J., and Pinch, T. (eds) (1995) *Handbook of Science and Technology Studies*, Thousand Oaks, CA: Sage.

Jasanoff, S. and Wynne, B. (1998) "Science and decisionmaking," in Rayner, S. and Malone, E. (eds) *Human Choice and Climate Change*, Columbus, OH: Battelle Press, pp. 1–87.

Jewitt, S. (1995) "Europe's 'others'? Forestry policy and practices in colonial and postcolonial India," *Environment and Planning D: Society and Space* 13: 67–90.

Jewitt, S. (2000) "Unequal knowledges in Jharkhand, India: de-romanticizing women's agroecological expertise," *Development and Change* 31: 961–985.

Joerges, B. (1999) "Do politics have artefacts?' *Social Studies of Science* 29: 3, 411–431.

Johnson, C. and Forsyth, T. (2002) "In the eyes of the state: negotiating a 'rights-based approach' to forest conservation in Thailand," *World Development* 30: 9, 1591–1605.

Johnson, M. (ed.) (1992) *Lore: Capturing Traditional Environmental Knowledge*, Dene Cultural Institute, Canada: International Development Research Center.

Joss, S. (1999) "Public participation in science and technology policy- and decision-making – ephemeral phenomenon or lasting change?" *Science and Public Policy* 25: 5, 290–293.

Kaiser, J. (2000) "Ecosystem assessment: ecologists hope to avoid the mistakes of previous assessment," *Science* 289: 5485, 1676–1677.

Kasperson, J. and Kasperson, R. (eds) (2001) *Global Environmental Risk*, London: Earthscan.

Kasperson, J., Kasperson, R., and Turner, B.L. (1995) *Regions at Risk: Comparisons of Threatened Environments*, Tokyo, New York, and Paris: United Nations University Press.

Kates, R. (2000a) "Population and consumption: what we know, what we need to know," *Environment* 42: 3, 10–19.

Kates, R. (2000b) "Cautionary tales: adaptation and the global poor," in Kanes, S. and Yohe, G. (eds) *Societal Adaptation to Climate Variability and Change*, Dordrecht: Kluwer, pp. 5–17.

Kates, R. and Clark, W. (1996) "Environmental surprise: expecting the unexpected," *Environment* 38: 6–11, 28–34.

Katyal, J. and Vlek, P. (2000) *Desertification – Concept, Causes and Amelioration*, Bonn: Zentrum für Entwicklungsforschung, Universität Bonn.

Kearns, G. (1998) "The virtuous circle of facts and values in the New Western History," *Annals of the Association of American Geographers* 88: 3, 377–387.

Keck, M. and Sikkink, K. (1998) *Activists Beyond Borders: Advocacy Networks in International Politics*, Ithaca, NY and London: Cornell University Press.

Keller, E. Fox (1995) "The origin, history, and politics of the subject called 'gender and science': a first person account," in Jasanoff, S., Markle, C., Petersen, J., and Pinch, T. (eds) *Handbook of Science and Technology Studies*, Thousand Oaks, CA: Sage, pp. 80–94.

Kelly, P. and Adger, W. (1999) *Assessing Vulnerability to Climate Change and Facilitating Adaptation*, CSERGE Working Paper GEC 99–07, CESERGE: University of East Anglia, Norwich.

Khor, M. (1997) "Is globalization undermining the prospects for sustainable development?" Fifth Annual Hopper Lecture presented at the University of Guelph, 21 October, and the University of Prince Edward Island, 23 October.

Kienholz, H., Schneider, G., Bichsel, M., Grunder, M., and Mool, P. (1984) "Mapping of mountain hazards and slope stability," *Mountain Research and Development* 4: 3, 247–266.

Kingdon, J. (1984) *Agendas, Alternatives, and Public Policies*, Boston, MA: Little, Brown.

Klooster, D. (2002) "Toward adaptive community forestry management: integrating local forest knowledge with scientific forestry," *Economic Geography* 78: 1, 43–70.

Knorr-Cetina, K. and Mulkay, M. (eds) (1983) *Science Observed: Perspectives on the Social Study of Science*, London: Sage.

Knowledge Center, The (2000) "Threat that never was" (http://www.biotech-knowledge.monsanto.com).

Koertge, N. (ed.) (1998) *A House Built on Sand: Exposing Postmodernist Myths about Science*, New York and Oxford: Oxford University Press.

Kuhn, T. (1962) *The Structure of Scientific Revolutions*, Chicago, IL: University of Chicago Press.

Kukla, A. (1993) "Epistemic boundedness," *International Studies in the Philosophy of Science* 7: 2, 121–126.

Kull, C. (2000) "Deforestation, erosion, and fire: degradation myths in the environmental history of Madagascar," *Environment and History* 6: 423–450.

Kurian, P. (2000) *Engendering the Environment? Gender in the World Bank's Environmental Policies*, Aldershot: Ashgate.

Kwa, C. (1987) "Representations of nature modeling between ecology and science policy: the case of the International Biological program," *Social Studies of Science* 17: 413–442.

Kwa, C. (2001) "The steering of the International Geosphere–Biosphere Programme (IGBP)," Draft BCSIA Discussion Paper, Environment and Natural Resources Program, Kennedy School of Government, Harvard University.

Lahsen, M. (1999) "The detection and attribution of conspiracies: the controversy over Chapter 8," in Marcus, G. (ed.) *Paranoia Within Reason: A Case Book on Conspiracy and Explanation*, Chicago, IL: Chicago University Press.

Lakatos, I. (1978) *The Methodology of Scientific Research Programs*, Worrall, J. and Currie, G. (eds), Cambridge: Cambridge University Press.

Lambin, E.F., Turner, B.L., Geist, H.J., *et al.* (2001) "The causes of land-use and land-cover change: moving beyond the myths," *Global Environmental Change* 11: 4, 261–269.

Lamprey, H. (1988 [1975]) "Report on the desert encroachment reconnaissance in northern Sudan, 21 October to 10 November, 1975," *Desertification Control Bulletin* 17: 1–7.

Landy, M. (1995) "The new politics of environmental policy," in Landy, M. and Levin, M. (eds) *The New Politics of Public Policy,* Baltimore, MA: Johns Hopkins Press, pp. 207–227.

Landy, M., Roberts, M., and Thomas, S. (1994) *The Environmental Protection Agency: Asking the Wrong Questions,* New York: Oxford University Press.

Lankester, E. (1914) "Nature reserves," *Nature* 93: 2315, 33–35.

Latour, B. (1983) "Give me a laboratory and I shall raise the world," in Knorr-Cetina, K. and Mulkay, M. (eds) *Science Observed,* London: Sage, pp. 141–170.

Latour, B. (1987) *Science in Action: How to Follow Scientists and Engineers Through Society,* Cambridge, MA: Harvard University Press.

Latour, B. (1988) *The Pasteurization of France,* Cambridge, MA: Harvard University Press.

Latour, B. (1993 [1991]) *We Have Never Been Modern,* Hemel Hempstead: Harvester Wheatsheaf.

Latour, B. and Woolgar, S., with Salk, J. (1979) *Laboratory Life: the Social Construction of Scientific Facts,* Beverly Hills, CA: Sage.

Laudan, L. (1977) *Progress and its Problems,* London: Sage.

Laudan, L. (1990) *Science and Relativism: Some Key Controversies in the Philosophy of Science,* Chicago, IL: University of Chicago Press.

Law, J. (ed.) (1991) *A Sociology of Networks: Essays on Power, Technology and Domination,* London: Routledge.

Law, J. and Hassard, J. (eds) (1999) *Actor Network Theory and After,* Oxford: Blackwell and The Sociological Review.

Layne, L. (1990) "Motherhood lost: cultural dimensions of miscarriage and still-birth in America," *Women and Health* 16: 3, 75–104.

Layne, L. (1997) "Breaking the silence: an agenda for a feminist discourse of pregnancy loss," Special edition of *Feminist Studies* 23: 2, 289–315.

Layne, L. (2001) "In search of community: tales of pregnancy loss in three toxically assaulted US communities," *Women's Studies Quarterly* 29: 1, 2, 25–50.

Leach, G. and Mearns, R. (1988) *Beyond the Fuelwood Crisis,* London: Earthscan.

Leach, M., Fairhead, J., and Amanor, K. (eds) (2002) "Science and the policy process: perspectives from the forest," Special edition of *IDS Bulletin* 33: 1.

Leach, M. and Mearns, R. (1991) *Poverty and Environment in Developing Countries, an Overview Study,* report presented to the Economic and Social Research Council and the Overseas Development Administration, Brighton: Institute of Development Studies.

Leach, M. and Mearns, R. (eds) (1996) *The Lie of the Land: Challenging Received Wisdom on the African Environment,* Oxford: James Currey.

Leach, M., Mearns, R., and Scoones, I. (1999) "Environmental entitlements: dynamics and institutions in community-based natural resource management," *World Development* 27: 2, 225–247.

Leach, M., Mearns, R., and Scoones, I. (eds) (1997) "Community-based sustainable development: consensus or conflict?' Special edition *IDS Bulletin* 28: 4.

Lederman, M. and Bartsch, I. (eds) (2001) *The Gender and Science Reader,* London and New York: Routledge.

Lee, N. and Hassard, J. (1999) "Organization unbound: actor-network theory, research strategy and institutional flexibility," *Organization* 6: 3, 391–404.

Leff, E. (1995) *Green Production: Toward an Environmental Rationality*, New York, London: Guilford.

Leggett, J. (ed.) (1990) *Global Warming: The Greenpeace Report*, Oxford: Oxford University Press.

Leiss, W. (1972) "Technological rationality: Marcuse and his critics," *Philosophy of the Social Sciences* 2: 34–35.

Leopold, A. (1933) *Game Management*, New York and London: C. Scribner's and Sons.

Leplin, J. (1984) "Truth and scientific progress," in Leplin, J. (ed.) *Scientific Realism*, Berkeley, CA, Los Angeles, CA, London: University of California Press, pp. 193–217.

Levidow, L. and Carr, S. (1997) "How biotechnology regulation sets a risk/ethics boundary," *Agriculture and Human Values* 14: 29–43.

Levin, M. (1984) "What kind of explanation is truth?" in Leplin, J. (ed.) *Scientific Realism*, Berkeley, CA, Los Angeles, CA, London: University of California Press, pp. 124–139.

Levitt, N. (1999) *Prometheus Bedeviled: Science and the Contradictions of Contemporary Culture*, New Brunswick, NJ: Rutgers University Press.

Lewis, J. (1990) "The vulnerability of small island states to sea level rise: the need for holistic strategies," *Disasters* 14: 3, 241–249.

Lewis, P. (1996) "Metaphor and critical realism," *Review of Social Economy* 54: 4, 487–506.

Light, A. and Katz, E. (eds) (1996) *Environmental Pragmatism*, London: Routledge.

Limerick, P. (1991) "What on earth is the new western history?" in Limerick, P., Milner, C., and Rankin, E. (eds) *Trails: Toward a New Western History*, Lawrence, KS: University of Kansas Press, pp. 81–88.

Lipietz, A. (1992) *Towards a New Economic Order: Postfordism, Ecology and Democracy,* translated by Malcolm Slater, Cambridge: Polity.

Lipietz, A. (2000) "Political ecology and the future of Marxism," *Capitalism, Nature, Socialism* 11: 1, 69–85.

List, M. and Rittberger, V. (1992) "Regime theory and international environmental management," in Hurrell, A. and Kingsbury, B. (eds) *The International Politics of the Environment*, Oxford: Clarendon, pp. 85–109.

Litfin, K. (1994) *Ozone Discourses: Science and Politics in Global Environmental Cooperation*, New York: Columbia University Press.

Little, C. (1995) *The Dying of the Trees: The Pandemic in America's Forests*, New York: Penguin.

Liverman, D. (1999) "Vulnerability and adaptation to drought in Mexico," *Natural Resources Journal* 39: 1, 99–115.

Lohmann, L. (1993) "Resisting green globalism," in Sachs, W. (ed.) *Global Ecology: a New Arena of Political Conflict*, London, Zed, Halifax, Novia Scotia: Fernwood, pp. 157–169.

Lohmann, L. (1995) "No rules of engagement: interest groups, centralization and the creative politics of 'environment' in Thailand," in Rigg, J. (ed.) *Counting the Costs: Economic Growth and Environmental Change in Thailand*, Singapore: ISEAS, pp. 211–234.

Lohmann, L. (1998) *Whose Voice is Speaking? How Opinion Polling and Cost–Benefit Analysis Synthesize New Publics*, Briefing Paper number 7, The Corner House, Sturminster Newton, UK (http://cornerhouse.icaap.org/).

Lohmann, L. (1999) *The Dyson Effect: Carbon "Offset" Forestry and the Privatisa-*

Bibliography 297

tion of the Atmosphere, Briefing number 15, The Corner House, Sturminster Newton, UK (http://cornerhouse.icaap.org/).

Lohmann, L. (2001) *Democracy or Carbocracy? Intellectual Corruption and the Future of the Climate Debate*, Briefing number 24, The Corner House, Sturminster Newton, UK (http://cornerhouse.icaap.org/).

Lomborg, B. (2001) *The Skeptical Environmentalist: Measuring the Real State of the World*, Cambridge: Cambridge University Press.

Long, N. and Long, A. (eds) (1992) *Battlefields of Knowledge: the Interlocking of Theory and Practice in Social Research and Development*, London: Routledge.

Longino, H. (1990) *Science as Social Knowledge: Values and Objectivity in Scientific Inquiry*, Princeton, NJ: Princeton University Press.

Lookingbill, B. (2001) *Dust Bowl, USA: Depression America and the Ecological Imagination, 1929–1941*, Athens, OH: Ohio University Press.

Losey, J., Raynor, L., and Carter, M. (1999) "Transgenic pollen harms monarch larvae," *Nature* 399: 214.

Low, N. and Gleeson, B. (1998) *Justice, Society and Nature: an Exploration of Political Ecology*, London: Routledge.

Lowe, P. and Rüdig, W. (1986) "Political ecology and the social sciences: the state of the art," *British Journal of Political Science* 16: 513–550.

Luhmann, N. (1989) *Ecological Communication*, Chicago, IL: University of Chicago Press.

Luke, T. (1999) *Capitalism, Democracy and Ecology: Departing from Marx*, Chicago and Urbana, IL: University of Illinois Press.

Luzzarder-Beach, S. and MacFarlane, A. (2000) "The environment of gender and science: status and perspectives of women and men in physical geography," *Professional Geographer* 52: 3, 407–424.

Lynch, M. and Woolgar, S. (eds) (1990) *Representation in Scientific Practice*, Cambridge, MA: MIT Press.

McComas, K. and Shanahan, J. (1999) "Telling stories about global climate change: measuring the impact of narratives on issue cycles," *Communication Research* 26: 1, 30–57.

McHenry, H. (2000) "Wild flowers in the wrong field are weeds! Examining farmers' constructions of conservation," *Environment and Planning A* 30: 1039–1053.

MacKenzie, D. (1990) *Inventing Accuracy: a Historical Sociology of Nuclear Missile Guidance*, Cambridge, MA: MIT Press.

McManus, P. (1999) "Histories of forestries: ideas, networks and silences," *Environment and History* 5: 2, 185–208.

McNaughton, P. and Urry, J. (1998) *Contested Natures*, London and Thousand Oaks, CA: Sage.

Malin, J. (1946) "Dust storms 1850–1900," *Kansas History Quarterly* 14: 129–144, 265–296.

Marcuse, H. (1964) *One Dimensional Man*, Boston, MA: Beacon Press.

Marcuse, H. (1969) *An Essay on Liberation*, Boston, MA: Beacon Press.

Marsh, G.P. (1864) *Man and Nature, or, Physical Geography as Modified by Human Action*, New York: Scribner.

Marshall, B. (1999) "Globalization, environmental degradation and Ulrich Beck's Risk Society," *Environmental Values* 8: 253–275.

Martens, P. (2000) "Malaria and global warming in perspective?" *Journal of Infectious Diseases* 6: 3, 313–314.

Martin, B. (1991) *Scientific Knowledge in Controversy: the Social Dynamics of the Fluoridation Debate*, Albany, NY: State University of New York Press.

Mason, M. (1999) *Environmental Democracy*, London: Earthscan.

Massey, D. (1999) "Space–time, 'science' and the relationship between physical geography and human geography," *Transactions of the Institute of British Geographers* NS 24: 261–276.

Mather, A. (1992) "The forest transition," *Area* 24: 367–379.

Mather, A. and Needle, C. (2000) "The relationships of population and forest trends," *The Geographical Journal* 166: 1, 2–13.

Matthews, J. and Ho, Mae-Wan (2001) "The new thought police," *Genetics Forum* September/October (http: //www.geneticsforum.org.uk/).

Maunder, W. (1992) *Dictionary of Global Climate Change,* London: UCL Press.

May, R. (1974) "Biological populations with non-overlapping generations: stable points, stable cycles and chaos," *Science* 186: 645–647.

Mayer, S. (2000) "Genetic engineering in agriculture," in Huxham, M. and Sumner, D. (eds) *Science and Environmental Decision Making*, Harlow: Prentice Hall, pp. 94–117.

Meadows, D., Meadows, D., Randers, J., and Behrens III, W. (1972) *The Limits to Growth: a Report for the Club of Rome's Project on the Predicament of Mankind*, New York: University Books.

Mehta, L. (2001) "The World Bank and its emerging knowledge empire," *Human Organisation* 60: 2, 189–196.

Mehta, L., Leach, M., Newell, P., Scoones, I., Sivaramakrishnan, K., and Way, S. (1999) *Exploring Understandings of Institutions and Uncertainty: New Directions in Natural Resource Management*, IDS Discussion Paper 372, Falmer: Institute of Development Studies.

Mehta, L., Leach, M., and Scoones, I. (eds) (2001) *Environmental governance in an uncertain world*, Special edition of *IDS Bulletin*, 32: 4.

Merchant, C. (1980) *The Death of Nature: Women, Ecology and the Scientific Revolution,* San Francisco, CA: Harper and Row.

Merton, R.K. (1973 [1942]) "The normative structure of science," in Merton, R.K., *The Sociology of Science*, Storer, N. (ed.), Chicago, IL: University of Chicago Press, pp. 267–278.

Metz, J. (1989) "Himalayan political economy: more myths in the closet?" *Mountain Research and Development* 9: 2, 175–181.

Metz, J. (1991) "A reassessment of the causes and severity of Nepal's environmental crisis," *World Development* 19: 7, 805–820.

Midgley, M. (1992) *Science as Salvation: a Modern Myth and its Meaning*, London: Routledge.

Miller, A. (1978) *A Planet to Choose: Value Studies in Political Ecology*, New York: Pilgrim Press.

Miller, C. (2001) "Hybrid management: boundary organizations, science policy, and environmental governance in the climate regime," *Science, Technology and Human Values* 26: 4, 478–500.

Millington, A. (1986) "Environmental degradation, soil conservation and agricultural policies in Sierra Leone, 1895–1984," in Anderson, D. and Grove, R. (eds) *Conservation in Africa,* Cambridge: Cambridge University Press, pp. 229–248.

Mol, A. (1996) "Ecological modernization and institutional reflexivity: environmental reform in the late modern age," *Environmental Politics* 5: 2, 302–323.

Mol, A. and Law, J. (1994) "Regions, networks and fluids: anaemia and social topology," *Social Studies of Science* 24: 641–671.

Monbiot, G. (2002) "Corporate phantoms: the web of deceit over GM food has now drawn in the PM's speechwriters," the *Guardian*, Wednesday 29 May.

Morgan, R. (1986) *Soil Erosion and Conservation*, Harlow: Longman Scientific and Technical Series.

Morris, A. and Mueller, C. (eds) (1992) *Frontiers in Social Movement Theory*, New Haven, CT: Yale University Press.

Morris, J. (1995) *The Political Economy of Land Degradation: Pressure Groups, Foreign Aid and the Myth of Man-Made Deserts*, London: Institute of Economic Affairs.

Morrow, R. (1994) *Critical Theory and Methodology* (Contemporary Social Theory, vol. 3), Thousand Oaks, CA, London, Delhi: Sage.

Morse, S. and Stocking, M. (eds) (1995) *People and Environment*, London: UCL Press.

Mortimore, M. and Adams, W. (1999) *Working the Sahel: Environment and Society in Northern Nigeria*, London: Routledge.

Mukta, P. and Hardiman, D. (2000) "The political ecology of nostalgia," *Capitalism, Nature, Socialism* 11: 1, 113–133.

Multinational Monitor (periodical) http: //www.essential.org/monitor/monitor.html.

Murdoch, J. (1997) "Inhuman/nonhuman/human: actor-network theory and the prospects for a nondualistic and symmetrical perspective on nature and society," *Environment and Planning D: Society and Space* 15: 6, 731–756.

Murdoch, J. and Clark, J. (1994) "Sustainable knowledge," *Geoforum* 25: 2, 115–132.

Murphy, R. (1994) *Rationality and Nature: a Sociological Inquiry into a Changing Relationship*, Boulder, CO: Westview Press.

Murphy, R. (1995) "Sociology as if nature did not matter: an ecological critique," *British Journal of Sociology* 46: 4, 688–707.

Murton, J. (1999) "Population growth and poverty in Machakos, Kenya," *Geographical Journal* 165: 1, 37–46.

Myers, N. (1984) *The Primary Source: Tropical Forests and our Future*, New York: Norton.

Myers, N. (1990) "Tropical Forests," in Leggett, J. (ed.) *Global Warming: The Greenpeace Report*, Oxford: Oxford University Press, pp. 372–399.

Nader, L. (ed.) (1996) *Naked Science: Anthropological Inquiry into Boundaries, Power, and Knowledge*, London: Routledge.

Naka, K., Hammett, A., and Stuart, W. (2000) "Forest certification: stakeholders, constraints and effects," *Local Environment* 5: 4, 475–481.

Nash, R. (1973) *Wilderness and the American Mind*, New Haven, CT: Yale University Press.

Netting, R. (1993) *Smallholders, Householders: Farm Families and the Ecology of Intensive, Sustainable Agriculture*, Stanford, CA: Stanford University Press.

Neumann, R. (1996) "Dukes, earls and ersatz Edens: aristocratic nature preservationists in colonial Africa," *Environment and Planning D: Society and Space* 14: 79–98.

Neumann, R. (1998) *Imposing Wilderness: Struggles over Livelihood and Nature Preservation in Africa*, Los Angeles and Berkeley, CA: University of California Press.

Nietzsche, F. (1979 [1873]) "On truth and lies in a non-moral sense," in Breazeale,

D. (ed.) *Philosophy and Truth: Selection from Nietzsche's Notebooks of the Early 1970s*, Sussex, Harvester Press, New Jersey: Humanities Press, pp. 79–80.

Nietzsche, F. (1967 [1901]) *The Will to Power*, translated by W. Kaufmann, edited by R. Hollingdale, New York: Vintage Books.

Norberg-Bohm, V., Clark, W., Bakshi, B., *et al.* (2001) "International comparisons of environmental hazards," in Kasperson, J. and Kasperson, R. (2001) *Global Environmental Risk*, London: Earthscan, pp. 55–147.

Norris, C. (1995) "Truth, science and the growth of knowledge," *New Left Review* 210: 105–123.

Nowotny, H. (1979) "Science and its critics: reflections on anti-science," in Nowotny, H. and Rose, H. (eds) *Counter Movements in the Sciences: the Sociology of the Alternatives to Big Science*, Dordrecht, Boston, MA, London: Reidel, pp. 1–26.

Nowotny, H. and Rose, H. (eds) (1979) *Counter Movements in the Sciences: the Sociology of the Alternatives to Big Science*, Dordrecht, Boston, MA, London: Reidel, pp. 1–26.

NRC (National Research Council, Board on Sustainable Development Policy Division) (1999) *Our Common Journey: A Transition Towards Sustainability*, Washington, DC: NRC.

O'Connor, J. (1996) "The second contradiction of capitalism," in Benton, T. (ed.) *The Greening of Marxism*, New York: Guilford Press, pp. 187–196.

O'Riordan, T. and Jordan, A. (1999) "Institutions, climate change and cultural theory: towards a common analytical framework," *Global Environmental Change* 9: 81–93.

Oba, G., Stenseth, N., and Lusigi, W. (2000) "New perspectives on sustainable grazing management in arid zones of sub-Saharan Africa," *BioScience* 50: 35–51.

Odum, E. (1964) "The new ecology," *Bioscience* 14: 7, 14–16.

Odum, E. (1969) "The strategy of ecosystem development," *Science* 164: 262.

Offe, C. (1985) "New social movements: challenging the boundaries of institutional politics," *Social Research* 52: 4, 817–868.

Olsson, L. and Ardö, J. (2002) "Soil carbon sequestration in degraded semi-arid ecosystems: perils and potentials," *Ambio* 31: 6, 471–477.

Openshaw, K. (1974) "Woodfuels in the developing world," *New Scientist* 31: 271–272.

Ortony, A. (1993) *Metaphor and Thought*, 2nd edn, Cambridge: Cambridge University Press.

Ostrom, E. (1990) *Governing the Commons: the Evolution of Institutions for Collective Action*, Cambridge: Cambridge University Press.

Pasuk Phongpaichit and Baker, C. (1995) *Thailand: Economy and Politics*, Oxford: Oxford University Press.

Patz, J., McGeehin, M., Bernard, S., *et al.* (2000) "The potential health impacts of climate variability and change for the United States: executive summary of the report of the health sector of the U.S. National Assessment," *Environmental Health Perspectives* 108: 367–376.

Peet, R. and Watts, M. (eds) (1996) *Liberation Ecologies: Environment, Development, Social Movements*, London: Routledge.

Pereira, H. (1989) *Policy and Practice in the Management of Tropical Watersheds*, Boulder, CO: Westview Press.

PERG (Political Ecology Research Group) (1979) *A First Report of the Work of the Political Ecology Research Group*, Oxford: PERG.

Perlman, M. (1994) *The Power of Trees: the Reforesting of the Soul*, Dallas, TX: Spring Publications.

Pickering, A. (1995) *The Mangle of Practice*, Chicago, IL: University of Chicago Press.

Pickett, S. and White, P. (1985) *The Ecology of Natural Disturbance and Patch Dynamics*, New York: Academic Press.

Pielke, R. (1999) "Nine fallacies of floods," *Climatic Change* 42: 2, 413–438.

Pirandello, L. (1998 [1922]) *Naked: a New Version by Nicholas Wright*, London: Nick Hern Books.

Plumwood, V. (1993) *Feminism and the Mastery of Nature*, London: Routledge.

Poincaré, H. (1958) *The Value of Science*, London: Dover.

Polanyi, M. (1962) "The republic of science: its political and economic theory," *Minerva* 1: 1, 54–73.

Popper, K. (1945) *The Open Society and its Enemies*, London: Routledge.

Popper, K. (1962) *Conjectures and Refutations: the Growth of Scientific Knowledge*, New York: Basic Books.

Popper, K. (1994) *The Myth of the Framework: in Defense of Science and Rationality*, edited by M.A. Notturno, New York: Routledge.

Porritt, J. (2000) *Playing Safe: Science and the Environment*, London: Thames and Hudson.

Postel, S. (1993) "Facing water scarcity," in Starke, L. (ed.) *State of the World 1993*, New York: W.W. Norton and Co., pp. 22–41.

Preston, D., Macklin, M., and Warburton, J. (1997) "Fewer people, less erosion: the twentieth century in southern Bolivia," *The Geographical Journal* 163: 2, 198–205.

Price, D. (1965) *The Scientific Estate*, Cambridge, MA: Harvard University Press.

Price, D. (1986) *Little Science, Big Science . . . and Beyond*, New York: Columbia University Press.

Price, M. and Thompson, M. (1997) "The complex life: human land uses in mountain ecosystems," *Global Ecology and Biogeography Letters* 6: 77–90.

Princen, T., Finger, M., and Manro, J. (1994) "Translational linkages," in Princen, T. and Finger, M. *Environmental NGOs in World Politics: Linking the Local and the Global*, London: Routledge, pp. 217–236.

Proctor, J. (1998) "The social construction of nature: relativist accusations, pragmatist and critical realist responses," *Annals of Association of American Geographers* 88: 3, 352–376.

Psillos, S. (1999) *Scientific Realism: How Science Tracks Truth*, London: Routledge.

Radder, H. (1998) "The Politics of STS," *Social Studies of Science* 28: 2, 328–332.

Rangan, H. (2000) *Of Myths and Movements: Rewriting Chipko into Himalayan History*, London: Verso.

Rappaport, R. (1968) *Pigs for the Ancestors*, New Haven, CT: Yale University Press.

Rasmussen, D. (ed.) (1996) *The Handbook of Critical Theory*, Oxford: Blackwell.

Rasmussen, K., Fog, B., and Madsen, J.E. (2001) "Desertification in reverse? Observations from northern Burkina Faso," *Global Environmental Change* 11: 4, 271–282.

Raustilia, K. and Victor, D. (1996) "Biodiversity since Rio: the future of Convention on Biological Diversity," *Environment* 38: 4, 17–20, 37–45.

Reich, R. (1990) *Public Management in a Democratic Society*, Englewood Cliffs, NJ: Prentice Hall.

Reid, W. (1997) "Strategies for conserving biodiversity," *Environment* 39: 7, 16–20.

Reij, C., Scoones, I., and Toulmin, C. (ed.) (1996) *Sustaining the Soil: Indigenous Soil and Water Conservation in Africa*, London: Earthscan.

Reiter, P. (1998) "Global warming and vector-borne disease," *Lancet* 351: 1738.

Reiter, P. (2000) "Malaria and global warming in perspective?' *Journal of Infectious Diseases* 6: 4, 438.

Revenga, C., Murray, S., Abramovitz, J., and Hammond, A. (1998) *Watersheds of the World: Ecological Value and Vulnerability*, Washington, DC: World Resources Institute, and Worldwatch Institute.

Ribot, J., Magalhaes, A., and Panagides, S. (eds) (1996) *Climate Variability, Climate Change and Social Vulnerability in the Semi-arid Tropics*, Cambridge: Cambridge University Press.

Rich, B. (1994) *Mortgaging the Earth: the World Bank, Environmental Impoverishment and the Crisis of Development*, London: Earthscan and Boston, MA: Beacon Press.

Richards, P. (1952) *The Tropical Rain Forest*, Cambridge: Cambridge University Press.

Richards, P., Slikkerveer, L., and Phillips, A. (1989) *Indigenous Knowledge Systems for Agriculture and Rural Development: The CIKARD inaugural lectures*, Studies in technology and social change no. 13, Technical and social change program, Iowa State University, Adams, IA.

Robbins, P. (1998) "Paper forests: imagining and deploying exogenous ecologies in arid India," *Geoforum* 29: 1, 69–86.

Robbins, P. (2000) "The practical politics of knowing: state environmental knowledge and local political economy," *Economic Geography* 76: 2, 126–144.

Robertson, R. (1992) *Globalization: Social Theory and Global Culture*, Thousand Oaks, CA: Sage.

Rocheleau, D. (1995) "Maps, numbers, text, and context: mixing methods in feminist political ecology," *Professional Geographer* 47: 4, 458–466.

Rocheleau, D. and Edmunds, D. (1997) "Women, men and trees: gender, power and property in forest and agrarian landscapes," *World Development* 25: 8, 1351–1371.

Rocheleau, D. and Ross, L. (1995) "Trees as tools, trees as text: struggles over resources in Zambrana-Chacuey, Dominican Republic," *Antipode* 27: 4, 407–428.

Rocheleau, D., Ross, L., Morrobel, J., Malaret, L., Hernandez, R., and Kominiak, T. (2001) "Complex communities and emergent ecologies in the regional agroforest of Zambrana-Chacuey, Dominican Republic," *Ecumene* 8: 4, 465–492.

Rocheleau, D., Thomas-Slayter, B., and Wangari, E. (eds) (1996) *Feminist Political Ecology: Global Issues and Local Experiences*, London: Routledge.

Roe, E. (1991) "'Development narratives' or making the best of blueprint development," *World Development* 19: 4, 287–300.

Roe, E. (1994) *Narrative Policy Analysis: Theory and Practice*, Chapel Hill, NC: Duke University Press.

Roe, E. (1995) "Except Africa: postscript to a special section on development narratives," *World Development* 23: 6, 1065–1069.

Roquelpo, P. (1995) "Scientific expertise among political powers, administrations and public opinion," *Science and Public Policy* 22: 3, 175–182.

Rorty, R. (1989a) *Contingency, Irony, and Solidarity*, Cambridge: Cambridge University Press.

Rorty, R. (1989b) "Solidarity or objectivity?" in Krausz, M. (ed.) *Relativism*, Notre Dame, IN: University of Notre Dame Press, pp. 35–50.

Rose, H. (1990) "Toward pragmatic realism in human geography," *Cahiers de Géographie du Québec* 34: 92, 161–179.

Ross, A. (1996) *Strange Weather: Culture, Science and Technology in the Age of Limits*, London: Verso.

Rosswall, T., Woodmansee, R., and Risser, P. (1988) *Scales and Global Change: Spatial and Temporal Variability in Biospheric and Geospheric Processes*, New York: John Wiley.

Rostand, E. (1985) *Cyrano de Bergerac*, translated and adapted by Anthony Burgess, London: Hutchinson.

Rouse, J. (1987) *Knowledge and Power: Toward a Political Philosophy of Science*, Ithaca, NY: Cornell University Press.

Royal Society (1992) *Risk: Analysis, Perception and Management*, London: The Royal Society.

Ruether, R. (1972) *Liberation Theology: Human Hope Confronts Christian History and American Power,* New York: Paulist Press.

Rundle, B. (1993) *Facts*, London: Duckworth.

Russell, B. (1940) *An Inquiry into Meaning and Truth*, London: Allen and Unwin.

Russett, B. (1967) *International Regions and the International System: a Study in Political Ecology*, Chicago, IL: Rand McNally.

Sabatier, P. (1987) "Knowledge, policy oriented learning and policy change," *Knowledge: Creation, Diffusion, Utilization* 8: 4, 649–662.

Sabatier, P. and Jenkins-Smith, H. (eds) (1993) *Policy Change and Learning: an Advocacy Coalition Approach,* Boulder, CO: Westview Press.

Saberwal, V. (1997) "Science and the desiccationist discourse of the 20th Century," *Environment and History* 3: 309–343.

Sachs, W. (ed.) (1993) *Global Ecology: a New Arena of Political Conflict*, London: Zed, Halifax, Novia Scotia: Fernwood.

Satterthwaite, D. (1997) "Environmental transformations in cities as they get larger, wealthier and better managed," *The Geographical Journal* 163: 2, 216–224.

Sayer, A. (1997) "Essentialism, social constructivism, and beyond," *Sociological Review* 45: 453–477.

Sayer, A. (2000) *Realism and Social Science*, Thousand Oaks, CA: Sage.

Schaefer, W. (ed.) (1983) *Finalization in Science*, Dordrecht: Reidel.

Schiebinger, L. (1993) *Nature's Body: Gender in the Making of Modern Science*, Boston, MA: Beacon Press.

Schmidt-Vogt, D. (1998) "Defining degradation: the impacts of swidden on forests in northern Thailand," *Mountain Research and Development*, 18: 2, 135–149.

Schön, D. and Rein, M. (1994) *Frame Reflection: Towards the Resolution of Intractable Policy Controversies*, New York: Basic Books.

Schumm, S. (1991) *To Interpret the Earth: Ten Ways to be Wrong*, Cambridge: Cambridge University Press.

Schuurman, F. (1993) "Introduction: development theory in the 1990s," in Schuurman, F. (ed.) *Beyond the Impasse: New Directions in Development Theory*, London: Zed, pp. 1–18.

Schwarz, M. and Thompson, M. (1990) *Divided We Stand: Redefining Politics, Technology and Social Choice*, Hemel Hempstead: Harvester Wheatsheaf.

Scoones, I. (1994) *Living with Uncertainty: New Directions in Pastoral Development in Africa*, London: Intermediate Technology Publications.

Scoones, I. (1998) *Sustainable Rural Livelihoods: a Framework for Analysis*, IDS Working Paper 72, Brighton, UK: IDS.

Scoones, I. and Thompson, J. (eds) (1994) *Beyond Farmer First: Rural People's Knowledge, Agricultural Research and Extension Practice,* London: Intermediate Technology Publications.

Scott, J. (1985) *Weapons of the Weak: Everyday Forms of Peasant Resistance*, New Haven, CT: Yale University Press.

Scott, J. (1998) *Seeing Like a State: How Certain Schemes to Improve the Human Condition have Failed*, New Haven, CT: Yale University Press.

Searle, J. (1995) *The Construction of Social Reality*, New York: Free Press.

Sears, P. (1959 [1935]) *Deserts on the March*, Norman, OK: University of Oklahoma Press.

Sears, P. (1964) "Ecology – a subversive subject," *Bioscience* 14: 7, 11–13.

Segerstråle, U. (ed.) (2000) *Beyond the Science Wars: the Missing Discourse About Science and Society*, Albany, NY: State University of New York Press.

Seligman, A. (1997) *The Problem of Trust*, Princeton, NJ: Princeton University Press.

Sen, A. (1981) *Poverty and Famines: an Essay on Entitlement and Deprivation*, Oxford: Clarendon Press.

Shackley, S. (1997) "The IPCC: consensual knowledge and global politics," *Global Environmental Change* 7: 1, 77–79.

Shackley, S. and Wynne, B. (1995) "Global climate change: the mutual construction of an emergent science–policy domain," *Science and Public Policy* 22: 218–230.

Shackley, S. and Wynne, B. (1996) "Representing uncertainty in global climate change science and policy: boundary-ordering devices and authority," *Science, Technology and Human Values* 21: 3, 275–302.

Shah, M. and Banerji, D. (2001) "India needs vigilance against discredited technologies: GM crops and the world market" *The Hindu*, Thursday, 20 December.

Shapin, S. (1994) *A Social History of Truth: Civility and Science in Seventeenth-Century England*, Chicago, IL: University of Chicago Press.

Shapin, S. (2001) "Proverbial economies: how an understanding of some linguistic and social features of common sense can throw light on more prestigious bodies of knowledge, science for example," *Social Studies of Science* 31: 5, 731–769.

Shapin, S. and Schaffer, S. (1985) *Leviathan and the Air Pump: Hobbes, Boyle and the Experimental Life*, Princeton, NJ: Princeton University Press.

Shepard, P. and McKinley, D. (eds) (1969) *The Subversive Science: Essays Toward an Ecology of Man*, Boston, MA: Houghton Mifflin.

Shiva, V. (1993) "The greening of the global reach," in Sachs, W. (ed.) *Global Ecology: a New Arena of Political Conflict*, London: Zed, Halifax, Novia Scotia: Fernwood, pp. 149–156.

Shiva, V. (1999) "The global campaign against biopiracy and changing the paradigm of agriculture," speech made to the *International Forum on Globalization*, Seattle IFG Teach-In, 26 November 1999 (http: //gos.sbc.edu/s/shiva3.html).

Shrader-Frechette, K. and McCoy, E. (1994) "How the tail wags the dog: how value judgments determine ecological science," *Environmental Values* 3: 107–120.

Shrum, W., Chompalov, I., and Genuth, J. (2001) "Trust, conflict and performance in scientific collaborations," *Social Studies of Science* 31: 5, 681–730.

Sillitoe, P. (1993) "Losing ground? Soil loss and erosion in the highlands of Papua New Guinea," *Land Degradation and Rehabilitation* 5: 3, 179–190.

Sillitoe, P. (1998) "It's all in the mound: fertility management under stationary shifting cultivation in the Papua New Guinea highlands," *Mountain Research and Development* 18: 2, 123–134.

Silverman, D. (1993) *Interpreting Qualitative Data: Methods for Analysing Talk, Text and Interaction*, London: Sage.

Simon, J. and Kahn, H. (eds) (1984) *The Resourceful Earth: a Response to Global 2000*, Oxford: Blackwell.

Simpson, A. (1998) "Can democracy cope with biotechnology?" *Genetics Forum* December 1998/January 1999 (http: //www.geneticsforum.org.uk/).

Simpson, G. (1960) "The world into which Darwin led us," *Science* 131: 966–974.

Sismondo, S. (1996) *Science Without Myth: On Constructions, Reality, and Social Knowledge*, Albany, NY: State University of New York Press.

Sivraramakrishnan, K. (2000) "State sciences and development histories: encoding local forest knowledge in Bengal," *Development and Change* 31: 61–89.

Sluijs, J. van der, Eijndhoven, J. van, Shackley, S., and Wynne, B. (1998) "Anchoring devices for science in policy: the case of consensus around climate sensitivity," *Social Studies of Science* 28: 2, 291–323.

Smith, P. (1988) *Discerning the Subject*, Minneapolis, MN: University of Minnesota Press.

Smith, R. (1996) "Sustainability and the rationalization of the environment," *Environmental Politics* 5: 1, 25–47.

Snow, C.P. (1961) *Science and Government*, Cambridge, MA: Harvard University Press.

Social Learning Group (2000a) *Learning to Manage Global Environmental Risks: vol. 1, A Comparative History of Social Responses to Climate Change, Ozone Depletion, and Acid Rain*, Cambridge, MA: MIT Press.

Social Learning Group (2000b) *Learning to Manage Global Environmental Risks: vol. 2, A Functional Analysis of Social Responses to Climate Change, Ozone Depletion, and Acid Rain*, Cambridge, MA: MIT Press.

Sokal, A. (1996) "Transgressing the boundaries: toward a transformative hermeneutics of quantum gravity," *Social Text* 46/47, 217–252.

Solbrig, O. (1993) "Ecological constraints to savanna land use," in Young, M. and Solbrig, O. (eds) *The World's Savannas: Economic Driving Forces, Ecological Constraints, and Policy Options for Sustainable Land Use*, Man and the Biosphere Series 12, Paris: UNESCO.

Soper, K. (1995) *What is Nature?* Oxford: Blackwell.

Soulé, M. (1995) "The social siege of nature," in Soulé, M. and Lease, G. (eds) *Reinventing Nature? Responses to Postmodern Deconstruction*, Washington, DC: Island Press, pp. 137–170.

Soulé, M. and Lease, G. (eds) (1995) *Reinventing Nature? Responses to Postmodern Deconstruction*, Washington, DC: Island Press.

Spaagaren, G. and Mol, A. (1992) "Sociology, environment and modernity," *Society and Natural Resources* 5: 4, 323–344.

Spivak, G.C. (1988) "Can the subaltern speak?," in Nelson, C. and Grossberg, L. (eds) *Marxism and the Interpretation of Culture*, London: Macmillan, pp. 271–313.

Star, S. and Griesemer, J. (1989) "Institutional ecology, 'translations,' and boundary objects: amateurs and professionals in Berkeley's Museum of Vertebrate Zoology, 1907–1939," *Social Studies of Science* 19: 3, 387–420.

Stebbing, E. (1937) "The threat of the Sahara," *Journal of the Royal African Society* 36: 1–35.

Stehr, N. (1994) *Knowledge Societies*, Thousand Oaks, CA: Sage.

Steinbeck, J. (1983 [1939]) *The Grapes of Wrath*, London: Landmark Heinemann.

Stocking, M. (1996) "Soil erosion: breaking new ground," in Leach, M. and Mearns, R. (eds) *The Lie of the Land: Challenging Received Wisdom on the African Environment*, Oxford: Currey, pp. 140–154.

Stott, P. (1999) *Tropical Rain Forest: a Political Ecology of Hegemonic Mythmaking*, London: Institute of Economic Affairs.

Stott, P. and Sullivan, S. (eds) (2000) *Political Ecology: Science, Myth and Power*, London: Arnold.

Swyngedouw, E. (1999) "Modernity and hybridity: nature, *regeneracionismo*, and the production of the Spanish Waterscape, 1890–1930," *Annals of the Association of American Geographers* 89: 3, 443–465.

Szerszynski, B., Lash, S., and Wynne, B. (1996) "Introduction: ecology, realism and the social sciences," in Lash, S., Szerszynski, B., and Wynne, B. (eds) *Risk, Environment and Modernity: Towards a New Ecology*, London: Sage, pp. 1–26.

Tandon, Y. (1995) "Grassroots resistance to dominant land-use patterns in Southern Africa," in Taylor, B. (ed.) *Ecological Resistance Movements: the Global Emergence of Radical and Popular Environmentalism*, New York: State University Press, pp. 161–176.

Taylor, B. (ed.) (1995) *Ecological Resistance Movements: the Global Emergence of Radical and Popular Environmentalism*, New York: State University Press.

Taylor, P. and Buttel, F. (1992) "How do we know we have Global Environmental Problems? Science and the globalization of environmental discourse," *Geoforum* 23: 3, 405–416.

Taylor, P. and Garcia-Barrios, R. (1995) "The social analysis of ecological change: from systems to intersecting processes," *Social Science Information* 35: 5–30.

Tennant, N. (1997) *The Taming of the True*, Oxford: Clarendon.

Tesh, S. (2000) *Uncertain Hazards: Environmental Activists and Scientific Proof*, Ithaca, NY and London: Cornell University Press.

Tewdwr-Jones, M. and Allmendinger, P. (1998) "Deconstructing communicative rationality: a critique of Habermasian collaborative planning," *Environment and Planning A* 30: 11, 1975–1989.

Thomas, D. and Middleton, N. (1994) *Desertification: Exploding the Myth*, Chichester: Wiley.

Thompson, G. (1985) "New faces, new opportunities: the environment movement goes to business school," *Environment* 27: 4, 6–11, 30.

Thompson, M. (1989) "Commentary: from myths as falsehoods to myths as repositories of experience and wisdom," *Mountain Research and Development* 9: 2, 182–186.

Thompson, M. (1993) "Good science for public policy," *Journal of International Development* 5: 6, 669–679.

Thompson, M. (2000) "Understanding the impacts of global networks on local social, political and cultural values," vol. 42 of Engel, C. and Keller, K. (eds) *First Symposium of the German American Academic Council's Project: "Global Networks and Local Values," Dresden, 18–20 February 1999*, Baden-Baden: Nomos Verlagsgesellschaft.

Thompson, M., Ellis, R., and Wildavsky, A. (1990) *Cultural Theory*, Boulder, CO: Westview Press.

Thompson, M. and Rayner, S. (1998a) "Risk and governance part 1: the discourses of climate change," *Government and Opposition* 33: 2, 139–166.

Thompson, M. and Rayner, S. (1998b) "Risk and governance part 2: the discourses of climate change," *Government and Opposition* 33: 3, 330–354.

Thompson, M., Warburton, M., and Hatley, T. (1986) *Uncertainty on a Himalayan Scale: an Institutional Theory of Environmental Perception and a Strategic Framework for the Sustainable Development of the Himalayas*, London: Ethnographica, Milton Ash Publications.

Tiffen, M. and Mortimore, M., with Gichuki, F. (1994) *More People, Less Erosion? Environmental Recovery in Kenya,* Chichester: John Wiley.

Tol, R. and Dowlatabadi, H. (2001) "Vector-borne diseases, development and climate change," *Integrated Assessment* 2: 171–183.

Touraine, A. (1981) *The Voice and the Eye*, Cambridge: Cambridge University Press.

Trimble, S. (1983) "A sediment budget for Coon Creek basin in the Driftless area, Wisconsin 1853–1977," *American Journal of Science* 283: 454–474.

Trudgill, S. and Richards, K. (1997) "Environmental science and policy: generalizations and context sensitivity," *Transactions of the Institute of British Geographers* NS 22: 5–12.

Turnbull, D. (1991) "Local knowledge and 'absolute standards': A reply to Daly," *Social Studies of Science* 21: 3, 571–573.

Turner, B.L. II (2002) "Contested identities: human–environment geography and disciplinary implications in a restructuring academy," *Annals of the Association of American Geographers* 92: 1, 52–74.

Turner, B.L. II, Clark, W., Kates, R., Richards, J., Mathews, J., and Meyer, W. (eds) (1990) *The Earth Transformed by Human Action: Global and Regional Changes in the Biosphere over the Past 300 Years,* Cambridge: Cambridge University Press.

Turner, M. (1993) "Overstocking the range: a critical analysis of the environmental science of Sahelian pastoralism," *Economic Geography* 69: 4, 402–421.

Turner, S. (2001) "What is the problem with experts?" *Social Studies of Science* 31: 1, 123–149.

UNCED (United Nations Conference on Environment and Development) (1992) *Agenda 21 and the UNCED proceedings*, edited by Robinson, N.A. with Hassan, P. and Burhenne-Guikmin, F., under the auspices of The Commission on Environmental Law of the World Conservation Union – The International Union for the Conservation of Nature and Natural Resources (IUCN), Third Series, International Protection of the Environment, New York, London and Rome: Oceania Publications (in six volumes).

UNEP (United Nations Environment Program) (2001) *Cultural and Spiritual Values of Biodiversity: a Complementary Contribution to the Global Biodiversity Assessment*, edited by Posey, D. and the Oxford Centre for the Environment, Ethics and Society, Oxford University, Nairobi: UNEP.

UNFCCC/SBSTA (2000) (United Nations Framework Convention on Climate Change/Subsidiary Body on Scientific and Technological Advice), *Methodological Issues: Land Use, Land Cover Change, and Forestry: Draft Conclusions by the Chairman of the Subsidiary Body on Scientific and Technological Advice*, UNFCCC/SBSTA/2000/CRP.11, The Hague, November.

USCEQ (United States Council on Environmental Quality) (1980) *The Global 2000 Report to the President: Entering the Twenty-first Century.* A report prepared by the Council on Environmental Quality and the Department of State: Washington, DC: Department of State.

USCSP (1999) *Climate Change: Mitigation, Vulnerability, and Adaptation in Developing and Transition Countries*, Washington, DC: USCSP.

USDA (US Department of Agricultural Research Service) (1961) *A Universal Equation for Predicting Rainfall-Erosion Losses,* USDA-ARS special report, pp. 22–26.

van Zwanenberg, P. and Millstone, E. (2000) "Beyond skeptical relativism: evaluating the social constructions of expert risk assessments," *Science, Technology and Human Values* 25: 3, 259–282.

Vayda, A. (1996) *Methods and Explanations in the Study of Human Actions and their Environmental Effects*, CIFOR/WWF Special Publication, Jakarta: World Wide Fund for Nature (Indonesia).

Vayda, A. and Walters, B. (1999) "Against political ecology," *Human Ecology* 27: 1, 167–179.

Vogel, S. (1996) *Against Nature: the Concept of Nature in Critical Theory*, Albany, NY: State University of New York Press.

Wade, W. (1997) "Greening the Bank: the struggle over the environment, 1970–1995," in Kapur, D., Lewis, J., and Webb, R. (eds), *The World Bank: its First Half Century*, vol. 2, Washington, DC: Brookings Institution, pp. 611–734.

Walker, P. (1962) "Terrace chronology and soil formation on the south coast of New South Wales," *Journal of Soil Science* 13: 178.

Wallerstein, I. (1999) "Ecology and capitalist costs of production: no exit," in Goldfrank, W., Goodman, D., and Szasz, A. (eds) *Ecology and the World System*, Westport, CT: Greenwood Press, pp. 3–12.

Wallner, H., Naraodoslawsky, M., and Moser, F. (1996) "Islands of sustainability: a bottom-up approach towards sustainable development," *Environment and Planning A* 28: 1763–1778.

Walters, B. (2001) "Event ecology in the Philippines: explaining mangrove planting and its environmental effects," paper presented at the 97th Annual Meeting of the American Association of Geographers, New York City, New York, 27 February–3 March.

Wapner, P. (1995) "Politics beyond the state: environmental activism and world civic politics," *World Politics* 47: 311–340.

Ward, B. and Dubos, R. (1972) *Only One Earth,* New York: Deutsch.

Ward, S. (1996) *Reconfiguring Truth: Postmodernism, Science Studies, and the Search for a New Model of Knowledge*, Lanham, BO, New York, London: Rowman and Littlefield.

Warren, A. and Agnew, C. (1988) *An Assessment of Desertification and Land Degradation in Arid and Semi-Arid Areas*, London: IIED and University College, London.

Watts, M. (1983) *Silent Violence: Food, Famine and Peasantry in Northern Nigeria*, Berkeley, CA: University of California Press.

Watts, M. and McCarthy, J. (1997) "Nature as artifice, nature as artefact: development, environment and modernity in the late twentieth century," in Lee, R. and Wills, J. (eds) *Geographies of Economies,* London: Arnold, pp. 71–86.

WCED (World Commission on Environment and Development) (1987) *Our Common Future,* New York: Oxford University Press.

Weinberg, A. (1963) "Criteria for scientific choice," *Minerva* 1: 159.

Weinberg, A. (1972) "Science and trans-science," *Minerva* X: 2, 209–222.

Weingart, P. (1999) "Scientific expertise and political accountability: paradoxes of science in politics," *Science and Public Policy* 26: 3, 151–161.

Wells, D. and Lynch, T. (2000) *The Political Ecologist*, Aldershot: Ashgate.

Westra, L. and Lawson, B. (2001) *Faces of Environmental Racism: Confronting Issues of Global Justice*, 2nd edn, Lanham, MD: Rowman and Littlefield.

White, R. (1980) *Land Use, Environment, and Social Change: the Shaping of Island Country, Washington*, Seattle, WA: University of Washington Press.

Whitmore, T. (1984) *Tropical Rain Forests of the Far East*, 2nd edn, Oxford: Clarendon Press.

Wiens, J. (1977) "On competition and variable environments," *American Scientist* 65: 590–597.

Wiesenthal, H. (1993) *Realism in Green Politics: Social Movements and Ecological Reform in Germany*, edited by J. Ferris, Manchester: Manchester University Press.

Wilbanks, T. and Kates, R. (1999) "Global change in local places: how scale matters," *Climatic Change* 43: 3, 601–628.

Wildavsky, A. (1995) *But Is It True? A Citizen's Guide to Environmental Health and Safety Issues*, Cambridge, MA: Harvard University Press.

Williams, J. (2001) *The Rise and Decline of Public Interest in Global Warming: Toward a Pragmatic Conception of Environmental Problems*, Huntingdon, NY: Nova.

Williams, W. (1958) "Conservation: is it important?" *Journal of the Institute of Biology* 5: 4, 86–88.

Wilson, A.N. (2000) "Sermons on science from a royal soapbox," *The Sunday Telegraph* 12 November, p. 42.

Wilson, W. (1896) "A commemorative address delivered on October 21, 1896," in Wilson, W. (1992) *Papers of Woodrow Wilson 1896–1898*, Princeton, NJ: Princeton University Press.

Wittfogel, K. (1956) "The hydraulic civilizations," in Thomas, W. (ed.) *Man's Role in Changing the Face of the Earth*, Chicago, IL: University of Chicago Press, pp. 152–164.

Wolf, E. (1972) "Ownership and political ecology," *Anthropological Quarterly* 45: 3, 201–205.

Woodgate, G. and Redclift, M. (1994) "Sociology and the environment: discordant discourse?" in Redclift, M. and Benton, T. (eds) *Social Theory and the Global Environment*, London: Routledge, pp. 51–66.

Woodgate, G. and Redclift, M. (1998) "From a 'sociology of nature' to environmental sociology: beyond social constructivism," *Environmental Values* 7: 3–24.

Worster, D. (1977) *Nature's Economy: a History of Ecological Ideas*, Cambridge: Cambridge University Press.

Worster, D. (1979) *Dust Bowl: the Southern Plains in the 1930s*, Oxford: Oxford University Press.

WRI (World Resources Institute) (1990) *A Guide to the Global Environment*, New York: Oxford University Press.

WRM (World Rainforest Movement) (2000) "Sinks that stink," *WRM Bulletin*, June.

Wu, J. and Loucks, O. (1995) "From balance of nature to hierarchical patch dynamics: a paradigm shift in ecology," *The Quarterly Review of Biology* 70: 4, 439–466.

Wynne, B. (1992a) "Misunderstood misunderstanding: social identities and public uptake of science," *Public Understanding of Science* 1: 3, 231–304.

Wynne, B. (1992b) "Uncertainty and environmental learning: reconceiving science and policy in the preventive paradigm," *Global Environmental Change* 2: 111–127.

Wynne, B. (1994) "Scientific knowledge and the global environment," in Redclift, M. and Benton, T. (eds) *Social Theory and the Global Environment*, London: Routledge, pp. 169–189.

Wynne, B. (1996a) "May the sheep safely graze? A reflexive view of the expert–lay knowledge divide," in Lash, S., Szerszynski, B., and Wynne, B. (eds) *Risk, Environment and Modernity*, London: Sage, pp. 44–83.

Wynne, B. (1996b) "SSK's identity parade: signing-up, off-and-on," *Social Studies of Science* 26: 2, 357–391.

Wynne, B. (1998) "Reply to Radder," *Social Studies of Science* 28: 2, 339–342.

Wynne, B. (2001) "Creating public alienation: expert cultures or risk and ethics on GMOs," *Science as Culture* 10: 4, 445–481.

Yearley, S. (1992) "Green ambivalence about science: legal–rational authority and the scientific legitimization of a social movement," *The British Journal of Sociology* 43: 4, 511–532.

Yearley, S. (1994) "Social movements and environmental change," in Redclift, M. and Benton, T. (eds) *Social Theory and the Global Environment*, London: Routledge, pp. 150–168.

Yearley, S. (1996) *Sociology, Environmentalism, Globalization*, London: Sage.

Zimmerer, K. (1994) "Human geography and the 'new ecology': the prospect and promise of integration," *Annals of the Association of American Geographers* 84: 1, 108–125.

Zimmerer, K. (1996a) "Discourses on soil loss in Bolivia," in Peet, R. and Watts, M. (eds) *Liberation Ecologies: Environment, Development, Social Movements*, London: Routledge, pp. 110–124.

Zimmerer, K. (1996b) "Ecology as cornerstone and chimera in human geography," in Earle, C., Matthewson, K., and Kenzer, M. (eds) *Concepts in Human Geography*, Lanham, MD: Rowman and Littlefield, pp. 161–188.

Zimmerer, K. (2000) "The reworking of conservation geographies: nonequilibrium landscape and nature–society hybrids," *Annals of the Association of American Geographers* 90: 2, 356–369.

Zimmerer, K. and Young, K. (eds) (1998) *Nature's Geography: New Lessons for Conservation in Developing Countries*, Madison, WI: University of Wisconsin Press.

Index